Y0-BUD-669

COMBINATORIAL STRUCTURES
AND THEIR APPLICATIONS

COMBINATORIAL STRUCTURES AND THEIR APPLICATIONS

Proceedings of the Calgary International Conference on Combinatorial Structures and Their Applications held at the University of Calgary, Calgary, Alberta, Canada, June, 1969.

Editorial Committee

RICHARD GUY
HAIM HANANI
NORBERT SAUER
JOHANAN SCHONHEIM

GORDON AND BREACH, SCIENCE PUBLISHERS
NEW YORK · LONDON · PARIS

150 Fifth Avenue, New York, N.Y. 10011

Library of Congress Catalog Card Number: 78 111302

Editorial office for Great Britain:
Gordon and Breach, Science Publishers Ltd.
12 Bloomsbury Way
London W.C.1

Editorial office for France:
Gordon & Breach
7-9 rue Emile Dubois
Paris 14e

Printed in the United States of America

FOREWORD

This volume contains papers and problems presented at, or submitted to, the Calgary International Conference on Combinatorial Structures and their Applications, held at the University of Calgary from June 2 to 14, 1969.

Of many international conferences in combinatorics held in recent years, this has probably been the largest so far, and the participants included many of the better known workers in this field. The conference was made possible by generous support from a number of institutions, including The University of Calgary. It was organized by the Department of Mathematics, Statistics and Computing Science of the university, the committee consisting of R.K. Guy, A.S.B. Holland, E.C. Milner, C. Nasim, N.W. Sauer, J. Schönheim, W. Vollmerhaus and representatives of the Division of Continuing Education. The program included twenty one-hour and twenty half-hour invited addresses, eighty contributed quarter-hour papers and three sessions devoted to unsolved problems.

The Organizing Committee and the Editors thank the many participants whose contributions made the conference such a success, and hope that this success is reflected in this volume.

The Editors

SONG OF THE SUBGRAPH HUNTER

Blanche Descartes

I'll pack my polyhedron
And I'll hit the Northern trail
With a plan that's algebraic,
With a plan that cannot fail.
Let the storm-cloud rise in fury
Or the sun still calmly shine,
I'll cross the wide Missouri
And the line of Forth-Nine
And still I'll journey onward
Where the prairie wheatfields lie
And Aurora's shining curtain drapes
The nights' Albertan sky.

With multifacial theory
I'll hit the Northern trail,
To tell this Guy in Calgary
My many-sided tale.

ENUMERATORS' SONG

Blanche Descartes

In the land of Denes König
By the blue Danubian water
Theorems rise for men who need them
Visions come to those that heed them

There I saw with eye seraphic
Figures many, figures graphic
Ranged in vast configurations
Piled beyong the constellations
Past the scope of Man to mind them
Yet I perceived a law to bind them
There's a C that so contains them
An ε thus constrains them.

I will fly across the ocean
From the land of Denes König
From the blue Danubian water
I will publicize my notion
Where the gathered graphmen call
In the Cicocsatan hall.

LIST OF PARTICIPANTS

Aczel, J.
Alspach, Brian
Anderson, William
Barnette, David
Bean, Don
Berge, Claude
Berlekamp, E.R.
Berman, Joel
Blackwell, Paul
Bondy, J. Adrian
Bosak, Juraj
Bridges, W.G.
Bryant, Peter R.
Bush, K.A.
Butler, Geoffrey, J.
Capobianco, M.F.
Chvatal, Vaclav
Cori, Robert
Crapo, Henry H.
Croft, Hallard T.
Csima, J.
Denes, Jozsef
DiPaola, Jane
Doyen, Jean
Driessel, K.
Eberlein, Mrs. P.J.
Edmonds, Jack
Edwards, C.S.
Erdös, Paul
Farmer, Frank
Fenyvesi, Marguerite
Foulkes, H.O.
Fox, Bennett
Fray, Robert
Fulkerson, D.R.
Gewirtz, Allan
Goldman, Jay R.
Gomory, R.E.
Graham, Ronald L.
Graver, Jack E.
Greene, Curtis
Greenwood, Robert E.
Grünbaum, Branko
Gryte, Daniel
Gupta, R.P.
Guy, Richard
Haff, Charles
Haggard, Gary
Hanani, Haim
Harary, Frank
Harborth, Heiko
Hare, William
Harper, Lawrence H.
Harris, Bernard
Harzheim, Egbert
Havel, J.
Hedetnieme, Stephen T.
Hell, Pavol
Hoffman, Allan J.
Hoffman, Kenneth R.
Horton, J.D.
James, Lee
Jenkyns, Thomas A.
Johnson, Diane
Johnson, Ellis
Johnson, Selmer M.
Jucovic, Ernest
Jung, H.A.
Katona, Gyula
Kelly, John B.
Kenyon, J.C.
Kleitman, Daniel J.
Kotzig, Anton
Kramer, Earl S.
Kreweras, Germain
Krolik, Max
Kronk, Hudson V.
Laver, Richard
Lawes, C. Peter
Lawler, Eugene L.
Leeb, Klaus
Lempel, Abraham
Lindström, Bernt
Lovasz, Laszlo
Malkevitch, Joseph
Mani, P.
Mason, John H.
Matheson, David
Matula, David
McMullen, Peter
McWhirter, I.P.
Melter, Robert A.
Mendelsohn, N.S.
Meyer, Walter
Miller, Donald J.
Milner, Eric C.
Monk, Donald
Moon, John W.
Moser, Leo
Moser, William
Murchland, J.D.
Murty, U.S.R.
Narayana, T.V.
Nash-Williams, C.J.A.
Nasim, Cyril
Nesetril, Jaroslav
Niven, Scott
Odeh, Robert

Osterweil, Leon
Ostrand, P.A.
Peck, J.E.L.
Perfect, Hazel
Petry, Francoise
Pfaltz, John L.
Picard, C.F.
Pless, Vera
Pultr, Aleš
Quintas, Louis
Ray-Chaudhuri, D.K.
Read, R.C.
Reay, John R.
Reid, K.B.
Ringel, Gerhardt
Renyi, A.
Riddell, J.
Roberts, Fred
Roberts, J.B.
Robertson, Neil
Rosa, Alexander

Roselle, David
Rosenfeld, Moshe
Rosenstiehl, Pierre
Sachs, Horst
Sellee, Thomas
Sauer, Norbert
Schönheim, J.
Sedlacek, Jiri
Seidel, J.J.
Sekanina, M.
Selfridge, J.L.
Sheehan, J.
Shimrat, M.
Sichler, J.
Simonoff, L.J.
Simonovits, M.
Smith, John H.
Spencer, Joel H.
Stanton, Ralph G.
Stone, A.H.
Storer, T.

Taylor, Walter
Treash, Mrs. Christine
Turan, Pal
Turan, Vera Sos
Turyn, R.J.
Tutte, William T.
Tverberg, Helge
Vollmerhaus, Walter
Walkup, David
Wallis, J.
Watkins, Mark
Williams, M.R.
Wollmer, Richard D.
White, A.T.
White, L.J.
Younger, Daniel
Youngs, J.W.T.
Zaks, Joseph
Zimmer, J.A.
Znam, Stefan

TABLE OF CONTENTS

DIAGRAMS AND SCHLEGEL DIAGRAMS

David Barnette

University of California, Davis, CA, U.S.A.

1. Introduction. A *cell complex* is a collection C of convex polytopes with the property that any face of a member of C is a member of C and any non-empty intersection of members of C is a member of C. Two cell complexes C_1 and C_2 are *combinatorially equivalent* provided there is a 1-1 function from the members of C_1 onto the members of C_2 that preserves dimension and incidences. The set of all proper faces of a d-polytope (convex d-dimensional polytope), P, is called the *boundary complex* of P and will be denoted by $\beta(P)$. If we are given a d-polytope P, we may construct a cell complex in $(d-1)$-space which closely resembles $\beta(P)$ in the following way: let p be a point not in P but "close enough" to an interior point of a facet ($(d-1)$-dimensional face) F of P so that any line segment from p to a vertex of P not on F, intersects the interior of F (See Fig. I for the case where P is the 3-cube). We project $\beta(P)$ onto F from the point p. The projected images of all faces of P, other than F, form a complex whose union is F. Any complex obtained in this way is called a *Schlegel diagram* of P. If C is combinatorially equivalent to a Schlegel diagram of P we shall also call C a Schlegel diagram of P.

A *d-diagram* is a cell complex C consisting of a collection of d-polytopes and their faces, satisfying:

(1) The union of the d-polytopes in C is a d-polytope Q.

(2) Any non-empty intersection of a member of C with the boundary of Q is a member of C.

It follows that a Schlegel diagram of a d-polytope is a $(d-1)$-diagram. Clearly a 1-diagram is a Schlegel diagram for some 2-polytope. A well-known theorem of Steinitz states that *any 2-diagram is the Schlegel diagram of a 3-polytope*. Until recently it was not known whether this result generalizes to higher dimensions; however, Grünbaum has constructed a 3-diagram which is not the Schlegel diagram of any 4-polytope [2, p. 218]. In this paper we construct another such 3-diagram. The advantage to this example is that the construction is simpler and the proof gives more insight into why a diagram may not be a Schlegel diagram.

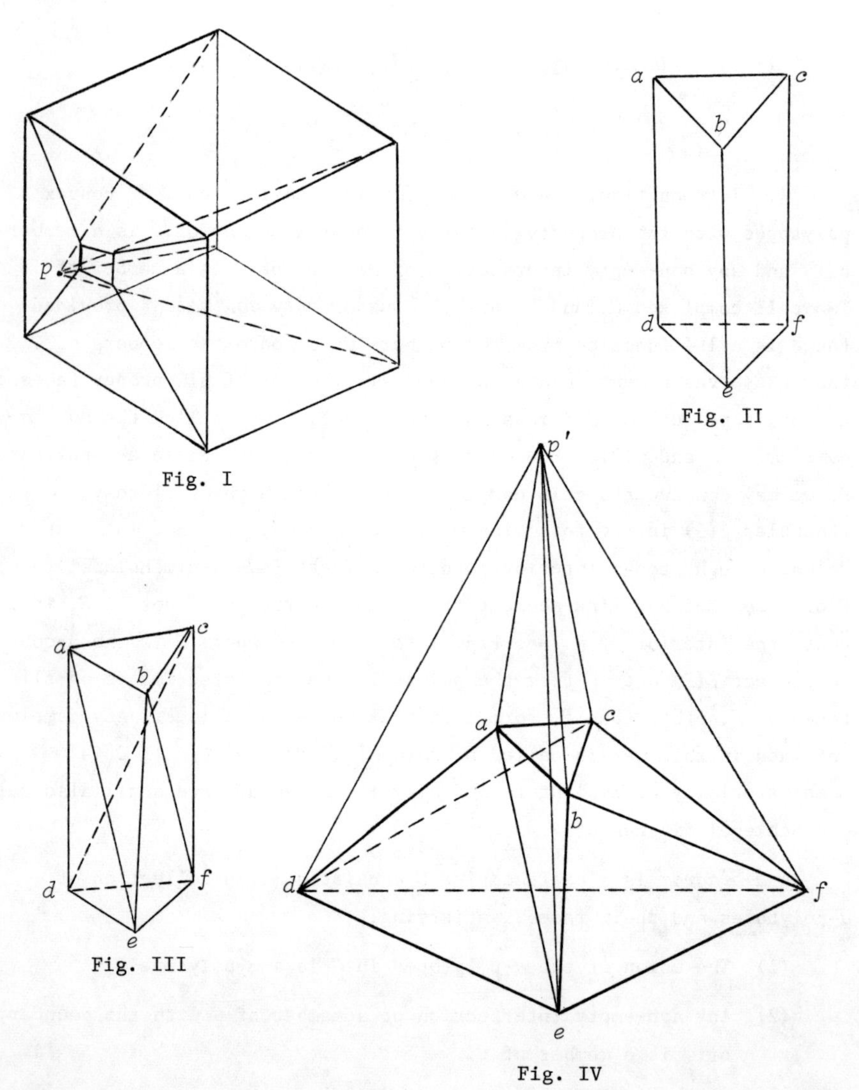

Fig. I

Fig. II

Fig. III

Fig. IV

2. The construction. We begin with a triangular prism (Fig. II) and obtain an octahedron by twisting the top face (a,b,c) slightly so that the vertices a, b, c, d, e and f are in general position. The convex hull of these vertices will be an octahedron F (Fig. III). Note that the segments $[b,d]$ $[c,e]$ and $[a,b]$ lie inside F. We shall take the following tetrahedra (and their faces) as members of the cell complex C which we are constructing: (a,b,d,e), (c,e,f,b), (a,f,c,d). Each of these tetrahedra has two 2-faces which lie inside F. These faces together with (a,b,c) and (d,e,f) form a topological sphere S. We place a point p inside S so that the line segments from p to the vertices of S lie in S. We now add the convex hulls of p with faces of S to our complex C. Next we apply a projective transformation to F so that there will exist a point p' with the property that the segments from P' to the vertices of F lie outside F (see Fig. IV). To complete our complex we add the convex hulls of p' and the faces of F other than the face (d,e,f). Note that the union of all of the tetrahedra in C is the tetrahedron (p',d,e,f).

Theorem. *C is not a Schlegel diagram.*

Proof: Suppose P were a polytope whose Schlegel diagram was C. We shall label the vertices of P with the same labels as the corresponding vertices in C. Since P is simplicial we may move any vertex an arbitrarily small amount without changing the combinatorial structure of P, thus we may assume that the vertices of P are in general position. We will look at the complex, Link (p), which consists of all faces of P which are incident to tetrahedra which meet p but which do not themselves meet p. This complex is a 2-sphere which separates p from all facets of P which do not meet p [1]. Let P' be the convex hull of all vertices of P other than p. The complex Link (p) is a subcomplex of the boundary of P' and it separates all facets of P' which were facets of P and did not meet p from the remaining facets of P', which are of the form convex hull of (x_1,x_2,x_3,x_4), where each x_i is a vertex of Link (p).

From the above we see that to construct a Schlegel diagram for P' we would take the complex in C corresponding to Link (p), remove p and fill the inside of this complex with tetrahedra whose vertices all lie on this complex. The complex corresponding to Link (p) will be the complex S. If we fill S with such tetrahedra we will have at least one edge of one of these tetrahedra lying inside S. However, any two vertices of S which are not joined by an edge on S are joined by an edge of C lying outside S, thus our diagram for P' must contain a double edge. Since a convex polytope cannot contain a double edge we have reached a contradiction.

3. Concluding remarks. The complex C is not the same as Grünbaum's example. The graph of Grünbaum's example is the complete graph on eight vertices, while in our example the vertices p and p' are not joined by an edge.

The example of Grunbaum has the property that it cannot be inverted, that is, it can be realized with some of its tetrahedra as the "outside" tetrahedron but not any tetrahedron can be the outside of a representation of this diagram. It can be shown that our example is also not invertible. One may conjecture that an invertible 3-diagram is a Schlegel diagram.

References

1. Barnette, D., A necessary condition for d-polyhedrality, *Pacific J. Math*, 23 (1967), 435-44.

2. Grünbaum, B., Convex Polytopes, Wiley, 1967.

LORENTZ TRANSFORMATIONS AND SYMMETRY PROPERTIES IN A GALOIS SPACE-TIME

E. G. Beltrametti

Instituto di Fisica dell'Università di Genova, Italy

Here we briefly summarize some results of a study which has been partially presented in previous papers [1]. Generally speaking, our study has to do with the question: how far can the physical space-time be thought of as a finite collection of points whose co-ordinates take values in a finite field $GF(k)$? First, the order k of the field is required to be of the form p^n, $p \equiv 3 \pmod 4$ and n odd, in order to meet some basic physical correspondence. We are mainly concerned with the properties displayed by standard relativity groups, namely three-dimensional rotations and Lorentz transformations acting onto the finite space-time. The structure of these groups is studied and their irreducible modular representations are classified, leading to the appearance of a number of symmetry properties caused by the finiteness of space and independent of the order of the co-ordinate field $GF(k)$. When dealing with the homogeneous Lorentz group including space reflection, one is met with the appearance of two kinds of Dirac spinors, say $\psi^{(e)}$, $e = 0,1$, which transform under inequivalent ray representation. We then look at the bispinor sesquilinear forms, i.e. the so called "currents" which are of basic importance in describing, e.g., weak interactions of elementary particles; such currents are found to fulfil covariance properties which are different according to whether they are built up with spinors of the kind $e = 0$ or $e = 1$. The choice $e = 0$ reproduces the usual pattern related to the continuous Lorentz group, but the choice $e = 1$ gives rise only to vector and axial-vector currents: this result looks relevant for physical interpretations since just these currents are known to be the only ones occurring in nature.

The previous results exhibit some connections with studies of different Authors [2]. We also mention that it has been pointed out [3] the opportunity of using Galois geometries as a tool for models of quantum field theory.

Finally we quote possible connections with the study of the Lorentz group in a p-adic space [4].

References

1. Beltrametti, E.G. and Blasi, A., *Journal Math. Phys.* 9, 1027 (1968); *Nuovo Cimento* 55A, 301 (1968); *Atti Acc. Naz. Lincei* 44, 384 (1968) and 46, (April 1969).

2. Coish, H.R., *Phys. Rev.* 114, 383 (1959); Shapiro, I.S., *Nucl. Phys.* 21, 474 (1960).

3. Joos, H., *Journal Math. Phys.* 5, 155 (1964); Ahmavaara, Y., *Journal Math. Phys.* 6, 87, 220 (1965); 7, 197, 201 (1966).

4. Everett, C.J. and Ulam, S.M., *Journal of Combinatorial Mathematics*, 1, 248 (1966).

UNE PROPRIÉTÉ DES GRAPHES k-STABLES-CRITIQUES

C. Berge

C.N.R.S., Paris, France

Soit $G = (X,\Gamma)$ un graphe non orienté sans arêtes multiples ni boucles; X est l'ensemble des sommets, et pour $x \varepsilon X$, l'ensemble $\Gamma(x)$ est l'ensemble des sommets adjacents à x.

$\alpha(G)$ dénote le *nombre de stabilité* de G, c'est à dire la cardinalité maximum d'un ensemble stable de G.

Un graphe G est dit *k-stable-critique* si $\alpha(G) = k$, et si tout graphe H déduit de G par élimination d'une arête vérifie $\alpha(H) = k+1$. Ce concept remonte à A. A. Zykov [1949]. L'étude des graphes *k-stable-critiques* fournit un grand nombre de résultats sur le nombre de stabilité d'un graphe quelconque.

Propriété 1. *Dans un graphe k-stable-critique, pour tout sommet x, il existe un ensemble stable* S_x *avec* $|S_x| = k-1$, *et qui forme avec le point x un ensemble stable maximum; on appellera "cellule" de x un tel ensemble* S_x.

1º Soit x un sommet; s'il existe un sommet a adjacent, la suppression de l'arête $[a,x]$ crée un ensemble stable S **à** k+1 éléments. On a $a,x \varepsilon S$, et l'ensemble $S - \{a,x\}$ est bien une cellule de x.

2º S'il n'existe pas de sommets adjacents à x, deux cas peuvent se produire: ou bien G consiste en un ensemble stable (et alors $X - \{x\}$, qui est stable et a k-1 éléments, est bien une cellule de x); ou bien il existe deux sommets b et c adjacents (et alors $b \neq x$, $c \neq x$).

Dans ce cas, la suppression de l'arête $[b,c]$ crée un ensemble stable S avec $b,c \varepsilon S$, $|S| = k+1$.

On a $x \varepsilon S$, et l'ensemble $S - \{b,x\}$, qui est stable et a k-1 éléments, est bien une cellule de x.

Propriété 2. *Dans un graphe k-stable-critique, la condition nécessaire et suffisante pour que deux sommets a et b aient une cellule en commun est qu'ils soient adjacents.*

1º Si a et b sont adjacents, la suppression de l'arête $[a,b]$ crée un ensemble stable S de k+1 éléments, donc $S - \{a,b\}$ est une cellule commune aux sommets a et b.

2º Si a et b sont non-adjacents, ils ne peuvent avoir une cellule commune S_{ab} car alors $S_{ab} \cup \{a,b\}$ serait un ensemble stable de k+1 éléments.

Propriété 3. *Dans un graphe k-stable-critique G, à deux arêtes adjacentes [a,b] et [b,x], on peut faire correspondre un ensemble stable maximum S_o tel que*

(1) $a,b \notin S_o$

(2) *b est relié à S_o par l'arête [b,x] et par celle-là seulement.*

Soit S_{bx} une cellule commune aux sommets b et x ; elle existe d'après la propriété 2.

Posons

$$S_o = S_{bx} \cup \{x\}$$

Comme $a \notin S_o$ (car a est adjacent à b et $\neq x$), et $b \notin S_o$, on a (1); on a aussi (2).

Donnons maintenant un énoncé, qui contient à la fois un résultat d'Andrásfai [1967] et un résultat de Beineke, Harary, Palmer [1967]:

Théorème. *Dans un graphe k-stable-critique, chaque paire de deux arêtes adjacentes est contenue dans un cycle élémentaire impair sans cordes.*

Soient $[a,b]$ et $[b,x_o]$ deux arêtes adjacentes du graphe k-*critique* G; d'après la propriété 3, il existe un ensemble stable maximum S_o, avec:

$$|S_o| = k; \quad a,b \notin S_o \ ; \quad \Gamma(b) \cap S_o = \{x_o\}$$

D'autre part, la suppression de l'arête $[a,b]$ crée un ensemble stable T de k+1 éléments, avec:

$$|T| = k+1, \quad a,b \in T$$

On va considérer les ensembles stables suivants de G:

$S = S_0 - T$

$A = T - \{b\} - S_0$

$B = T - \{a\} - S_0$

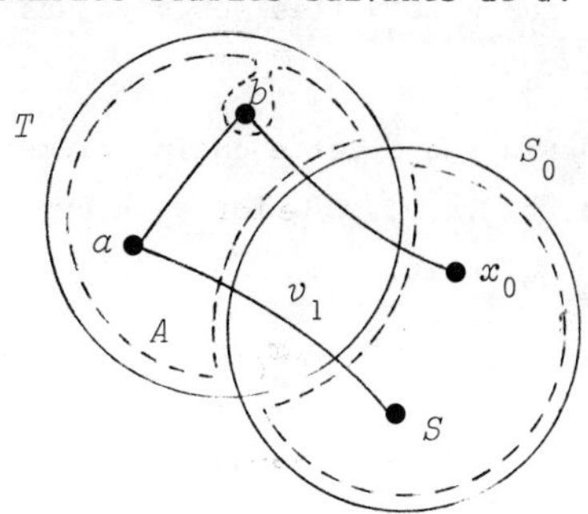

Le sous-graphe $G_{A \cup S}$ consiste en deux ensembles disjoints A et S, avec $|A| = |S|$, et en $m(A,S)$ arêtes reliant A et S. C'est donc un graphe bipartite qui admet S pour ensemble stable maximum (car s'il admettait un ensemble stable E avec $|E| > |S|$, l'ensemble $E \cup (S_0 \cap T)$ serait stable dans G avec $|E \cup (S_0 \cap T)| > |S \cup (S_0 \cap T)| = |S_0|$). Il admet donc l'ensemble complémentaire de S, c'est à dire A, comme ensemble transversal minimum; donc d'après le théorème de König, il admet un couplage maximum V avec

$$|V| = |A| = |S|.$$

De même, le sous-graphe $G_{B \cup S}$ est bipartite et admet un couplage maximum W avec

$$|W| = |B| = |S|.$$

Dans le sous-graphe engendré par $S \cup A \cup B$, le couplage V admet un seul sommet insaturé, b , et le couplage W admet un seul sommet insaturé, a; la composante connexe du graphe partiel engendré par $(V-W) \cup (W-V)$ et qui contient a ne peut être que

1º un point isolé;

2º un cycle élémentaire pair dont les arêtes appartiennent respectivement à V et à W;

3º une chaîne élémentaire dont les arêtes appartiennent respectivement à V et à W, et dont les extrêmités sont deux sommets distincts insaturés pour un des deux couplages V ou W.

Le cas 1º est impossible (car a est saturé dans $V-W$); le cas 2º aussi (car a est insaturé dans le couplage W).

C'est donc une chaîne élémentaire, dont les deux extrêmités ne peuvent être que a et b (les seuls points insaturés dans les couplages V et W).

Il s'ensuit que cette chaîne élémentaire est de longueur paire, et, avec l'arête $[a,b]$, elle forme un cycle élémentaire impair du sous-graphe $G_{S \cup A \cup B}$, soit:

$$\mu = [a,b,x_0,x_1,x_2,\ldots,x_{2l},x_{2l+1} = a]$$

avec

$$x_0,x_2,\ldots,x_{2l} \in S$$

$$a,b,x_1,\ldots,x_{2l-1} \in A \cup B$$

Une corde du cycle μ ne peut avoir une extrêmité en b (car b est incident seulement aux arêtes $[a,b]$ et $[b,x_o]$); elle ne peut avoir ses deux extrêmités dans S, ni ses deux extrêmités dans $A \cup B$.

Elle a donc une extrêmité dans A, et l'autre dans S; elle détermine donc avec μ un cycle plus court contenant les arêtes $[a,b]$ et $[b,x_o]$ et de longueur impaire.

Si ce nouveau cycle admet encore une corde, on peut répéter la même opération pour le raccourcir, etc.; au bout d'un nombre fini d'opérations, on obtiendra un cycle élémentaire impair, sans cordes, contenant les arêtes $[a,b]$ et $[b,x_o]$.

Corollaire 1. *Un graphe k-stable-critique connexe n'a pas de points d'articulation.*

S'il existait un point d'articulation a, soient B,C deux composantes connexes du graphe obtenu par suppression du point a, et soient deux arêtes $[a,b]$, $[a,c]$, avec $b \in B$, $c \in C$.

Par ces deux arêtes, il ne peut passer un cycle élémentaire, ce qui contredit le théorème.

Corollaire 2. *Un graphe k-stable-critique connexe n'admet pas une clique pour ensemble d'articulation.*

Soit A_o une clique qui serait un ensemble d'articulation, et montrons que le graphe ne peut être k-critique. Soit $A \subset A_o$ un ensemble d'articulation minimal contenu dans A_o; on peut supposer $|A| > 1$ (car sinon, d'après le Corollaire 1, le graphe n'est pas k-critique). Soient B,C deux composantes connexes du graphe obtenu par suppression de l'ensemble d'articulation A_o.

Tout sommet $a \varepsilon A$ est adjacent à B et à C, car sinon, $A - \{a\}$ serait un ensemble d'articulation, ce qui contredit la minimalité de A.

Soient $[a,b]$, $[a,c]$ deux arêtes avec $b \varepsilon B$, $c \varepsilon C$. Tout cycle élémentaire qui contiendrait ces deux arêtes passe par un sommet $a' \varepsilon A$ et différent de a; mais comme A est une clique, ce cycle admettrait la corde $[a,a']$.

Donc, d'après le théorème, le graphe n'est pas k-stable-critique.

References

1. Andrásfai, B. Über ein Extremalproblem der Graphentheorie. *Acta Math. Acad. Sc. Hung.* 13(1962), 443-455.

2. Andrásfai, B. On critical graphs. International Symposium Theory of Graphs, Gordon and Breach (1967), 9-19.

3. Beineke, L. W., Harary, F., and Plummer, M. D. On the critical lines of a graph. *Pacific J. Math.* 22(1967), 205-212.

4. Berge, C. Problèmes de colorations en theorie des graphes. *Publ. Inst. Stat. Université de Paris.* 9(1960), 123-160.

5. Berge, C. Sur une propriété du nombre de stabilité. Miméographe du Séminaire sur les Problèmes Combinatoires, nr. 3, IHP, 1962.

6. Erdös, P., and Gallai, T. On the minimal number of vertices representing the edges of a graph. *Mag. Tud. Akad.,* 6(1961), 181-203.

7. Plummer, M. P. On a family of line-critical graphs. *Monatsh. Math.,* 71(1967), 40-48.

8. Zykov, A. A. On some properties of linear complexes. *Mat. Sb.,* 24(1949), 163-188. *Am. Math. Soc. translations nr.* 79.

THE EQUIVALENCE PROBLEM FOR POST LOGIC FUNCTIONS

Paul Blackwell

University of Missouri, Columbia, MO, U.S.A.

Let J_k denote the ring of integers modulo k and $V_n[J_k]$ denote the n-dimensional vector space over J_k. A function $f:V_n[J_k] \to J_k$ is called a k-valued logic function of n variables. The function f is often called Boolean if $k = 2$ and Post if $k > 2$. Let $\oplus$ denote addition in J_k. For $x \varepsilon J_k$, and $i = 1, \ldots, k-1$, $x \oplus 1$ will be called the i-th complement of x. Then $f_1(x_1,\ldots,x_n)$ is said to be equivalent to $f_2(x_2,\ldots,x_n)$, denoted $f_1 \sim f_2$, if there exists a set of constants $C_1,\ldots,C_{n+1}$ and a permutation p of the integers $1,\ldots,n$ such that for all $(x_1,\ldots,x_n) \varepsilon V_n[J_k]$

$$f_1(x_1,\ldots,x_n) = f_2(x_{p(1)} \oplus C_1,\ldots,X_{p(n)} \oplus C_n) \oplus C_{n+1}$$

S. W. Golomb [2] gave an algorithm for deciding whether two given Boolean functions are equivalent. By the use of the finite Fourier transform, it is possible to give an algorithm for the general case, by which a given function can be reduced to a form that is canonical for its equivalence class.

Let ε be a primitive k^{th} root of unity and let $\boxdot$ denote the inner product in $V_n[J_k]$. Then by the n-dimensional k-point Fourier transform we mean the matrix

$$M = (\varepsilon^{i \boxdot j}) \qquad i,j = 0,1,\ldots,k^n-1$$

where the row and column indices i and j are to be expressed in base k and identified in the natural way with elements of $V_n[J_k]$.

For a given function f, let $f\#$ denote the function in exponentiated form, i.e., $f\#\ (x_1,\ldots,x_n) = \varepsilon^{f(x_1,\ldots,x_n)}$. The spectrum of f is defined as $S = Mf\#$. The weight of f is an n-tuple $(w_0,w_1,\ldots,w_{n-1})$ where w_i is the number of occurences of the symbol i among the functional values of f. The component S_o of S is the oth order component while any S_i for which

$i = k^m$ $(m = o, 1, \ldots, n-1)$ is called a first order component of S. A permutation P of the elements of $V_n[J_k]$ is called weight-preserving if v and $P(v)$ have the same weight n-tuple and is called linear if

$$P(u \boxplus v) = P(u) \boxplus P(v)$$

for all $u, v \in V_n[J_k]$ where $\boxplus$ denotes addition in $V_n[J_k]$. Then it can be shown [1] that complementation of variables is equivalent to component by component multiplication of the spectrum of f by a suitable column of the matrix M, that permutation of variables is equivalent to a suitable linear, weight-preserving permutation of the spectrum and that a complementation of the function f is equivalent to multiplication of each component of the spectrum by a power of ε.

The algorithm may then be stated as follows:

1. Compute the spectrum $S = Mf\#$.
2. Let $r_1, \ldots, r_n$ be the values of r $(0 \leq r \leq k-1)$ for which $\varepsilon^r S_o$ is lexicographically minimal. Form $\varepsilon^{r_1} S, \ldots, \varepsilon^{r_n} S$.
3. For each spectrum found in 2, multiply by the column of M that makes each first order spectral component lexicographically minimal. If more than one power of ε will minimize a given first order component, then use all appropriate columns.
4. For each spectrum in 3, permute the spectrum by that linear, weight-preserving permutation that puts the moduli of the first order spectral components into increasing order. If some first order components are equal, use linear, weight-preserving permutations to permute equal components in all possible ways.
5. Choose the lexicographically minimal spectrum in 4.

The significance of the spectrum of a logic function is shown, in part at least, by the fact that in the Boolean case, the celebrated Chow parameters are essentially the spectral components of orders zero and one.

References

1. Blackwell, P. K., The Classification of Logic Functions, Ph.D. Dissertation, Syracuse University, 1968.

2. Golomb, S. W., Shift Register Sequences, Holden Day, 1967.

CYCLES IN GRAPHS

J. A. Bondy

University of Waterloo, Ont., Canada

The two main results that we present in this paper are both descendants of a well-known theorem that Dirac proved in 1952.

Let G be a finite undirected graph of order $n \geq 3$ with no loops or multiple edges. $V(G) \equiv \{v_i\}_1^n$ is the vertex set of G, and $E(G)$ the edge set of G. Let d_i be the degree of vertex v_i, and assume the vertices ordered so that $d_1 \leq d_2 \leq \ldots \leq d_n$.

1. Hamiltonian graphs.

Definition: G is *Hamiltonian* if there is a cycle of length $|V(G)| = n$ in G.

The following is Dirac's theorem, together with successive generalizations by Ore and then Pósa.

G is Hamiltonian if:

$$d_1 \geq \frac{n}{2} \qquad \text{(Dirac [4])}$$

$$(v_i, v_j) \notin E(G) \Rightarrow d_i + d_j \geq n \qquad \text{(Ore [9])}$$

$$\left.\begin{array}{ll} \text{(i)} & d_i \leq i \Rightarrow i \geq \frac{n-1}{2} \\ \text{(ii)} & \text{when } n \text{ is odd, } d_{\frac{n+1}{2}} \geq \frac{n+1}{2} \end{array}\right\} \qquad \text{(Pósa [10])}$$

The first result is a slight strengthening of Pósa's theorem.

Theorem 1. *G is Hamiltonian if*

$$d_i \leq i,\ d_j \leq j\ (i \neq j) \Rightarrow d_i + d_j \geq n.$$

The proof will appear in [2]. Here I shall just give an example to show that Theorem 1 is indeed stronger than Pósa's theorem.

Example. Suppose G has degrees $\{d_i\}$ where $d_1 = d_2 = 2$, $d_3 = 4$, $d_4 = d_5 = d_6 = d_7 = 5$. Then, by Theorem 1, G is Hamiltonian. However the $\{d_i\}$ do not satisfy condition (i) of Pósa's theorem. There is in fact only one graph, up to isomorphism, having these degrees.

If it is known that G is bipartite, with the same number of vertices in each part, we can improve upon Theorem 1 in the following way.

Theorem 1a. *Let G be bipartite, with the same number of vertices in each part. Then G is Hamiltonian if*

$$d_i \leq i,\ d_j \leq j\ (i \neq j) \Rightarrow d_i + d_j \geq \frac{n}{2}.$$

When in addition we know that G has degrees $\{d_i\}_1^{n/2}$ in one part, and degrees $\{d_i'\}_1^{n/2}$ in the other part, we obtain a generalization of a result of Moon and Moser [8].

Theorem 1b. *Let G be bipartite, with vertices $\{v_i\}_1^{n/2}$ in one part, and vertices $\{v_i'\}_1^{n/2}$ in the other part. Then G is Hamiltonian if*

$$d_i \leq i,\ d_j' \leq j \rightarrow d_i + d_j' \geq \frac{n}{2},$$

where d_i, d_i' respectively are the degrees of vertices v_i, v_i' and $d_1 \leq d_2 \leq \ldots \leq d_{n/2}$, $d_1' \leq d_2' \leq \ldots \leq d_{n/2}'$.

2. Pancyclic Graphs.

Definition: G is *pancyclic* if there are cycles of all lengths l, $3 \leq l \leq n$, in G.

Pancyclic digraphs--that is, digraphs with directed cycles (circuits) of all lengths--have been studied for the particular case of tournaments. Harary and Moser [6] proved that strong tournaments are pancyclic, and Moon [7] showed further that every vertex of a strong tournament is in a circuit of each length. Alspach [1] proved that in a regular tournament every edge features in a circuit of each length.

From the definition it follows that a pancyclic graph is Hamiltonian. In the next theorem we give a condition under which the converse holds.

Theorem 2. *Let G be Hamiltonian, with $|E(G)| \geq \frac{n^2}{4}$. Then either G is pancyclic or else G is the complete bipartite graph $K_{\frac{n}{2},\frac{n}{2}}$.*

For a proof see [3].

Corollary 1. If G is Hamiltonian and $|E(G)| > \frac{n^2}{4}$, then G is pancyclic.

<u>Corollary 2</u>. If

$$(v_i, v_j) \notin E(G) \Rightarrow d_i + d_j \geq n \tag{1}$$

then either G is pancyclic or else $G \simeq K_{\frac{n}{2},\frac{n}{2}}$.

The proof of Corollary 2 follows from Ore's theorem and Theorem 2, since the condition (1) implies that G has at least $\frac{n^2}{4}$ edges.

We conclude with two conjectures.

<u>Conjecture 1</u> -- Digraph analogue of Theorem 2.

Let G be a Hamiltonian digraph with no loops or multiple edges and $|E(G)| \geq \frac{n^2}{2}$. Then either G is pancyclic or else $G \simeq \overrightarrow{K_{\frac{n}{2},\frac{n}{2}}}$, where $\overrightarrow{K_{\frac{n}{2},\frac{n}{2}}}$ is the symmetric digraph associated with $K_{\frac{n}{2},\frac{n}{2}}$.

<u>Conjecture 2</u> -- Planar analogue of Theorem 2.

Let G be a planar Hamiltonian graph with $|E(G)| \geq 2n - 2$. Then G is pancyclic.

<u>Notes</u>. 1. A special case of Conjecture 1 is when each in-degree and each out-degree of G is at least $\frac{n}{2}$. It is then known [5] that G has a Hamiltonian circuit.

2. The bound of $2n - 2$ in Conjecture 2 is necessary. The planar Hamiltonian graph in Figure 1 has $2n - 3$ edges but no cycle of length $n - 1$.

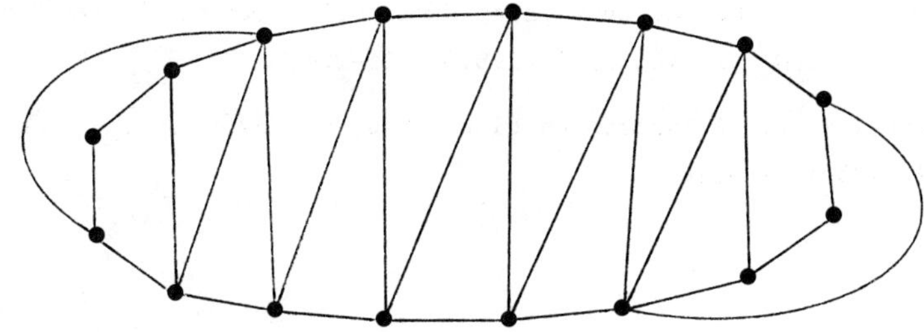

Figure 1

3. Tutte [11] has shown that all four-connected planar graphs are Hamiltonian. Since every vertex of a four-connected graph has degree at least four, such a graph must have at least $2n$ edges. Conjecture 2 would then imply that every four-connected planar graph was pancyclic.

References

1. Alspach, B., Cycles of Each Length in Regular Tournaments, *Canad. Math. Bull.*, 10 (1967), 283-286.

2. Bondy, J. A., Properties of Graphs with Constraints on Degrees, to appear in Studia Scientiarum Mathematicum Hungarica.

3. Bondy, J. A., Pancyclic Graphs, to appear.

4. Dirac, G. A., Some Theorems on Abstract Graphs, *Proc. London Math. Soc.*, 2 (1952), 69-81.

5. Ghouila-Houri, A., Une Condition Suffisante d'Existence d'un Circuit Hamiltonien, *C. R. Acad. Sci. Paris*, 251 (1960), 495-497.

6. Harary, F. and Moser, L., The Theory of Round Robin Tournaments, *Amer. Math. Monthly*, 73 (1966), 231-246.

7. Moon, J. W., On Subtournaments of a Tournament, *Canad. Math. Bull.*, 9 (1966).

8. Moon, J. W. and Moser, L., On Hamiltonian Bipartite Graphs, *Israel J. Math.*, 1 (1963), 163-165.

9. Ore, O., Note on Hamilton Circuits, *Amer. Math. Monthly*, 67 (1960), 55.

10. Pósa, L., A Theorem concerning Hamiltonian Lines, *Magyar Tud. Akad. Mat. Kutató Int. Kozl.*, 7 (1962), 225-226.

11. Tutte, W. T., A Theorem on Planar Graphs, *Trans. Amer. Math. Soc.*, 82 (1956), 99-116.

ON THE ITERATION OF A GRAPH TRANSFORMATION

Juraj Bosák

Bratislava, Czechoslovakia

Let an integer $k > 1$ and a (non-oriented) graph G be given. Denote by $f^k(G)$ the graph constructed from G in such a way that we connect every pair of vertices of G whose distance is greater than k by an arc (= simple path) consisting of k new edges and containing $k - 1$ new inner vertices, no two of these new arcs having a common inner vertex. Put $G_1^k = G$, $G_{n+1}^k = f^k(G_n)$ for every natural number n. Thus we obtain a sequence $\{G_1^k, G_2^k, \ldots, G_n^k, \ldots\}$ of graphs. By the limit of this sequence we understand the graph G^k whose vertex [edge] set is the union of vertex [edge] sets of all graphs G_n^k $(n = 1, 2. 3, \ldots)$; the incidence in G^k is defined in a natural way. By a k-index of G we mean the least natural number n such that $G_n^k = G_{n+1}^k$; if such an n does not exist, we define k-index of G as ∞. For example, a snake of length k(*) has k-index 1, a snake of length $k + 1$ has k-index 2 and a snake of length $k + 2$ has k-index ∞. The graph of Fig. 1 has k-index 3.

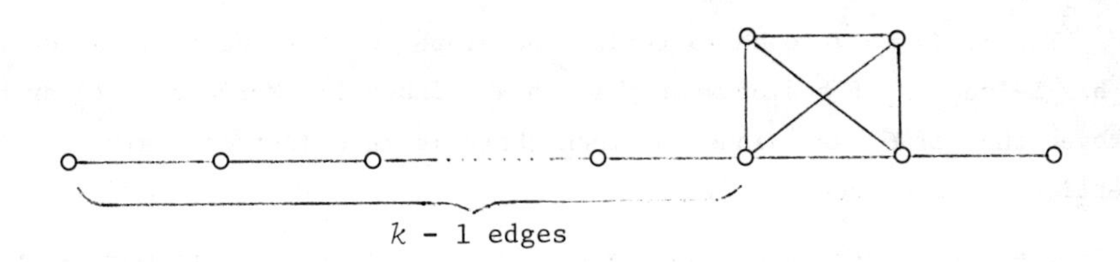

(*) By a snake of length k we understand a graph consisting of the vertices and edges of an arc with k edges.

Theorem 1. *k-index of G can be equal only to 1, 2, 3, or ∞.*

Theorem 2. *If the k-index of G is finite, then for the diameter κ_G of G we have either $\kappa_G \leq k + 1$ or $\kappa_G = \infty$.*

Let s be a non-zero cardinal number. Denote by G_{1s} the graph with two components, the first consisting of a single vertex and the second of a complete graph with s vertices.

Theorem 3. *Let G have diameter $\kappa_G = \infty$. The following three assertions are equivalent:*

(I) *k-index of G is finite;*

(II) *k-index of G is 2;*

(III) *G is isomorphic to one of the graphs C_{1s}.*

Our next aim is to characterize the graphs with a given finite k-index. Evidently, G has k-index 1 if and only if its diameger $\kappa_G \leq \infty$. Since the case of $\kappa_G = \infty$ was settled by Theorem 3, according to Theorems 2 and 1 it remains to investigate graphs G with a diameter $\kappa_G = k + 1$ and a k-index 2 or 3. In the following theorem, $\rho(a,b)$ denotes the distance of vertices a and b in G.

Theorem 4. *A graph G with diameter $\kappa_G = k + 1$ has k-index 2 if and only if for any of its vertices a, b, c and d we have:*

(1) *If $\rho(a,b) = \rho(c,d) = k + 1$, then a, b, c and d are not mutually different;*

(2) *If $\rho(a,b) = \rho(a,c) = k + 1$, then $\rho(b,c) \leq 1$;*

(3) *If $\rho(a,c) = k + 1$, but $\rho(a,b) < k + 1$ and $\rho(b,c) < k + 1$, then $\rho(a,c) = \rho(a,b) + \rho(b,c)$.*

Now it is easy to characterize the graphs with k-index 3; evidently G has k-index 3 if and only if $f^k(G)$ has k-index 2. Moreover, it can be proved that if G has k-index 3, then there is in G just one pair of vertices at distance $k + 1$.

The problem to determine the k-index of a graph will be more simple if we consider only graphs without circuits of length $\leq 2k$.

Theorem 5. *The k-index of a graph G containing no circuits of length $\leq 2k$ is equal to*

(i) *1, if its diameter $\kappa_G \leq k$;*

(ii) *2, if G is isomorphic to C_{11}, C_{12} or H_{k+1} (snake of length* *(iii)* ∞, *otherwise.*

Let a cardinal number d and an integer $k > 1$ be given. By a tied graph of type (d,k) we mean a regular graph of degree d with diameter containing no circuits of length $\leq 2k$. It is an open problem (cf. [1] and [2]) whether there exist two non-isomorphic tied graphs of the same type (d,k). We shall indicate a method, which, as we hope, will lead do an affirmative solution of the above problem in the case of infinite d.

From Theorem 5 we easily obtain (δ_G denotes the degree of G, i.e. the supremum of degrees of its vertices).

Theorem 6. *Let an integer $k > 1$ and a graph G with a diameter $\kappa_G > k$ containing no circuits of length $\leq 2k$ be given. If G is not isomorphic to any of graphs C_{11}, C_{12} and H_{k+1}, then the limit of the sequence $\{G_1^k, G_2^k, \ldots, G_n^k \ldots\}$ is a tied graph of type (d,k), where*

$$d = \begin{cases} \aleph_0, & \text{if } \delta_G < \aleph_0; \\ \delta_G, & \text{if } \delta_G \geq \aleph_0. \end{cases}$$

Note that Theorem 6 does not yield a solution of our problem of isomorphism of tied graphs immediately, because non-isomorphic initial graphs G and H may lead to isomorphic limit graphs G^k and H^k.

References

1. Bosák, J., Kotzig, A., Znám, S., Strongly geodetic graphs, *J. Comb. Theory* 5(1968), 170-176.

2. Bosák, J., Cubic Moore graphs, *Matematický casopis*, to appear.

INTRODUCTION TO JORDAN CIRCUITS OF A GRAPH*

J. Richard Büchi

Purdue University, Lafayette, IN, U.S.A.

and

Gary Haggard

University of California, Santa Cruz, CA, U.S.A.

A graph G will be denoted by $\langle V,E,I\rangle$ where V is the set of vertices of G, E is the set of edges of G and $I \subseteq V \times E$ defines the incidence of vertices and edges in G. $|V|$ and $|E|$ will always be finite. If $S \subseteq E$, then $G \cdot S$ is that subgraph of G whose edges are the members of S and whose vertices are the ends in G of the members of S.

In the remainder of the note let $G = \langle V,E,I\rangle$ be a fixed connected graph.

Let $H = \langle V',E',I'\rangle$ be a connected subgraph of G. A way $W = (v_0,e_0,v_1,\ldots,v_n,e_n,v_{n+1})$ in G is *H-avoiding* if and only if (1) $\{e_0,e_1,\ldots,e_n\} \cap E' = \emptyset$ and (2) $\{v_1,\ldots,v_n\} \cap V' = \emptyset$. Let $E' \subseteq E$ such that $G \cdot E'$ is connected. For each $e_1,e_2 \in E$ $e_1 \approx e_2$ if and only if there is a $G \cdot E'$ avoiding way in G which contains both e_1 and e_2. $\approx$ is an equivalence relation. If $A \in (E-E')/\approx$, then $G \cdot A$ is a *bridge of* $G \cdot E'$. Now let $C = \langle V_c,E_c,I_c\rangle$ be a circuit of G and let $B_1 = \langle V_1,E_1,I_1\rangle$ and $B_2 = \langle V_2,E_2,I_2\rangle$ be bridges of C. B_1 and B_2 are *strongly equivalent 3-bridges* if and only if (1) $|V_1 \cap V_c| = |V_2 \cap V_c| = 3$ and (2) there are connected subgraphs $P_1 = \langle V_{11},E_{11},I_{11}\rangle \subseteq B_1$ and $P_2 = \langle V_{22},E_{22},I_{22}\rangle \subseteq B_2$ such that (a) $E_{11} \cap E_{22} = \emptyset$ and (b) $V_{11} \cap V_{22} = V_c \cap V_1$. A pair of points $\langle a,b\rangle$ on C *separate* a pair of points $\langle a_1,b_1\rangle$ on C if and only if the four points a,b,a_1,b_1 are distinct and these points occur on C in the cyclic order a,b,a_1,b_1. Let W and W_1 be distinct ways in G such that W meets C in exactly two vertices a,b and W_1 meets C in exactly two vertices a_1,b_1. $\langle W,W_1\rangle$ *separate* each other on C if and only if

*This work was supported in part by N.S.F. grant No. CJ-120.

$\langle a,b \rangle$ separates $\langle a_1,b_1 \rangle$. Two bridges B_1 and B_2 of C *separate each other on C* if and only if there is a way $P_i \subseteq B_i$ such that P_i meets C at its end points a_i and b_i for $i = 1,2$ and such that $\langle P_1,P_2 \rangle$ separate each other on C. Two bridges B_1 and B_2 of C *overlap on C* if and only if (1) B_1 and B_2 separate each other on C, or (2) B_1 and B_2 are strongly equivalent 3-bridges. For a circuit C of G $\overline{\beta}(C)$, *the weak bridge graph of C in G,* is defined as follows: (1) the vertices of $\overline{\beta}(C)$ are the bridges of C in G and (2) there is an edge in $\overline{\beta}(C)$ joining vertices B_1 and B_2 if and only if B_1 and B_2 overlap on C. C is *weakly Jordan* if and only if $\overline{\beta}(C)$ is bipartite. G is *weakly Jordan* if and only if $\overline{\beta}(C)$ is bipartite for each circuit C in G. For a circuit C of G $\beta(C)$, *the strong bridge graph of C in G,* is defined as follows: (1) the vertices of $\beta(C)$ are the bridges of C in G and (2) there is an edge in $\beta(C)$ joining vertices B_1 and B_2 if and only if B_1 and B_2 separate each other on C. C is *Jordan* if and only if $\beta(C)$ is bipartite. G is Jordan if and only if $\beta(C)$ is Jordan for each circuit C in G.

The following seven statements are equivalent for G:

(i) $\gamma(G) = 0$, i.e., G is planar.

(ii) For any circuit C in G (a) $\gamma(G \cdot E(B \cup C)) = 0$ for each bridge B of C and (b) C is weakly Jordan.

(iii) G is weakly Jordan.

(iv) G is Jordan.

(v) For no circuit C of G does $\beta(C)$ contain an odd circuit.

(vi) For no circuit C of G does $\beta(C)$ contain a loop or a triangle.

(vii) G does not contain any subgraph which is isomorphic to $K_{3,3}$ or K_5, up to replacing suspended chains by single edges.

(ii) corrects a statement found in [1]. At (iii) the planarity of a graph becomes equivalent to a purely combinatorial condition. At (vi) it becomes apparent that the class of 1-irreducible graphs is finite. The equivalence of (i) - (vii) is proven in [2].

It is hoped that this new proof of the Kuratowski Theorem may contain ideas which generalize to surfaces of higher genus. As there are many 2-irreducible graphs, nobody will want to list all of them. In case of a surface of genus $n \geq 1$, an adaptation of our proof could stop at stage (vi), yielding the finiteness of the class of

n-irreducible graphs without actually listing them all which (vii) does.

References

1. Auslander, L., Parter, S.V., On Imbedding Graphs in the Sphere. *J. Math. Mech.*, 10 (1961), No. 3, 517-524.

2. Büchi, J.R., Haggard, G., Jordan Circuits of a Graph, *J. Comb. Th.* (to appear).

SOME PROPERTIES OF TENSOR COMPOSITE GRAPHS

Michael F. Capobianco

St.John's University, Jamaica, NY, U.S.A.

A graph (digraph) G (no loops or multiple edges) is said to be tensor composite if there exist graphs (digraphs) G_u and G_v such that $G \simeq G_u \otimes G_v$. Here $G_u \otimes G_v$ is the tensor product of G_u and G_v. See, for example, [1], for the definition. If two such graphs G_u and G_v do not exist, then G is said to be tensor prime.

We define the distance between two lines l_1 and l_2 of a graph, $d(l_1,l_2)$, as the length of the shortest path joining any vertex of l_1 with any vertex of l_2.

Theorem 1. *Let G be a tensor composite graph with n vertices and l lines, $(l > 0)$ then*

(1) $n = n_u n_v$ where n_u and n_v are integers greater than one such that for all v of G, $d(v) \leq (n_u-1)$ where $d(v)$ is the degree of v.

(2) $l = 2k$, $k = 1,2,\ldots, \binom{n_u}{2}\binom{n_v}{2}$ and these l lines can be listed in k pairs such that for each such pair l_1, l_2, $d(l_1,l_2) > 1$.

The proof is almost immediate.

If $\frac{l}{2}$ is a prime, we can say a bit more.

Theorem 2. *Let G be a tensor composite graph with n vertices and l lines $(l > 0)$ such that $\frac{l}{2}$ is prime, then*

(1) $n = n_u n_v$ $(n_u > 1,\ n_v > 1)$ such that $d(v) \leq n_u-1$

(2) $l = 2k$, $k = 1,2,\ldots, \binom{n_u}{2}$ and these l lines can be listed in k pairs such that for each pair l_1, l_2 $d(l_1,l_2) > 1$.

(3) G is bicolorable.

These conditions are not sufficient. The graph

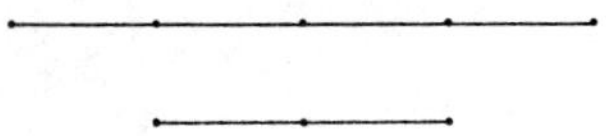

provides a counterexample.

One of our earlier results states that no tree is tensor composite. However, tensor composite digraphs which are (unoriented) trees do exist. On the other hand, one has the following:

Theorem 3. *Except for the trivial graph, any arborescence (oriented tree is tensor prime.*

The proof follows readily from the fact that an arborescence must have exactly one point with indegree 0. No tensor composite graph can have exactly one point with indegree 0, or exactly one point with out-degree 0, except the trivial one.

Theorem 3 can be applied to a situation in which two sets of players compete in tournaments. We represent each set by a digraph in which the vertices are the players and u is adjacent to v if and only if u defeats v. In the case of a tie we let u and v be adjacent to each other. Now form all possible teams or coalitions of two players, one from the first set and one from the second. Assuming that the strengths of the players are additive, the tensor product will represent the "victory-defeat" pattern for these coalitions. From the remarks following theorem 3 we see that there can not be exactly one undefeated coalition. Either there is none, or there are more than one. Both of these alternatives are possible.

References

1. Capobianco, M. F., Tournaments and tensor products of digraphs, *SIAM J. App. Math.* 15 (1967) 624-626.

2. Harary, F. and Trauth, C., Connectedness of products of two directed graphs, *SIAM J. App. Math.* 14 (1966) 250-254.

3. Miller, D., The categorical product of graphs, *Can. J. Math.* 20 (1968) 1511-1521.

SOME REMARKS ON BALLOT-TYPE SEQUENCES OF POSITIVE INTEGERS

L. Carlitz and R.A. Scoville
Duke University, Durham, NC, U.S.A.

D.P. Roselle
Louisiana State University, Baton Rouge, LA, U.S.A.

Recently various authors have considered special cases of the following problem: Let $(f(1), f(2),\ldots)$ be a non-decreasing sequence of positive integers and let $T_f(n)$ denote the number of n-tuples of positive integers $(\alpha_1,\ldots,\alpha_n)$ satisfying $\alpha_{i-1} \leq \alpha_i \leq f(i)$, $i = 1(1)n$, where for convenience we write $\alpha_0 = 1$.

In this paper we include a discussion of the properties of $T_f(n)$ although our main concern is with the relationship between this problem and a problem posed by Elwyn Berlekamp at the May 1968 Combinatorics Conference held at the University of Waterloo. The Berlekamp problem is as follows: The "roof" determined by a non-decreasing sequence of positive integers $(f(1), f(2),\ldots)$ will mean the collection of lattice-points

$$\{(x,y)\,|\,f(x-1) \leq y \leq f(x)\} \cup \{(1,y)\,|\,1 \leq y \leq f(1)\}.$$

Let us assign the number 1 to each of the cells which comprise the roof. Can we assign positive integers to the cells under the roof in such a way that every square subarray, comprised of adjacent rows and columns and containing at least one 1, will have determinant 1? If so, what are these numbers? It is shown here that the lattice-point $(n,1)$ should be assigned the entry $T_f(n-1)$. The proof of this and more general results is scheduled to appear in the Journal of Combinatorial Theory.

SOME PROBLEMS [illegible] SEQUENCES OF INTEGER SYSTEMS

[illegible] and E.A. [illegible]

[illegible]

[illegible]

[illegible]

Recently [illegible] special case of the following problem: Let $f(1), f(2), \ldots$ be a non-decreasing sequence of positive integers, and let [illegible] positive integers [illegible] where for every [illegible].

In this paper we find a discussion [illegible] properties of [illegible]. Although our main concern is with the [illegible] of this problem [illegible] Conference held at the [illegible]. The [illegible] problem is as follows: [illegible] a non-decreasing sequence of positive integers [illegible] the collection of all the points

$$[illegible]$$

Let us assign the number 1 to each of the cells [illegible] in such a way that [illegible] rows and columns and containing [illegible]. [illegible] what are these numbers? [illegible] should be assigned the [illegible]. The proof of the [illegible] result is attempted to appear [illegible] of Combinatorial Theory.

SOME UNKNOWN VAN DER WAERDEN NUMBERS

V. Chvátal

University of New Brunswick,
Fredericton, NB, Canada

Van der Waerden's theorem [1] may be formulated as follows:
Given any positive integer k and positive integers $t_1, t_2, \ldots, t_k$ there is an integer m such that given any partition

$$\{1,2,\ldots, m\} = V_1 \cup V_2 \cup \ldots \cup V_k \tag{1}$$

there is always a class V_j containing an arithmetic progression of $t_j + 1$ terms. Let us denote the least m with this property by $W(k;t_1,t_2,\ldots t_k)$. Using the computing facilities of the University of New Brunswick, I found $W(3;2,2,2) = 27$ and the following numbers $W(2;t_1,t_2)$:

$t_1 \backslash t_2$	2	3	4	5	6
2	9	18	22	32	46
3	18	35	55		
4	22	55			
5	32				
6	46				

By a *good partition*, we shall mean a partition of the form (1) such that no V_j contains an arithmetic progression at $t_j + 1$ terms. Obviously, if (1) is a good partition, then a partition

$$\{1,2,\ldots,m\} = X_1 \cup X_2 \cup \ldots \cup X_k \tag{2}$$

obtained by "reflection" (i.e. defined by $X_j = \{p; m + 1 - p \in V_j\}$) is also good.

If (1) is a good partition and $t_1 = t_2 = \ldots = t_k$, then a partition (2) obtained by "relabelling" (i.e. defined by $X_j = V_{\pi(j)}$ where π is a fixed permutation of the subscripts) is also good.

All the good partitions corresponding to the numbers $W(k;t_1,t_2,\ldots,t_k)-1$ are as follows:

W(2;2,2):

11221 122 (i.e. $V_1 = \{1,2,5,6\}$, $V_2 = \{3,4,7,8\}$)

12122 121

12211 221

and the other three obtained by relabelling.

W(2;3,2):

11121 22111 21221 11

and another one obtained by reflection.

W(2;4,2):

11212 11112 21211 11212 1

121A1 12211 11212 2111B C

where BC ≠ 11,

and seven others obtained by reflection.

W(2;5,2):

21111 12211 11121 12211 11212 11112 2

11111 221A1 11221 11211 11122 1BC11 D

where ABCD is 1112, 1211, 1212, 1221 or 2112,

and six others obtained by reflection.

W(2;6,2):

21111 12112 11111 21112 21211 11211 11121 12111 12121

11111 12121 11121 22111 11121 11221 11112 A1211 11B12

where AB ≠ 22,

and four others obtained by reflection.

W(2;3,3):

A2122 21112 1B212 22111 21C21 22211 121D

where $|\{A,B,C,D\}| = 2$,

and 14 others obtained by relabelling.

W(2;4,3):

11121 12111 12221 22122 21111 21121 11122 21221 22211 11211 2111
A2B12 212C2 11112 11211 11222 12212 22111 12112 11112 D2122 1E2F

where ABCDEF is arbitrary.

W(3;2,2,2):

11221 12323 31311 21223 13323 2
C2112 12332 23321 21123 31131 3
112AB 11313 22332 23131 1BA21 1

where AB = 23 or 32, C = 1 or 3,
and 43 others obtained by reflection and relabelling.

Incidentally, a similar algorithm was developed and some of the above results obtained also by A. Rosa and Š. Znám.

References

1. van der Waerden, B.L., Beweis einer Baudetschen Veermutung, *Nieuw Archief voor Wiskunde* 15(1927), 212-216.

PLANAR MAPS AND BRACKETING SYSTEMS

Robert Cori

Centre National de la Recherche Scientifique, Paris, France

In his "Census" series of papers Professor W. T. Tutte [1] has obtained a theorem which may be restated as follows:

Let $p = (p_0, p_1, \ldots, p_k)$ be a vector with strictly positive integral coordinates and $t(p)$ be the number of rooted planar maps with fixed vertices $s_0, s_1, \ldots, s_k$ whose respective degrees are: $p_0, p_1, \ldots, p_k$, the root being incident with the vertex s_0. If the numbers p_i are even for all i, $p_i = 2q_i$, then $t(p)$ is given by:

$$t(p) = \frac{(n-1)!}{(n-k+1)!} \cdot 2q_0 \cdot \prod_{i=0,k} \frac{(2q_i-1)!}{q_i!(q_i-1)!} \tag{1}$$

It is also stated in this paper that if $p' = (2q_0-1, 2q_1+1, 2q_2, \ldots, 2q_k)$ one has:

$$t(p') = \frac{(n-1)!}{(n-k+1)!} \quad \frac{(2q_0-1)!}{(q_0-1)!(q_0-1)!} \quad \frac{(2q_1)!}{q_1!q_1!} \quad \prod_{i=2,k} \frac{(2q_i-1)!}{q_i!(q_i-1)!} \tag{2}$$

The formula (1) is proved by using techniques of differential calculus on generating functions.

We have shown with Professor M. P. Schützenberger that these two results are equivalent with the two other ones:

$$t(2n-2k, 2, \ldots, 2) = \frac{(n-1)!(2n-2k)!}{(n-k+1)!(n-k)!(n-k-1)!} \tag{3}$$

$$t(p) = t(p'). \tag{4}$$

(3) may be proved by a recursion formula stated in [1].

To obtain (4) we have established a one to one correspondence between rooted Planar maps with vertices $s_0, s_1, \ldots, s_k$ and the words of a language L_k on the free monoïd generated by the set:
$\{x_0, x_1, x_2, \ldots, x_k\} \cup \{\bar{x}_0, \bar{x}_1, \ldots, \bar{x}_k\}$. This language may be considered as a

generalization of the well known "Bracketing systems" or Dyck language on two letters. If in a rooted planar map the vertex s_i has degree p_i, then in the corresponding word the letters x_i or $\overline{x}_i$ appear p_i times.

And so (4) is a direct consequence of the building of a bijection between the sets L_k^p (of the words of L_k having p_i occurrences of x_i or $\overline{x}_i$ for all i) and $L_k^{p'}$.

We have thus obtained a purely Combinatorial proof of (1) and (2).

Reference

1. Tutte, W. T., A Census of Slicings, *Can. J. Math.* 14, (1962), 402-417.

GEOMETRIC DUALITY AND THE DILWORTH COMPLETION

Henry H. Crapo

University of Waterloo, Ont., Canada

A number of interesting geometric constructions arise as solutions to categorical problems, such as how to realize a given general mapping as a composite of special types of mappings. With this in mind, we shall look at three constructions: lift, join, and the Dilworth completion.

Consider two classes of maps defined on geometric lattices (lattices of flats of combinatorial geometries [2]). *Strong maps* preserve the lattice operation supremum and the covering relation $\downarrow$:

> $y \downarrow x$ iff the flat y contains or equals the flat x, but there are no intermediate flats.

Typical strong maps are injections of subgeometries and contractions $x \to x \vee z$ for some fixed flat z. These are the two operations referred to in Tutte's work [7] as "taking minors". *Comaps* preserve the operation infimum on modular pairs of flats, and the covering relation $\downarrow$. Typical comaps are injections of subgeometries and intersections $x \to x \wedge z$ for some fixed modular flat z.

Every strong map $\sigma\colon P \to Q$ may be factored as an injection $P \to L$ followed by a contraction $L \to Q$. Higgs [5] invented the *lift* construction for making up the required lattice L. Assume $\sigma\colon P \to Q$ is onto Q. The lift lattice L has a flat z such that the interval $[z,1]$ is isomorphic to the image lattice Q and such that $x \to x \vee z$ restricted to flats x in the subgeometry $P \subseteq L$, agrees with the strong map $\sigma\colon P \to Q$. The choice of an interval lattice $[0,z]$ is not at all critical: it may be any geometric lattice of the correct rank. Figure 1 provides an example of the lift construction.

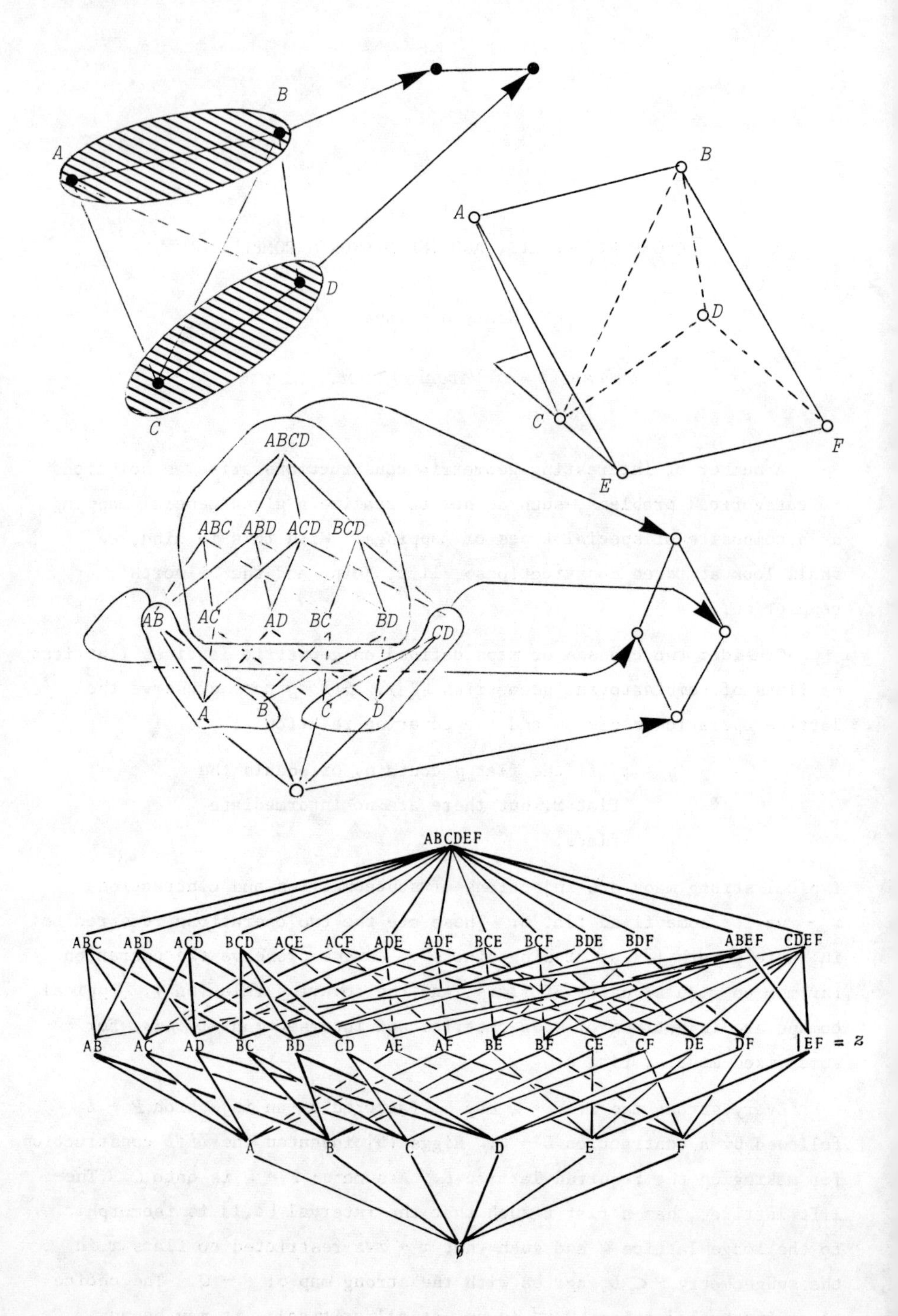
A
B
C
D
A
B
C
D
E
F
ABCD
ABC ABD ACD BCD
AB AC AD BC BD CD
A B C D
ABCDEF
ABC ABD ACD BCD ACE ACF ADE ADF BCE BCF BDE BDF ABEF CDEF
AB AC AD BC BD CD AE AF BE BF CE CF DE DF EF = z
A B C D E F
∅

Figure 1

Every comap $\tau: P \to R$ may be factored as an injection $P \to J$ followed by a contraction $J \to R$. The join construction [1] furnishes the required lattice J. The join lattice has a modular flat z such that the interval $[0,z]$ is isomorphic to the image lattice R, and such that $x \to x \wedge z$, restricted to flats in the subgeometry $P \subseteq J$, agrees with the comap τ. The interval $[z,1]$ cannot be freely chosen, as we shall now see.

The embedding $P \to J$ preserves the covering relation, so if a flat y covers a flat x in P, there are perhaps four possibilities for $x \vee z$, $y \vee z$, $x \wedge z$, $y \wedge z$:

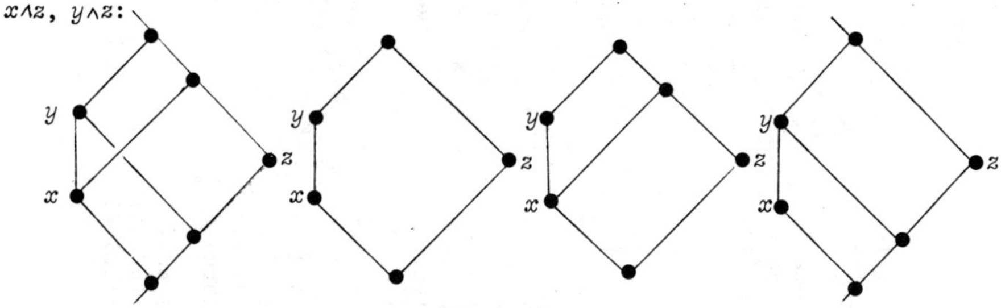

Figure 2

In the first instance, we have

$$\left.\begin{matrix} x \\ y \wedge z \end{matrix}\right\} \leq \left\{\begin{matrix} y \\ x \vee z \end{matrix}\right.,$$

so there must be an intermediate element. This is impossible in any lattice, if y covers x, so there are actually only three possibilities. If the flat z is modular, then the second instance is also excluded, and *exactly one* of the equalities

$$x \vee z = y \vee z$$

$$x \wedge z = y \wedge z$$

holds.

Back on the subgeometry P we find two closure operators, one going up

$x \to \overline{x}$ = greatest flat in P having the same supremum as x with z

and one going down,

$x \to \dot{x}$ = least flat in P having the same infimum as x with z,

related on each covering pair $x < y$ by

$$\overline{x} = \overline{y} \qquad \text{or} \qquad \dot{x} = \dot{y}$$

(bot not both)

If P is a Boolean algebra, this is precisely Whitney duality (orthogonality) of pregeometries. In Figure 3 (an example of the join operation) the associated closure operators are shown by arrows between the covering pairs in the lattice P.

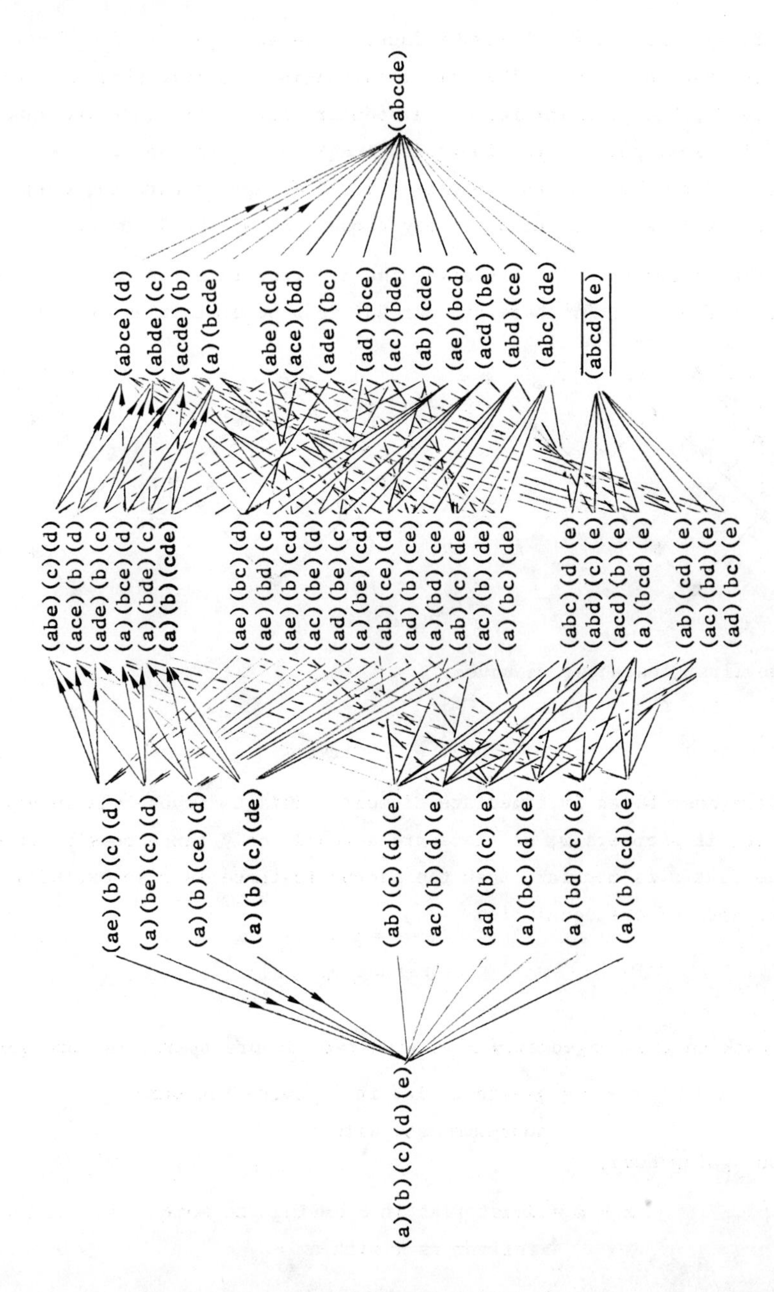

Figure 3

Recall, for instance, that if a set B of edges of an arbitrary graph covers a subset A, of its edges, the Whitney dual has the edge $e = B - A$ in the dual closure of the set $\overset{c}{-}B$ if and only if e is *not* in the closure of the set A (i.e., if and only if e is an isthmus in the subgraph B).

The question naturally arises,

I When can an orthogonal pair of pregeometries on a set X be realized by an embedding of the Boolean algebra $B(X)$ in a geometric lattice J with a modular flat z, such that the associated closure operators are realized by supremum and infimum with z?

Whenever this is accomplished, a peculiar thing happens. If Q and Q^* are the lattices of flats of the orthogonal pair, then the lattice Q^* is embedded *upside-down* in the interval $[0,z]$, which is some geometric lattice, say R, of rank equal to that of Q^*. This raises problem II, surely one of the most interesting unsolved problems in combinatorial geometry.

II Given a geometric lattice S, embed the inverted lattice $\tilde{S}$ in a geometric lattice R, preserving infimum and cover.

Theorem. *A solution to problem I yields a solution to problem II and conversely.*

Proof. If we have solved problem I, and we are given a geometric lattice Q to invert and complete, we realize Q by a strong map $\sigma: B \to Q$ on a Boolean algebra, and represent the pair (σ^*,σ) by the solution to I, where σ^* is the orthogonal map. Then the interval $[0,z]$ is the required completion of the inverted lattice $\tilde{Q}$. Conversely, if we have solved problem II, and wish to represent an orthogonal pair of closure operators, we represent the downward closure operator as a strong map $\tau: \tilde{B} \to S$, we complete $\tilde{S}$ to a geometric lattice R of equal rank, and join the Boolean algebra B to the geometric lattice R across the pair of comaps $(\tau$, identity).

Once problem I is solved, the various concepts relevant to pregeometries take on a lattice-theoretic character. For instance, the bases of the pregeometry become the lattice complements of the flat z in the lattice J. Geometrically speaking, two orthogonal pregeometries on an n-element set X are realized by placing the points of X freely in space, then *projecting from* and *intersecting with* a fixed flat z. The flat z is characterized as follows: it has rank $|X| - r(X)$, equal to the rank of the orthogonal pregeometry and it is in general position relative to the free geometry X, in the same space of rank n.

For vector spaces, this construction was mentioned as an example of orthogonality by Whitney in his 1935 paper [8] on abstract linear dependence.

A number of special cases of problem II are solved.

(a) A projective geometry needs no completion, because the inverted lattice is already geometric.

(b) The eleven planes of the 6-point prism (figure 4), when regarded as points in the inverted lattice form a triple-projection of a cube. (Higgs)

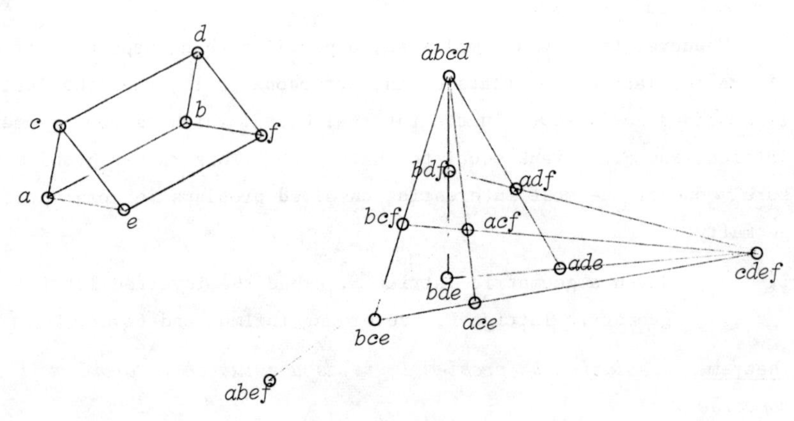

Figure 4

(c) The inverted partition lattice $\tilde{P}_n$ has 2^n-1 copoints, and may be embedded in the modular lattice of subspaces of the vector space $GF(2^{n-1})$.

(d) David Sachs [6] has studied a number of partial inversions.

One case of problems I and II has been completely solved. On a geometric lattice P, let the empty flat, together with the flats of rank $k+1$ and higher, be closed relative to a downward closure. Let's call the lattice of closed flats $L_k(P)$, the *lower truncation*. In the orthogonal upward closure, the entire flat, together with those flats of rank $k-1$ and lower are closed. This yields the upper truncation $T_k(P)$, which is geometric. To solve problem I we must extend the geometry P of rank n to include a flat z in general position, rank $n-k$. Dilworth constructed the geometry needed for the interval $[0,z]$ in 1944 [3]. Let us call it $D_k(P)$ the *Dilworth completion* of the lower k-truncation. Recently we have found a direct lattice-theoretic construction of this lattice.

Each flat $x \in L_k(P)$ has a rank

$$f(x) = \lambda_P(x) - k,$$

and may be associated with the set $\binom{x}{k+1}$ of flats of rank $k+1$ in P which it contains.

Theorem. *A set of $(k+1)$-flats of P is a geometric flat of the Dilworth completion $D_k(P)$ if and only if it is a (necessarily disjoint) union*

$$\binom{x_1}{k+1} \cup \ldots \cup \binom{x_m}{k+1}$$

for a collection $\{x_1, \ldots, x_m\}$ *of flats of* $L_k(P)$ *satisfying, for all sub-collections* $\{x_j\}$ *having at least two members*

$$(*) \qquad f(\bigvee x_j) > \sum f(x_j).$$

If a Boolean algebra B is truncated at $k = 1$, a new geometry is formed on the set of pairs. Condition (*) requires only that the sets X_i be disjoint, so the flats $\binom{X_1}{2} \ldots \binom{X_m}{2}$ are the *partitions* with blocks X_i. The solution to problem I in this case is the larger partition lattice P_{n+1} (as in figure 3; figures 5 and 6 show this geometrically.)

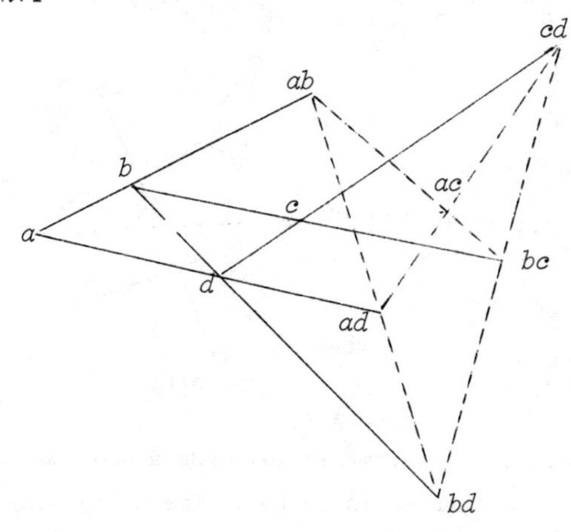

Figure 5

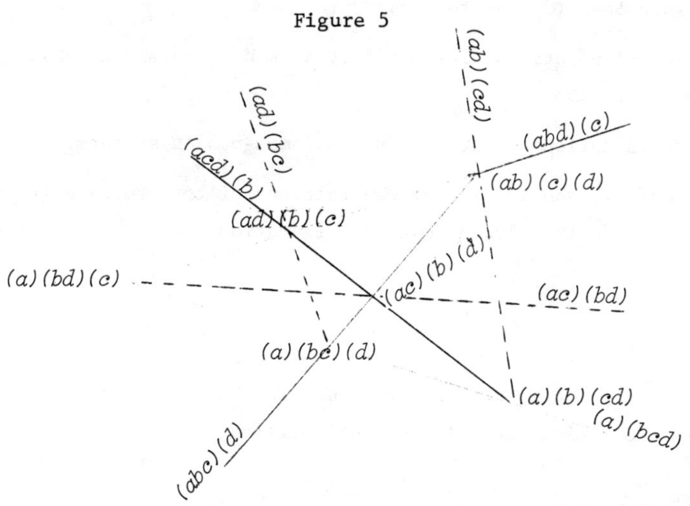

Figure 6

The Dilworth completions $D_k(B_n)$ are appropriate generalizations of the partition lattices $D_1(B_n)$. Geometrically, they are the geometries of *n flats of rank n-k-1 in general position in rank n-k.*

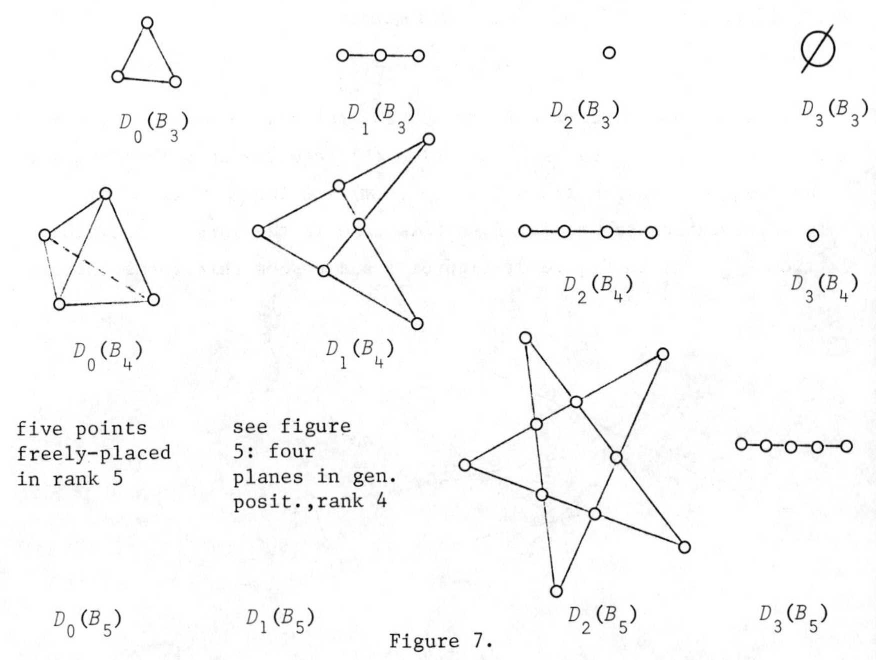

Figure 7.

It is clear now that Juris Hartmanis [4] made a poor choice when he defined some "partitions of type k" in 1956-9. The difference can be seen as follows. Let us find those minimal configurations N_i which define the closure in $D_k(B_n)$ as follows:

(1) A set of points is closed iff it does not contain a subset isomorphic to one of the N_i.

These configurations N_i are the *minimal non-closed* subsets.

In $D_1(B_n)$, the lattice of contractions of the complete graph on vertices, there is only one such configuration

N_2 =

In $D_2(B_n)$ the points are triples. The Hartmanis closure on triples has only one minimal non-closed configuration

N_2 =

In $D_2(B_n)$, however, there are infinitely many different minimal non-closed subsets. We indicate all those consisting of p triples, for $p = 2,\ldots,7$, in figure 8.

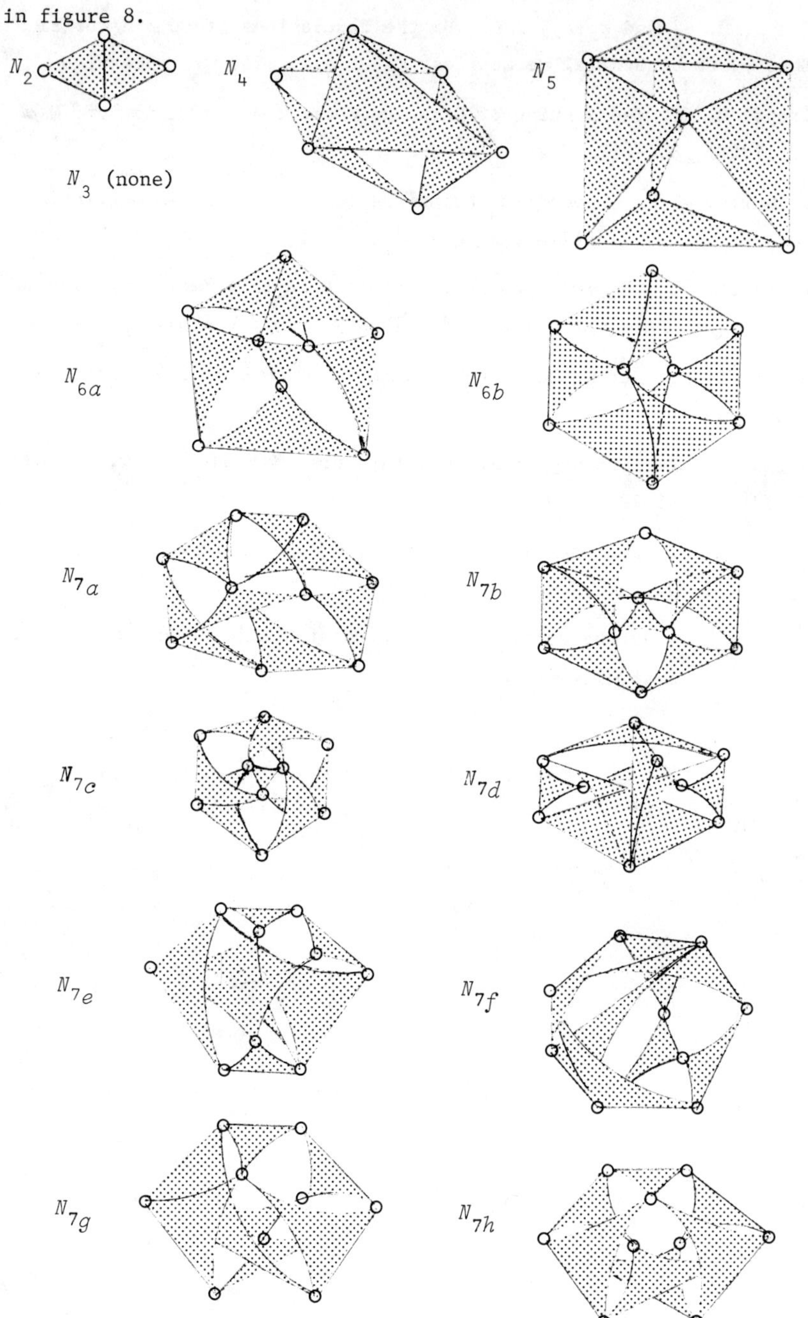

Figure 8

References

1. Crapo, H. H. and Rota, G. C., On the Foundations of Combinatorial Theory, Combinatorial Geometries, MIT Press, 1969, to appear.

2. Crapo, H. H., The Joining of Exchange Geometries, *J. Math. and Mech.* 17 (1968), 837-852.

3. Dilworth, R. P., Dependence Relations in a Semi-modular Lattice, *Duke Math. J.*, 11 (1944), 575-587.

4. Hartmanis, J., Generalized Partitions and Lattice Embedding Theorems, Proc. Symp. Pure Math. 2 (Lattice Theory) A.M.S., 1961.

5. Higgs, D. A., Strong Maps of Geometries, *J. Comb. Th.* 5 (1968), 185-191.

6. Sachs, D., Reciprocity in Matroid Lattices, *Sem. Mat. Univ. di Padova* 36 (1966), 55-79.

7. Tutte, W. T., Lectures on Matroids, *J. Res. Nat. Bur. Stand.* 69B (1965), 1-47.

8. Whitney, H., On the Abstract Properties of Linear Dependence, *Amer. J. Math.* 57 (1935), 509-533.

SOME PROBLEMS OF COMBINATORIAL GEOMETRY

H.T. Croft

Peterhouse, Cambridge, England

1. [Erdös]. A plane set S of infinite Lebesque measure necessarily has the property that it contains the vertices of a triangle of area 1; find a constant c (ultimately as small as possible) such that any set of plane measure $> c$ has the same property. [The proof for infinite measure is elegant: There exist 2 perpendicular lines--call them (x,y)-axes--that cut S in sets of positive linear measure. Since difference sets of sets of positive measure contain a neighbourhood of the origin, we have that $\exists h_1, h_2$ such that any point with $|x| \geq h_1$ or $|y| \geq h_2$, together with some 2 points on the appropriate axis furnishes such a triangle. This proof extends to any unbounded set (comment by Davies).]

2. [Steinhaus]. Let D be a set of infinite measure in the plane. Is it true that we can always (shift and) rotate it so as to cover an infinite number of lattice points? If no, we may add conditions on D e.g. closed and simply-connected.

Again, suppose a set is such that after any (or perhaps 'almost any') such motion it covers an infinite number of lattice points. Must it have positive (or even infinite) outer measure? Many variants e.g. consider just translations, or just rotations about one fixed point.

3. Let S be a planar set of points; let C be the set of points each of which is unit distance from some point of S, i.e. C is the union of unit circles with centres in S.

Now, assuming measurability of S and C, is the following true:

$$m(S) = \pi r^2 \Rightarrow m(C) \geq \begin{cases} 4\pi r & (r \leq 1) \\ \pi(1+r)^2 & (r \geq 1) \end{cases} \quad ?$$

If the linear measure of the plane set S is ℓ, can we say anything about the upper and lower bounds of the plane measure of C?

Define 'circular measure' $m_c(C)$ of set C to be $\int_0^\infty f(r)dr$, where $f(r)$ is the plane measure of the set S of centres of circles of radius r, whose circumference lies wholly in C.

If $m(C) = \pi R^2$, is $m_c(C) \leq \int_0^R \pi(R-r)^2\, dr = \frac{\pi}{3} R^3$?

4. [Fejes Toth]. If C is a plane convex set of constant width, can 8 congruent non-overlapping sets touch it. 7 can, if C is a Realeaux triangle. See [Fejes Toth, Stud. Sci. Math. Hung. 2 (1967), 363-7].

5. It is well-known that only 12 non-overlapping spheres can touch one sphere, all of the same size. The maximum number of non-overlapping translates of a convex body C that can touch C is 26 (when C is a parallelepiped). But what is this maximum if we restrict consideration to bodies with a unique tangent plane at each point of the boundary (or weaker, so that at each point of contact of C with any of its translates in the given position there is a unique separating plane)? Or again, if we re-interpret 'touch' to mean touch along a set of positive area? Similar questions restricted to centrally symmetric C. Perhaps the answers to all are 14 (as they are for cubes). See [Grünbaum, Proc. J. Math., 11 (1961), 215-219].

How many congruent convex bounded bodies in 3-D can touch pairwise, with no overlapping? Certainly 7 (right circular cylinders, not too long). Presumably the answer is finite. Is there any difference if 'bounded' is removed?

6. (a) Find the shortest curve in the plane that cuts every chord (possibly produced) of a fixed, say unit, circle;

(b) Find the shortest curve in the plane that cuts every tangent to a unit circle;

where we interpret 'curve' as either

(i) a continuous curve i.e. homeomorph of [0,1], or

(ii) an enumerable collection of such rectifiable sections.

Extremals exist (and are the same) for (a) (i) and (b) (i); do they exist for (a) (ii)? There are similar problems on the surface of a sphere. See my paper [to appear, JLMS]. Closely connected are the interesting problems of Burr: see [Ogilvy, Tomorrow's Math., Oxford 1962, pp. 23-24].

7. Kelly has found a set of 8 points such that the perpendicular bisector defined by each pair of points contains at least 2 points of the set. Are there such (finite) sets with more than 8 points? Sets with at least 3 points on every perpendicular bisector?

8. Given 8 points 'in general position' in 3 dimensions, can one partition them by 3 separating planes such that 1 point lies in each compartment? Same problem if the points are constrained to lie on the surface of a sphere, and the planes to be diametral. And similar problems for 2^n points in n dimensions. Also, how many planes do we need to separate N points in 3 dimensions, 1 into each compartment?

9. Two variants on Kakeya's needle turning problem (solved originally by Besicovitch):

(i) due to Wirsing (written communication). Let A, B be 2 congruent arcs of different unit circles in the plane, each of length α. Can one move one continuously to the other, only covering a set of arbitrarily small measure on the way? Perhaps this is true iff $\alpha < \pi$.

(ii) if a straight unit segment is moved through angle π, what is the least measure of the set covered if this is simply-connected? See [Cunningham and Schoenberg, Can. Math. J. 17 (1965) 946-956].

10. [Conway]. Can one embed the complete 7-graph in 3-space so that no closed circuit is knotted? Is there any graph that is essentially knotted in this sense? Perhaps there is a similar theorem to Kuratowski's, with just a few basic 'knotted types'. Conway has proved: given any 7 types of knot, we can embed a complete 4-graph in 3-space so that its 7 circuits (4 triangular, 3 quadrangular) are of the specified types.

11. [L. Moser]. Estimate the size of the largest circle which can be placed in the square lattice so as not to be nearer than ε to any lattice point.

Given a closed curve of bounded curvature round the origin, show that, if it is magnified radially, then as it tends to infinity, the distance of the curve to the closest lattice point $\to 0$. How fast?

12. [Erdös]. It is known that there exist sequences of arbitrary length composed of the digits 0, 1, 2, with the property that no two equal adjacent blocks of any length are the same. If we have k digits, how long a sequence, $N(k)$, can we form if no two adjacent blocks may be rearrangements of one another? Does $N(4) = \infty$?

Erdös informs me that recent computation gives $N(4) > 1{,}678$.

References for the known results on 0, 1, 2 are: Morse and Hedlund, Duke Math. J. 11 (1944), 6; Leech, Math. Gaz. XLI (1957), 277; Hawkins and Mientka, Math. Student XXIV (1956), 85.

13. [Conway]. A "thrackle" is a finite planar set of points with simple arcs joining some pairs of them, such that every pair of such arcs "cut" just once in the sense defined below. Prove that there are at least as many points as arcs. We define "cut" as satisfying just one or other of (i), (ii) below:

(i) the arcs intersect just once and cross over at the point of intersection;

(ii) the arcs have just one common end-point;

(iii) no arc may cut itself.

The conjecture has been verified for small numbers of points (up to 10).

14. Given n points P_j in a plane, there are $\frac{1}{2}n(n-1)(n-2)$ angles of the type $P_i\widehat{P_j}P_k$ formed. What proportion of them must necessarily be obtuse or right angles? Guy has shown that this proportion has a limit; and that it lies between $\frac{1}{9}$ and $\frac{4}{27}$, by most elegant arguments.

As Erdös has pointed out, we may more generally ask what proportion of angles are $\geqslant \alpha$; or $\leqslant \alpha$. The arguments for the existence of a limit, and lower bound, can apparently still be applied, and perhaps in some cases the upper bound one too. If $f(\alpha)$ is this proportion (of angles $\geqslant \alpha$), when is $f(\alpha)$ continuous? Erdös says that f is discontinuous at $\alpha = 60°$.

15. [Steinhaus]. Let C be a convex, differentiable, plane curve. Take an inscribed $\Delta X_1Y_1Z_1$; keep X_1Y_1 fixed, and move Z_1 to a point Z_2 on C such that $\Delta X_1Y_1Z_2$ has maximum perimeter; now keep Z_2, X_1 fixed and

similarly move Y_1 to Y_2; and so on, dealing with one vertex at a time in rotation. Do the triangles necessarily tend to a limit; and if so, is this limit necessarily a 'P-triangle', i.e. a triangle whose sides are equally inclined to the tangents of C at each vertex? We may vary the question by maximizing instead of the perimeter the area of the triangle, and expecting the limit-triangle to be a 'Q-triangle' i.e. to have the property that each side is parallel to the tangent to C at the opposite vertex. For the proof that 2 P-triangles and 2 Q-triangles exist in any convex curve, see [Croft and Swinnerton-Dyer, P.C.P.S. 59 (1963), 37-41].

Other possibilities are suggested by dropping the condition of convexity, or by asking for 3-dimension analogues, for example, how many P-triangles (i.e. triangles round which a ball will bounce) must there be in a (sufficiently smooth) 3-dimensional convex body?

16. [Besicovitch; Steinhaus]. The 'balls'. Given n distinguishable balls, of which we know that no two have the same weight, we wish to arrange them in order of magnitude as regards weight by weighing them one against one. What is the least number of weighings $k(n)$ that will always suffice, and what strategy does one adopt? (Applications to tennis tournaments.) There is also a conjecture that the best strategy as defined above will also minimize the expected number of weighings (which is a function of the original ordering), assuming all original orderings equally likely.

For the present state of knowledge see [Ford and Johnson, AMM 66 (1959), 387-9]. The situation is as follows: we cannot hope to do better than $k(n) = [\log_2(n!)] + 1 = K(n)$ say, where [ ] denotes integral part, for we cannot get more than a certain amount of information from each operation and there are $n!$ possible configurations. There are now two extreme conjectures:

(A) We can always find the most 'economical' strategy, so that $k(n) = K(n)$;

(B) We can never do better than the scheme described in the reference.

Critical cases occur when 2^m is very close to, and larger than $n!$ e.g., 5 is a critical case, $120 < 128$. It seems fairly clear that cases of arbitrary 'criticalness' will occur as $n \to \infty$. Typical values are $k(5) = K(5) = 7$, $k(10) = K(10) = 22$, $K(12) = 29$. (This raises the interesting and difficult question of the distribution of the fractional part of $\log_2(n!)$; in particular how small can it get, in terms of n? Perhaps $o(1/n)$?

Now the first case for which conjectures (A), (B) diverge is for $n = 12$; remarkably they re-converge, temporarily, for $n = 20$ and 21; asymptotically the limits are very narrow. It would therefore be of great interest to investigate completely $n = 12$. I think here (B) is wrong; it would be the easier alternative to prove, but the less interesting result. I think it quite likely that both conjectures are wrong. The only hope of tackling (A) (short of the luck of disproving it for some small n), seems to be to construct a general scheme for indicating mechanically which weighing to make next. There seems some connection between the configuration at any stage and the factors of the number of configurations still available. Information theory rules out some strategies: for example, for $n = 12$, we cannot use twice, during the procedure, the technique for arranging 5 balls in order (itself 15/16 efficient), for we have then lost too much information.

Recent computation by Paterson suggests strongly that $k(12) = 30$, not 29.

17. [Swinnerton-Dyer]. Several scorpions chase a beetle along the edges of a given regular (or semi-regular) solid. The beetle has speed at most 1. Find the minimal conditions on the scorpions' speeds to enable them to capture him. (for 2,3,4 etc. scorpions). Solved for the tetrahedron, cube, and octahedron, but not for the larger solids. Similar questions when the scorpions are replaced by spiders which can move on the faces of the solid, not just along the edges.

PATTERNS, COINCIDENT WITH THEIR CROSSPATTERN

J. Csima

McMaster University, Hamilton, Ont., Canada

Let J_{dn} be the set of all d-tuples $(i_1, i_2, \ldots, i_d)$ in which the components are integers between 1 and n inclusive. An element of J_{dn} is called a *point*. The subsets of J_{dn} are called *patterns*. If in $(i_1, i_2, \ldots, i_d)$ we keep $d - e$ components fixed and let the rest take up all integer values from 1 to n then the pattern (set) of the n^e points so obtained is called an *e-flat*. The 1-flats are called lines. Two e-flats are *parallel* if they have common variable components. We assume that d, n and e are fixed integers such that $d \geqslant 2$, $n \geqslant 2$ and $1 \leqslant e \leqslant d - 1$. A set C of n^{d-e} e-flats is an *e-cover* of the pattern S if $S \subseteq \bigcup_{f \in C} f$. A point $x \in J_{dn}$ is an *e-crosspoint* with respect to S if there exists an e-cover of S and two distinct e-flats f and g in this cover such that $x \in f \cap g$. The set of all e-crosspoints with respect to S is called the *e-crosspattern* of S and is denoted by S_e^x. Two points are *e-independent* if they are not in a common e-flat. An *e-transversal* (if exists) is a set of n^{d-e} pairwise e-independent points.

In a seminar, given by the author, B. Banaschewski asked if it is possible that $S = S_e^x$ for some d, e, n and S. For the planar case we can say the following.

THEOREM 1. *Let $S \subseteq J_{2n}$. Then $S = S_1^x$ if and only if S is the union of $n - 1$ distinct lines.*

A proof can be based on König's theorem. The characterization problem appears to be more difficult when $d \geqslant 3$. In this case we can prove the following.

<u>THEOREM 2</u>. *If S is the union of $n^{d-e} - 1$ distinct parallel e-flats and there exists an e-transversal in J_{dn} then $S = S_e^x$.*

The proof is based on the fact that if T is an e-transversal then $T \cap T_e^x = \emptyset$. We remark that an e-transversal is equivalent to (is the pattern of) a permutation matrix of dimension d, degree e and order n. Existence problems regarding such matrices are discussed by Jurkat and Ryser in (4). We mention here only that it is quite simple to demonstrate the existence of 1-transversals and $(n-1)$-transversals for all values of d and n.

Combinatorial properties of patterns for which $S \cap S_e^x = \emptyset$ are dealt with in (1), (2) and (3).

References

1. Csima, J., Investigations on a time-table problem, Ph.D. Thesis, University of Toronto, 1965.

2. Csima, J., Multidimensional Stochastic Matrices and patterns, *J. of Algebra* (to appear).

3. Csima, J., Restricted patterns, Mathematical Report, Vol. 1, No. 9, May 1969, McMaster University.

4. Jurkat, W.B. and H.J. Ryser, Extremal configurations and Decomposition Theorems I, *J. of Algebra*, 8 (1968), 194-22.

ON GRAPH REPRESENTATION OF SEMIGROUPS

József Dénes

Central Research Institute for Physics,
Budapest, Hungary

The graph representation of transformations has been introduced by A. Suschkewitsch. His graph representation can be formulated as follows: to every transformation of degree n there corresponds uniquely a directed graph having n labelled vertices in such a way that the vertices are labelled by the natural numbers 1, 2, ..., n and if the transformation maps i to j then the graph has a directed edge from i to j (see [6]).

In some of the author's papers (see [1], [2], [3], [4]) the combinatorial and algebraic characterization of transformations and transformation semi-groups has been studied with the help of graph representation.

It is easy to see that a directed graph corresponds to a transformation if and only if each of its connected components contains a single cyclically directed circuit and directed rooted trees. Such graphs with n vertices will be called $F(n)$ graphs. A transformation corresponding to a component of an $F(n)$ graph is conveniently called generalized cycle. Deleting the trees (apart from their roots) from an $F(n)$ graph one can obtain a special $F(k)$ graph ($k \leq n$), containing circuits only. The transformation corresponding to the $F(k)$ graph is a permutation: it is called the main-permutation of the original transformation. It is denoted by $f(\alpha)$. The maximum heights of trees in the transformation graph corresponding to α is denoted by $h(\alpha)$. If α denotes an arbitrary transformation of degree n, then the quasi-inverse will be defined as a power of α say α^s with the least exponent s whose main permutation is the inverse of the main permutation of α. Obviously every α has a unique quasi-inverse which always exists. Further if $\{\alpha\}$

the cyclic semigroup generated by α is a group then the inverse of α is equal to its quasi-inverse. Vagner introduced an inverse of an element a of a semigroup S. b is a Vagner inverse of a if $aba = a$ and $bab = b$ hold. It is easy to prove that it may happen for $x \varepsilon S$ that the Vagner inverse doesn't exist or that x has more than one Vagner inverse. If for every $x \varepsilon S$, x has exactly one inverse, then S is called inverse semigroup, if x has at least one inverse then S is called regular semigroup. Since all the transformations of degree n form a semigroup it is called the symmetric semigroup of degree n and for arbitrary abstract semigroups S of order n there exists a subsemigroup S' of F_{n+1} such that $S \cong S'$, we may consider transformation semigroups only without loss of generality.

Using the definitions above and not very complicated arguments we can prove Lemma 1. $\{\alpha\}$ is a cyclic group if and only if $h(\alpha) \leqslant 1$.

Lemma 2. T transformation semigroups is the union of its disjoint subgroups if and only if for arbitrary $\alpha \varepsilon T$, $h(\alpha) \leqslant 1$ holds.

Lemma 3. The quasiinverse of α is one of its Vagner inverses if and only if $h(\alpha) \leqslant 1$.

Theorem. *If the transformation graph corresponding to* α, *i.e.* Γ_α *contains an edge* $\vec{ab}$ *then* $\Gamma_{\alpha-1}$ *has an edge* $\vec{ba}$, *excepted if* $b \varepsilon f(\alpha)$ *and* α *is an end point. In the later case* $\Gamma_{\alpha-1}$ *has an edge* $\vec{ac}$ *where* a *is an end point and* $c \varepsilon f(\alpha^{-1})$.

Corollary. F_n is regular. (see also Doss [5]).

Two applications of the theorem were studies by the author.

(a) Let S be an arbitrary semigroup defined by its multiplication table then it is easy to construct, especially by computer, such a T transformation semigroup, that $S \cong T$. By using the theorem one can spare time to decide whether S is an inverse semigroup or not.

(b) Let $Ax = y$ be a system of linear equations (A is a matrix x, y are vectors) over a finite field. If A is known and non-singular then the solution is $x = A^{-1}y$. Let us restrict ourselves to the case when $Ax = y$ is over $GF(2)$. The best solution of $A^*x' = y$ is defined such that the Hamming distance of x and x' is the minimum. By using the theorem one can prove that the best solution can be obtained if A^* is a Vagner inverse of A. The above result can be generalized to $GF(p^m)$

and by the theorem and by other arguments one can create the algorithm of the solution. Such problems are raised when A is determined by devices or when A is transmitted over a noisy channel.

The proofs lacking here and the detailed description of the applications will be published elsewhere (see [4]).

References

1. Dénes, J., Connection between transformation semigroups and graphs. *Actes des Journees Internationales d'etude sur la theorie de graphes*, Rome (1966) 298-303.

2. Dénes, J., On transformations, transformation semigroups and graphs. *Theory of graphs*. Proc. of the Colloquium held at Tihany, Hungary 1966, 65-75.

3. Dénes, J., Some combinatorial properties of transformations and their connections with the theory of graphs. *J. of Combinatorial Theory*, (to appear).

4. Dénes, J., Leképezések és leképezesfélcsoportok. (Transformations and transformation semigroups) I, II, III to appear in Hungarian.

5. Doss, C.G., Certain equivalence relations in transformation semigroups. M.A. Thesis, University of Tennessee, 1955.

6. Suschkewitsch, A., Untersuchengen über verallgemeinerte substitutionen. *Atti del congresso Internazionale dei Matemaici Bologna*, 1928 147-157.

7. Vagner, V.V., Obobsenije gruppi. *Dokl. Akad. Nauk S.S.S.R.* 84(1952) 1119-1122.

STEINER TRIPLES AND TOTALLY NON-SYMMETRIC LOOPS

Jane W. Di Paola

New York University, University Heights, NY, U.S.A.

Totally non-symmetric loops are characterized by ($\underline{i}$) if $a \neq b$ then $ab = ba$ implies at least one of a and b is the identity, I; and ($\underline{ii}$) $(xy)x = x(yx) = y$. These conditions imply $a^2 = I$ for all a; and if $a,b \neq I$ then $a(ab) \neq b$ and $(ba)a \neq I$. Totally symmetric loops are characterized by commutativity and the requirement that $a(ab) = b$. (Bruck [3]). A totally non-symmetric loop, therefore, has the property that no pair of distinct elements a, b with $a \neq I$ and $b \neq I$ satisfies either condition for total symmetry. We construct a class of totally non-symmetric loops from an infinite set of disjoint Steiner triple systems. A pair of Steiner triple systems on V elements, $S(V)$ and $S'(V)$ are *disjoint* if no triple of one appears in the other.

Theorem 1. *If V is a power of a prime and $V = 6t + 1$, there exists a pair of disjoint Steiner triple systems on V elements.*

We sketch the proof briefly. A construction by Bose [2] for $V = 6t + 1 = p^n$, p a prime, provides a partitioned difference set which, without loss of generality, we write in the form

$$(0,b_0,c_0),\ (0,b_1,c_1),\ \ldots,\ (0,b_{t-1},c_{t-1}).$$

We use these triples as base blocks and for each $i \in A$, the additive group of $GF(p^n)$, we write, with addition in A, $(i,\ b_j+i,\ c_j+i)$ for the i-th block in the j-th class. Designate the triple system so obtained by $S(V)$ (cf. Hall [4]).

To obtain $S'(V)$, let α be a mapping of the points of $S(V)$ defined by $x\alpha = -x$, where $-x$ is the additive inverse of x in A. Write as the j-th base block in $S'(V)$ the triple $(0,b_j\alpha,c_j\alpha)$. For the r-th block, $r \in A$, in the j-th class write $(r,\ b_j\alpha+r,\ c_j\alpha+r)$ with addition in A. It is easily verified algebraically that $S(V)$ and $S'(V)$ have no triple in common.

Theorem 2. *If V is a power of a prime and $V = 6t + 1$ there is a totally non-symmetric loop of order $V + 1$.*

To prove this we consider the triples of $S(V)$ as ordered triples for each triple $(a,b,c) \in S(V)$ write the products $ab = c$, $bc = a$, and $ca = b$. Obtain the remaining products ba, cb, ac from $S'(V)$. Since α reverses the sign of the differences in the base blocks, we find that if ab appears in $S(V)$ in the order (a,b) then ba appears in $S'(V)$ in the order (b,a). It follows that a triple, say (x,y,z), in either $S(V)$ or $S'(V)$ determines three products $xy = z$, $yz = x$ and $zx = y$ so that the loop satisfies (*i*) and (*ii*).

That the class of totally non-symmetric loops is not coextensive with the class of disjoint Steiner triple systems is shown in the following example which appears in Stein [6] in a different context.

$$\begin{array}{ccccc} I & a & b & c & d \\ a & I & c & d & b \\ b & d & I & a & c \\ c & b & d & I & a \\ d & c & a & b & I \end{array}$$

That there are disjoint Steiner triple systems for V not a power of a prime is shown in a construction by Assmus and Mattson [1] who obtained such systems for $V = 2^k - 1$ with k odd.

From a loop in which each element is of order 2 there is obtained an idempotent quasigroup by removing the identity and defining $xx = x$. (cf. Bruck [3]). Theorem 2, therefore, yields a class of idempotent non-commutative quasigroups satisfying the identity $(xy)x = x(yx) = y$. Quasigroups satisfying this identity have been studied by Sade [5] and Stein [6].

References

1. Assmus, E. F., Jr. and Mattson, H. F., Jr., Steiner Systems and Perfect Codes, Inst. of Stat. Mimeo Series No. 484.1, The Institute of Statistics, University of North Carolina, 1966.

2. Bose, R. C., On the Construction of Balanced Incomplete Block Designs, *Ann. Eugenics*, 9 (1939) 353-399.

3. Bruck, R. H., What is a Loop?, MAA Studies in Modern Algebra, New York, 1963.

4. Hall, M., Jr., Combinatorial Theory, New York, 1968.

5. Sade, A., Quasi Groupe Demi-symétrique, *Ann. Soc. Sci. Bruxelles*, Ser. I, 79 (1965) 133-143 and 225-232.

6. Stein, S. K., On the Foundations of Quasigroups, *Trans. Amer. Math. Soc.* 85 (1957) 228-256.

ON THE NUMBER OF NON-ISOMORPHIC STEINER SYSTEMS $S(2,m,n)$

Jean Doyen*

University of Brussels, Belgium

A *Steiner system* $S(l,m,n)$ is a set S of n elements (called *points*) together with a family of subsets of S(called *blocks*) having m points each, such that every subset of S having l points is contained in exactly one block. Let $N(l,m,n)$ denote the number of non-isomorphic $S(l,m,n)$.

a) Recently, E.F. Assmus, Jr. and H.F. Mattson, Jr. [1] showed that $N(2,3,2^k-1)$ goes to infinity with k. We were able to generalise their result by proving:

Theorem 1. *Let n be an arbitrary positive integer satisfying the necessary and sufficient conditions for the existence of a $S(2,3,n)$, namely $n \equiv 1$ or $3 \pmod 6$. Then $N(2,3,n)$ goes to infinity with n.*

In fact, we produce a lower bound of the form

$N(2,3,n) > 2^{\log_3 n/17}$ for every $n \geq 15$.

The complete proof is to appear elsewhere [2].

b) A first step towards the extension of the latter result to other Steiner systems is given by:

Theorem 2. *Let p be a fixed prime number greater than 2. The function $k \to N(2,p,p^k)$ goes to infinity with k. Moreover, this function is convex, i.e.*

$$N(2,p,p^{k+2}) - N(2,p,p^{k+1}) \geq N(2,p,p^{k+1}) - N(2,p,p^k)$$

for every $k \geq 0$.

The proof is based on a seemingly new method of construction of transversal systems $T[p,p^k]$ (in H. Hanani's notations):

---

*Aspirant du Fonds National belge de la Rechersche Scientifique

$G_1, G_2, \ldots, G_p$ being p mutually disjoint copies of an additive abelian group G of order p^k and g_i denoting an arbitrary element of $G_i (i = 1,\ldots,p)$, we consider the following system of $p - 2$ equations in p unknowns $g_1, \ldots, g_p$:

$$\left[\frac{j-1}{2}\right] g_1 + (-1)^{j-1} g_2 + g_j = 0 \qquad (j = 3,\ldots,p)$$

(where [ ] is the greatest integer function).

It can be easily checked that the set T of all solutions $\{g_1, g_2, \ldots, g_p\}$ of this system is a transversal system $T[p,p^k]$ over $G_1, \ldots, G_p$.

<u>Lemma 1</u>. *If T' is a transversal system over $G_1', \ldots, G_p'$, where $G_i' \subseteq G_i$ $(i = 1,\ldots,p)$ and $T' \subseteq T$, then G_1' is a coset of a subgroup in G_1.*

<u>Lemma 2</u>. *If S is a Steiner system $S(2,p,p^k)$ with $k \geq 2$, the set S can be partitioned in p subsets $X_1, \ldots, X_p$ of cardinality p^{k-1} in such a way that none of the X_i $(i = 1,\ldots,p)$ is a subsystem of S.*

Let $S_1, \ldots, S_p$ be p Steiner systems $S(2,p,p^k)$, isomorphic or not, constructed on p mutually disjoint sets $S_1, \ldots, S_p$ of cardinality p^k. With the help of lemmas 1 and 2 (in the particular case of a cyclic group G), we prove that a Steiner system $S(2,p,p^{k+1})$, in which $S_1, \ldots, S_p$ are the only subsystems of order p^k, can be constructed on the set $S = S_1 \cup \ldots \cup S_p$. The inequality

$$N(2,p,p^{k+1}) \geq \frac{1}{p!} N(2,p,p^k) \left[N(2,p,p^k) + 1\right] \ldots \left[N(2,p,p^k) + p - 1\right]$$

is then deduced, and theorem 2 obviously follows, as $N(2,p,p^3) \geq 2$ for every odd prime p.

<u>References</u>

1. Assmus, Jr. E.F., and Mattson, Jr. H.F., On the number of inequivalent Steiner triple systems, *J. Combinatorial Theory* 1(1966), 301-305.

2. Doyen,J., Sur la croissance du nombre de systèmes triples de Steiner non isomorphes, *J. Combinatorial Theory* (to appear).

ON ORBITS IN RELATIONAL STRUCTURES

Kenneth R. Driessel

University of Colorado, Boulder, CO, U.S.A.

Let $\mathfrak{A}$ be a relational structure - a set A together with a collection of finitary relations on A - and let $Aut\,\mathfrak{A}$ be the set of automorphisms of $\mathfrak{A}$. An *orbit* of a point $a \in A$, denoted by $[a]$, is defined by $[a] = \{f(a): f \in Aut\,\mathfrak{A}\}$. We shall prove the following:

<u>Theorem</u>. *Let $\mathfrak{A}$ be a relational structure on a set A that satisfies:*

(1) Every point in $\mathfrak{A}$ has a finite orbit.

Then $\mathfrak{A}$ also satisfies:

(2) For every subset B of A, the set $\{f|B: f \in Aut\,\mathfrak{A}\}$ is either finite or uncountable.
(The restriction of f to B is denoted by $f|B$.)

For any sequence x of points from , the mapping from the orbit $\left\{\left(f(x_0),\ f(x_1),\ldots\right): f \in Aut\,\mathfrak{A}\right\}$ into the set $\{f|\text{range } x: f \in Aut\,\mathfrak{A}\}$ given by $\left(f(x_0),\ f(x_1),\ldots\right) \to f|\text{range } x$ is one-one and onto. Consequently, if $\mathfrak{A}$ satisfies (1) then no sequence in $\mathfrak{A}$ has a countably infinite orbit. This consequence of the theorem answers a question raised by J. Mycielski in 1967. From the theorem it also follows that if $\mathfrak{A}$ satisfies (1) then $Aut\,\mathfrak{A}$ is either finite or uncountable.

Let us consider another application. Let G be a graph. A sequence of vertices and edges from G may be called a *configuration* in G. Two configurations of G are *congruent* if there is an automorphism of G that carries one onto the other. If G is locally finite, connected and at least one vertex or edge has a finite orbit then it follows that every vertex and every edge has a finite orbit.

Hence, by the theorem, every congruence class of configuration in G is either finite or uncountable.

The proof of the theorem uses some topological results which we now review. (For undefined terminology see J.L. Kelly, General Topology, D. Van Nostrand Company, Inc., 1955.) A topological space is a *Baire* space if every co-meager subset is dense. Then the Baire category theorem may be stated as follows:

T.1. Every locally compact regular space is a Baire space.
It is also well known (and easy to prove) that:

T.2. The open continuous image of a Baire space is a Baire space.
A topological space is *homogeneous* if for every pair of points there is a homeomorphism of the space with itself that carries one point onto the other. It is not hard to prove:

T.3. Let G be a topological group and let H be a subgroup of G. Then $G/H = \{x H: x \in G\}$ with the quotient topology is a homogeneous space and the canonical map $x \mapsto x H$ is open and continuous. Furthermore, if H is closed then G/H is a T_1-space.
We also have:

T.4. If X is a compact, homogeneous, Baire T_1-space then X is either finite or uncountable.
The proof of T.4 splits into two cases.

Case 1: All the points of X are open. Then X is finite since X is compact and $X = \bigcup \left\{\{p\}: p \in X\right\}$.

Case 2: The points of X are not open. Then the points are nowhere dense and X is uncountable since X is a Baire space and $X = \bigcup\left\{\{p\}: p \in X\right\}$.

Now let $\mathfrak{A}$ be any relational structure on a set A satisfying (1). Let each $[a]$ have the discrete topology. Then the cartesian product $X\{[a];a \in A\}$ with the product topology is a compact regular Hausdorff space since each component space has these properties. Note Aut $\mathfrak{A}$ is a closed subset of this product. It follows that Aut $\mathfrak{A}$ with the relative topology is a topological group that is compact, regular, Hausdorff and (by T.1) a Baire space.

For B a subset of A, let $H(B) = \{f \in \text{Aut}\,\mathfrak{A}: f$ is the identity on $B\}$. Then $H(B)$ is a closed subgroup of Aut $\mathfrak{A}$ and (using T.2 and T.3) Aut $\mathfrak{A}/H(B)$ with the quotient topology is a compact, homogeneous, Baire T_1-space. Now the cardinality of $\{f|B: f \in \text{Aut}\,\mathfrak{A}\}$ is the same as the

cardinality of $\mathrm{Aut}\mathfrak{A}/H(B)$ since $f|B = g|B$ if and only if $f\,H(B) = g\,H(B)$.
Finally (2) follows from T.4.

SUBMODULAR FUNCTIONS, MATROIDS, AND CERTAIN POLYHEDRA*

Jack Edmonds

National Bureau of Standards, Washington, D.C., U.S.A.

I.

The viewpoint of the subject of matroids, and related areas of lattice theory, has always been, in one way or another, abstraction of algebraic dependence or, equivalently, abstraction of the incidence relations in geometric representations of algebra. Often one of the main derived facts is that all bases have the same cardinality. (See Van der Waerden, Section 33.)

From the viewpoint of mathematical programming, the equal cardinality of all bases has special meaning -- namely, that every basis is an optimum-cardinality basis. We are thus prompted to study this simple property in the context of linear programming.

It turns out to be useful to regard "pure matroid theory", which is only incidentally related to the aspects of algebra which it abstracts, as the study of certain classes of convex polyhedra.

(1) A *matroid* $M = (E,F)$ can be defined as a finite set E and a nonempty family F of so-called *independent* subsets of E such that

(a) Every subset of an independent set is independent, and

(b) For every $A \subseteq E$, every maximal independent subset of A, i.e., every *basis* of A, has the same cardinality, called the *rank*, $r(A)$, of A (with respect to M).

(This definition is not standard. It is prompted by the present interest).

---

*Synopsis for the Instructional Series of Lectures, "Polyhedral Combinatorics".

(2) Let R_E denote the space of real-valued vectors $x = [x_j]$, $j \varepsilon E$. Let $R_E^+ = \{x : 0 \leqslant x \varepsilon R_E\}$.

(3) A *polymatroid* P in the space R_E is a compact non-empty subset of R_E^+ such that

(a) $0 \leqslant x^0 \leqslant x^1 \varepsilon P \Longrightarrow x^0 \varepsilon P$.

(b) For every $a \varepsilon R_E^+$, every maximal $x \varepsilon P$ such that $x \leqslant a$, i.e., every *basis* x of a, has the same sum $\sum_{j \varepsilon E} x_j$, called the rank, $r(a)$, of a (with respect to P).

Here *maximal* x means that there is no $x' > x$ having the properties of x.

(4) A polymatroid is called *integral* if (b) holds also when a and x are restricted to being integer-valued, i.e., for every integer-valued vector $a \varepsilon R_E^+$, every maximal integer-valued x, such that $x \varepsilon P$ and $x \leqslant a$, has the same sum $\sum_{j \varepsilon E} x_j = r(a)$.

(Sometimes it may be convenient to regard an *integral polymatroid* as consisting only of its integer-valued members).

(5) Clearly, the 0-1 valued vectors in an integral polymatroid are the "incidence vectors" of the sets $J \varepsilon F$ of a matroid $M = (E,F)$.

II

(6) Let f be a real-valued function on a lattice L. Call it a β_0-function if

(a) $f(a) \geqslant 0$ for every $a \varepsilon K = L - \{\phi\}$;

(b) it is non-decreasing: $a \leqslant b \Longrightarrow f(a) \leqslant f(b)$; and

(c) submodular:
$f(a \vee b) + f(a \wedge b) \leqslant f(a) + f(b)$
for every $a \varepsilon L$ and $b \varepsilon L$.

(d) Call it a β-function if, also, $f(\phi) = 0$. In this case, f is also subadditive, i.e., $f(a \vee b) \leqslant f(a) + f(b)$.

(We take the liberty of using the prefixes *sub* and *super* rather than "upper semi" and "lower semi". *Semi* refers to either. The term *semimodular* is taken from lattice theory where it refers to a type of lattice on which there exists a semimodular function f such that if a is a maximal element less than element b then $f(a) + 1 = f(b)$ See [1].)

(7) For any $x = [x_j] \in R_E$, and any $A \subseteq E$, let $x(A)$ denote $\sum_{j \in A} x_j$.

(8) Theorem. *Let L be a family of subsets of E, containing E and* ϕ, *and closed under intersections,* $A \cap B = A \wedge B$. *Let f be a* β_0-*function on L. Then the following polyhedron is a polymatroid:*

$$P(E,f) = \{x \in R_E^+ : x(A) \leq f(A) \text{ for every } A \in L - \{\phi\} = K\}.$$

Its rank function r is, for any $a = [a_j] \in R_E^+$,

$$r(a) = \min \left(\sum_{j \in E} a_j z_j + \sum_{A \in K} f(A)\, y_A \right)$$

where the z_j*'s and* y_A*'s are* 0*'s and* 1*'s such that for every* $j \in E$,

$$z_j + \sum_{j \in A \in K} y_A \geq 1.$$

Where $f(\phi) \geq 0$, *only one non-zero* y_A *is needed.*
Where f is integer-valued, $P(E,f)$ *is an integral polymatroid.*

(9) Theorem. *A function f of all sets* $A \subseteq E$ *is itself the rank function of a matroid* $M = (E,F)$ *iff it is an integral* β-*function such that* $f(\{j\}) = 1$ *or* 0 *for every* $j \in E$. *Such an f determines M by:*
$J \in F \Longleftrightarrow J \subseteq E$ *and* $|J| = f(J)$.

(10) For any $a = [a_j] \in R_E^+$ and $b = [b_j] \in R_E^+$, let $a \vee b = [u_j] \in R_E^+$ and $a \wedge b = [v_j] \in R_E^+$, where

$$u_j = \max(a_j, b_j) \text{ and } v_j = \min(a_j, b_j).$$

(11) Theorem. *The rank function* $r(a)$, $a \in R_E^+$, *for any polymatroid* $P \subset R_E^+$, *is a* β-*function on* R_E^+ *relative to the above* $\vee$ *and* $\wedge$.

(12) For any $x = [x_j] \in R_E^+$ and any $A \subseteq E$, let $x/A = [(x/A)_j] \in R_E^+$ denote the vector such that $(x/A)_j = x_j$ for $j \in A$, and $(x/A)_j = 0$ for $j \notin A$.

(13) Given a polymatroid $P \subset R_E^+$, let $\alpha \in R_E^+$ be an integer-valued vector such that $x < \alpha$ for every $x \in P$. Where r is the rank function of P, let $f_P(A) = r(\alpha/A)$ for every $A \subseteq E$.

Let $L_E = \{A : A \subseteq E\}$. Clearly, by (11), f_P is a β-function on L_E. Furthermore, if P is integral, then f is integral.

(14) Theorem. *For any polymatroid* $P \subset R_E^+$,

$$P = P(E, f_P).$$

Thus, all polymatroids $P \subset R_E^+$ are polyhedra, and they correspond to certain β-functions on L_E.

Theorem 8 provides a useful way of constructing matroids which is quite different from the usual algebraic constructions.

(15) For any given integral β_0-function f as in (8), let a set $J \subseteq E$ be a member of F iff for every $A \in K = L - \{\phi\}$, $|J \cap A| \leq f(A)$. In particular, where $L = L_E$, let a set $J \subseteq E$ be a member of F when for every $\phi \neq A \subseteq J$, $|A| \leq f(A)$. Then (8) implies that $M = (E,F)$ is a matroid, and gives a formula for its rank function in terms of f. (This generalizes a construction given by Dilworth [1]).

III.

In this section, K will denote $L_E - \{\phi\} = \{A : \phi \neq A \subseteq E\}$.

(16) Given any $c = [c_j] \in R_E$, and given a β-function f on L_E, we show how to solve the linear program:

$$\text{maximize } c \cdot x = \sum_{j \in E} c_j x_j \text{ over } x \in P(E,f).$$

(17) Let $j(1), j(2), \ldots$ be an ordering of E such that

$$c_{j(1)} \geq c_{j(2)} \geq \ldots \geq c_{j(k)} > 0 \geq c_{j(k+1)} \geq \ldots$$

(18) For each integer i, $1 \leq i \leq k$, let

$$A_i = \{j(1), j(2), \ldots, j(i)\}.$$

(19) <u>Theorem</u>. *(The Greedy Algorithm).*
$c \cdot x$ is maximized over $x \in P(E,f)$ by the following vector x^0:

$$x^0_{j(1)} = f(A_1);$$

$$x^0_{j(i)} = f(A_i) - f(A_{i-1}) \quad \text{for } 2 \leq i \leq k;$$

$$x^0_{j(i)} = 0 \quad \text{for } k < i \leq |E|.$$

(There is a well-known non-polyhedral version of this for graphs, given by Kruskal [9]. A related theorem for matroids is given by Rado [15]).

The dual $\ell.p.$ is to minimize

$$f \cdot y = \sum_{A \in K} f(A)\, y(A) \text{ where}$$

(20) $y(A) \geq 0$; and for every $j \in E$, $\sum_{j \in A} y(A) \geq c_j$.

(21) Theorem. *An optimum solution, $y^0 = [y^0(A)]$, $A \in K$, to the dual l.p. is*

$$y^0(A_i) = c_{j(i)} - c_{j(i+1)} \quad \textit{for } 1 \leq i \leq k\text{-}1;$$

$$y^0(A_k) = c_{j(k)}; \text{ and } \; y^0(A) = 0 \textit{ for all other } A \in K.$$

(22) Theorem. Corollary to (19). *The vertices of the polyhedron $P(E,f)$ are precisely the vectors of the form x^0 in* (19) *for some sequence $j(1), j(2), \ldots, j(k)$.*

(23) Where f is the rank function of a matroid $M = (E,F)$, (9) and (22) imply that the vertices of $P(E,f)$ are precisely the incidence vectors of the members of F, i.e., the independent sets of M. Such a $P(E,f)$ is called a *matroid polyhedron.*

(24) Let f be a β-function on L_E. A set $A \in L_E$ is called *f-closed* or an *f-flat*, when, for any $C \in L_E$ which properly contains A, $f(A) < f(C)$.

(25) Theorem. *If A and B are f-closed then $A \cap B$ is f-closed.*

(In particular, for the f of (9), the f-flats form a "geometric" or "matroid" lattice.)

Proof. Suppose that C properly contains $A \cap B$. Then either $C \not\subseteq A$ or $C \not\subseteq B$. Since f is non-decreasing we have $f(A \cap B) \leq f(A \cap C)$ and $f(A \cap B) \leq f(B \cap C)$. Thus, since f is submodular, we have either

$$0 < f(A \cup C) - f(A) \leq f(C) - f(A \cap C) \leq f(C) - f(A \cap B), \quad \text{or}$$

$$0 < f(B \cup C) - f(B) \leq f(C) - f(B \cap C) \leq f(C) - f(A \cap B).$$

(26) A set $A \in K$ is called f-separable when

$$f(A) = f(A_1) + f(A_2)$$

for some partition of A into non-empty subsets A_1 and A_2. Otherwise A is called f-inseparable.

(27) Theorem. *Any $A \in K$ partitions in only one way into a family of f-inseparable sets A_i such that $f(A) = \Sigma f(A_i)$.* The A_i's are called the *f-blocks* of A.

If a polyhedron $P \subset R_E$ has dimension equal to $|E|$ hen there is a unique minimal system of linear inequalities having P as its set of solutions. These inequalities are called the *faces* of P.

It is obvious that a polymatroid $P \subset R_E^+$ has dimension $|E|$ if and only if, where f is the β-function which determines it, and set ϕ is f-closed. It is obvious that inequality $x(A) \leq f(A)$, $A \in K$, is a face of $P(E,f)$ only if A is f-closed and f-inseparable.

(28) Theorem. *Where f is a β-function on L_E such that the empty set is f-closed, the faces of polymatroid $P(E,f)$ are: $x_j \geq 0$ for every $j \in E$; and $x(A) \leq f(A)$ for every $A \in K$ which is f-closed and f-inseparable.*

IV.

(29) Let each V_p, $p = 1$ and 2, be a family of disjoint subsets of E. Where $[a_{ij}]$, $i \in H$, $j \in E$, is the 0-1 incidence matrix of $V_1 \cup V_2 = H$, the following ℓ.p. is known as the Hitchcock problem.

(30) Maximize $c \cdot x = \sum_{j \in E} c_j x_j$, where

(31) $x_j \geq 0$ for every $j \in E$, and $\sum_{j \in E} a_{ij} x_j \leq b_i$ for every $i \in H$.

The dual ℓ.p. is

(32) Minimize $b \cdot y = \Sigma\, b_i y_i$, where

(33) $y_i \geq 0$ for every $i \in H$, and $\sum_{i \in H} a_{ij} y_i \geq c_j$ for every $j \in E$.

Denote the polyhedron of solutions of a system Q by $P[Q]$.

The following properties of the Hitchcock problem are important in its combinatorial use.

(34) Theorem. (a) *Where the b_i's are integers, the vertices of $P[(31)]$ are integer-valued.* (b) *Where the c_j's are integers, the vertices of $P[(33)]$ are integer-valued.*

Theorem (34a) generalizes to the following.

(35) Theorem. *For any two integral polymatroids P_1 and P_2 in R_E^+, the vertices of $P_1 \cap P_2$ are integer-valued.*

The following technique for proving theorems like (34) is due to Alan Hoffman [7].

(36) Theorem. *The matrix $[a_{ij}]$ of the Hitchcock problem is totally unimodular -- that is, the determinant of every square submatrix has value 0, 1, or -1.*

(37) Theorem. *Theorem* (34) *holds whenever* $[a_{ij}]$ *is totally unimodular.*

(38) Let each V_p, $p = 1$ and 2, be a family of subsets of E such that any two members of V_p are either disjoint or else one is a subset of the other.

(39) Theorem. *The incidence matrix of the* $V_1 \cup V_2$ *of* (38) *is totally unimodular.*

Property (29) is a special case of (38). Property (38) is a special case of the following.

(40) Let each V_p, $p = 1$ and 2, be a family of subsets of E such that for any $R \in V_p$ and $S \in V_p$ either $R \cap S = \phi$ or $R \cap S \in V_p$.

The incidence matrix of the $V_1 \cup V_2$ of (40) is generally not totally unimodular. However,

(41) Theorem. *From the incidence matrix of each* V_p *of* (40), *one can obtain, by subtracting certain rows from others, the incidence matrix of a family of mutually disjoint subsets of* E. *Thus, in the same way, one can obtain from the incidence matrix of the* $V_1 \cup V_2$ *of* (40), *a matrix of the Hitchcock type.*

(42) Theorem. *For any polymatroid* $P(E,f)$ *and any* $x \in P(E,f)$, *if* $x(A) = f(A)$ *and* $x(B) = f(B)$ *then either* $A \cap B = \phi$ *or* $x(A \cap B) = f(A \cap B)$.

Theorems (42), (41), and (34a) imply (35).

(43) Assuming that each V_p of (38) contains the set E, $L_p = V_p \cup \{\phi\}$ is a particularly simple lattice. For any non-negative non-decreasing function $f(i) = b_i$, $i \in V_p$, let $f(\phi) = -f(E)$. Then f is a β_0-function on L_p.

(44) The only integer vectors in a matroid polyhedron P are the vectors of the independent sets of the matroid, and these vectors are all vertices of P. Thus, (35) implies:

(45) Theorem. *Where* P_1 *and* P_2 *are the polyhedra of any two matroids* M_1 *and* M_2 *on* E, *the vertices of* $P_1 \cap P_2$ *are precisely the vectors which are vertices of both* P_1 *and* P_2 -- *namely, the incidence vectors of sets which are independent in both* M_1 *and* M_2.

Where P_1, P_2, and P_3 are the polyhedra of three matroids on E, polyhedron $P_1 \cap P_2 \cap P_3$ generally has many vertices besides those which are vertices of P_1, P_2, and P_3.

Let $c = [c_j]$, $j \in E$, be any numerical weighting of the elements of E. In view of (45), the problem:

(46) Find a set J, independent in both M_1 and M_2, that has maximum weight-sum, $\sum_{j \in J} c_j$, is equivalent to the $\ell.p.$ problem:

(47) Find a vertex x of $P_1 \cap P_2$ that maximizes $c \cdot x$.

(48) Assuming there is a good algorithm for recognizing whether or not a set $J \subseteq E$ is independent in M_1 or in M_2, there is a good algorithm for problem (46). This seems remarkable in view of the apparent complexity of matroid polyhedra in other respects. For example, a good algorithm is not known for the problem:

(49) Given a matroid $M_1 = (E, F_1)$ and given an element $e \in E$, minimize $|D|$, $D \subseteq E$, where $e \in D \notin F_1$;

Or the problems:

(50) Given three matroids M_1, M_2, and M_3, on E, and given an objective vector $c \in R_E$, maximize $c \cdot x$ where $x \in P_1 \cap P_2 \cap P_3$.

Or maximize $\sum_{j \in J} c_j$ where $J \in F_1 \cap F_2 \cap F_3$.

V.

Where f_1 and f_2 are β-functions on L_E, the dual of the $\ell.p.$:

(51) Maximize $c \cdot x = \sum_{j \in E} c_j x_j$, where

(52) For every $j \in E$, $x_j \geq 0$; and for every $A \in K$, $x(A) \leq f_1(A)$ and $x(A) \leq f_2(A)$; is the $\ell.p.$:

(53) Minimize $f \cdot y = \sum_{A \in K} [f_1(A) y_1(A) + f_2(A) y_2(A)]$ where

(54) For every $A \in K$, $y_1(A) \geq 0$ and $y_2(A) \geq 0$; and for every $j \in E$, $\sum_{j \in A \in K} [y_1(A) + y_2(A)] \geq c_j$.

Combining systems (52) and (54) we get,

(55) $\sum_{j\varepsilon E} x_j \ (\sum_{j\varepsilon A\varepsilon K} [y_1(A) + y_2(A)] - c_j)$

$+ \sum_{A\varepsilon K} y_1(A) \ [f_1(A) - \sum_{j\varepsilon A} x_j]$

$+ \sum_{A\varepsilon K} y_2(A)[f_2(A) - \sum_{j\varepsilon A} x_j] \geq 0.$

Expanding and cancelling we get

(56) $c \cdot x \leq f \cdot y$

for any x satisfying (52) and any $y = (y_1, y_2)$ satisfying (54).

(57) Equality holds in (56) if and only if equality holds in (55).

The $\ell.p.$ duality theorem says that

(58) If there is an x^0, a vertex of $P[(52)]$, which maximizes $c \cdot x$, then there is a $y^0 = (y_1^0, y_2^0)$, a vertex of $P[(54)]$, such that

(59) $c \cdot x^0 = f \cdot y^0$,

and hence such that y^0 minimizes $f \cdot y$.

For the present problem obviously there is such an x^0.

The vertices of (54) are not generally all integer-valued when the c_j's are. However,

(60) Theorem. *If the c_j's are all integers, then, regardless of whether f_1 and f_2 are integral, there is an integer-valued solution $y^4 = (y_1^4, y_2^4)$ of (54) which minimizes $f \cdot y$.*

Let $y^3 = (y_1^3, y_2^3)$ be any solution of (54) which minimizes $f \cdot y$.

(61) For every $j \varepsilon E$, and $p = 1,2$ let $c_j^p = \sum_{j\varepsilon A\varepsilon K} y_p^3$

For each $p = 1,2$ consider the problem,

(62) Minimize $f_p \cdot y_p = \sum_{A\varepsilon K} f_p(A) \ y_p(A)$ where

(63) for every $A \varepsilon k$, $y_p(A) \geq 0$; and for every $j \varepsilon E$,

$\sum_{j\varepsilon A\varepsilon K} y_p(A) \geq c_j^p.$

(64) By (21), for each p, there is an optimum solution, say y_p^4, to (62) having the following form:

(65) The sets $A \in K$, such that $y_p^4(A) > 0$, form a nested sequence, $A_1 \subset A_2 \subset A_3 \subset \dots$.

Since y_p^3 is a solution of (63), we have $f_p y_p^4 \leq f_p y_p^3$, for each p, and thus $f \cdot y^4 \leq f \cdot y^3$. Since $c_j^1 + c_j^2 \geq c_j$ for every $j \in E$, y^4 is a solution of (54), and hence y^4 is an optimum solution of (54). Thus, we have that

(66) <u>Theorem</u>. *There exists a solution* y^4 *of* (54) *which minimizes* $f \cdot y$ *and which has property* (65) *for each* $p = 1,2$.

The problem, minimize $f \cdot y$ subject to (54) and also subject to $y_p(A) = 0$ for every $y_p^4(A) = 0$, has the form [(32, (33)] where $[a_{ij}]$ is the incidence matrix of a $V_1 \cup V_2$ as in (38). Thus, by (39) and (37), we have:

(67) <u>Theorem</u>. *If the* c_j*'s are all integers then the* y^4 *of* (66) *can be taken to be integer-valued.*

In particular this proves (60).

An immediate consequence of (35), (60), and the *ℓ.p.* duality theorem is

(68) <u>Theorem</u>. $\max c \cdot x = \min f \cdot y$ *where* $x \in P[(52)]$ *and* $y \in P[(54)]$.

If f *is integral,* x *can be integral.*
If c *is integral,* y *can be integral.*

In particular, where f_1 and f_2 are the rank functions, r_1 and r_2, of any two matroids, $M_1 = (E,F_1)$ and $M = (E,F_2)$, and where every $c_j = 1$, (68) implies:

(69) <u>Theorem</u>. $\max|J| = \min\,[r_1(S) + r_2(E-S)]$, *where* $J \in F_1 \cap F_2$, *and where* $S \subseteq E$.

(A related result is given by Tutte [16]).

VI.

(70) <u>Theorem</u>. *For each* $i \in E'$, let Q_i be a subset of E. For each $A' \subseteq E'$, let $u(A') = \bigcup_{i \in A'} Q_i$. Let f be any integral β-function on L_E.

Then $f'(A') = f(u(A'))$ *is an integral* β*-function on* $L_{E'} = \{A' : A' \subseteq E$

(71) This follows from the relations

$$u(A' \cup B') = u(A') \cup u(B') \text{ and}$$

$$u(A' \cap B') \subseteq u(A') \cap u(B').$$

(72) Applying (15) to f' yields a matroid on E'.

(73) In particular, taking f to mean cardinality, if we let $J' \subseteq E'$ be a member of F' iff $|A'| \leq |u(A')|$ for every $A' \subseteq J'$, then $M' = (E', F')$ is a matroid.

(74) Hall's SDR theorem says that: $|A'| \leq |u(A')|$ *for every* $A' \subseteq J'$ *iff the family* $\{Q_i\}$, $i \in J'$, *has a system of distinct representatives, i.e., a transversal.* A *transversal* of a family $\{Q_i\}$, $i \in J'$ is a set $\{j_i\}$, $i \in J'$, of distinct elements such that $j_i \in Q_i$. Thus,

(75) Theorem. *For any finite family* $\{Q_i\}$, $i \in E'$, *of subsets of* E, *the sets* $J' \subseteq E'$ *such that* $\{Q_i\}$, $i \in J'$, *has a transversal are the independent sets of a matroid on* E' (called a *transversal matroid).*

There are a number of interesting ways to derive (75). Some others are in [2], [3], [5], and [12]. The present derivation is the way (75) was first obtained and communicated.

The following is the same result with the roles of elements and sets interchanged.

(76) *Let* $J \in F_0$ *iff, for some* $J' \subseteq E'$, J *is a transversal of* $\{Q_i\}$, $i \in J'$. That is, let $J \in F_0$ iff J is a *partial transversal* of $\{Q_i\}$, $i \in E'$. *Then* $M_0 = (E, F_0)$ *is a matroid.*

(77) Thus, where P_0 is the polyhedron of M_0 and where P is the polyhedron of any other matroid, $M = (E, F)$, on E, the vertices of $P_0 \cap P$ are the incidence vectors of the M-independent partial transversals of $\{Q_i\}$, $i \in E'$.

By (8), the rank function r_0 of M_0 is, for each $A \subseteq E$,

(78) $$r_0(A) = \min[|A_0| + |\{i : (A - A_0) \cap Q_i \neq \phi\}|]$$

where $A_0 \subseteq A$.

Combining (69) and (78), we get

(79) $\max |J| = \min[r(A_1) + |A_0| + |E'| - |\{i : Q_i \subseteq A_1 \cup A_0\}|]$

$= \min [r(u(A')) + |E'| - |A'|]$, where $J \in F_0 \cap F$, $A_0 \cup A_1 \subseteq E$, $A_0 \cap A_1 = \phi$, and $A' \subseteq E'$.

In particular, (79) implies the following theorem of Rado [14], given in 1942.

(80) *For any matroid M on E, a family* $\{Q_i\}$, $i \in E'$, *of subsets of E, has a transversal which is independent in M iff* $|A'| \leq r(u(A'))$ *for every* $A' \subseteq E'$.

Taking the f of (70) to be r, (70), (15), and (80) imply:

(81) Theorem. *For any matroid M on E, and any family* $\{Q_i\}$, $i \in E'$, *of subsets of E, the sets* $J' \subseteq E'$ *such that* $\{Q_i\}$, $i \in J'$, *has an M-independent transversal are the independent sets of a matroid on* E'.

(82) A *bipartite graph G* consists of two disjoint finite sets, V_1 and V_2, of nodes and a finite set $E(G)$ of edges such that each member of $E(G)$ meets one node in V_1 and one node in V_2.

The following theorem of Konig is a prototype of (69).

(83) Theorem. *For any bipartite graph G,* $\max |J|$, $J \subseteq E(G)$, *such that*

(a) *no two members of J meet the same node in* V_1, *and*

(b) *no two members of J meet the same node in* V_2,

equals $\min (|T_1| + |T_2|)$, $T_1 \subseteq V_1$ *and* $T_2 \subseteq V_2$, *such that every member of* $E(G)$ *meets a node in* T_1 *or a node in* T_2.

(84) To get the Hall theorem, (74), from (83), let V_1 be the E' of (70), let V_2 be the E of (70), and let there be an edge in $E(G)$ which meets $i \in V_1$ and $j \in V_2$ iff $j \in Q_i$.

Clearly, if the family $\{Q_i\}$, $i \in E'$, has no transversal then, in (83), $\max |J| < |V_1|$. If the latter holds, then by (83), the T_1 of $\min(|T_1| + |T_2|)$, in (83), is such that

$$|V_1 - T_1| > |u(V_1 - T_1)|.$$

(85) For the Konig-theorem instance, (83) of (69), the matroids $M_1 = (E,F_1)$ and $M_2 = (E,F_2)$ are particularly simple: Let $E = E(G)$. For $p = 1$ and $p = 2$, let $J \subseteq E(G)$ be a member of F_p iff no two members of J meet the same node in V_p.

(86) Where P_1 and P_2 are the polyhedra of these two matroids, finding a vertex x of $P_1 \cap P_2$ which maximizes $c \cdot x$ is essentially the *optimum assignment problem* That is, the Hitchcock problem where every $b_i = 1$.

(87) Clearly, the inequality $x(A) \leq r_p(A)$ is a face of P_p, that is, A is r_p-closed and r_p-inseparable, iff, for some node $v \in V_p$, A is the set of edges which meet v.

VII.

(88) Let $\{M_i\}$, $i \in I$, be a family of matroids, $M_i = (E,F_i)$, having rank functions r_i. Let $J \subseteq E$ be a member of F iff:

(89) $|A| \leq \sum_i r_i(A)$ for every $A \subseteq J$.

Since $f(A) = \sum_i r_i(A)$ is a β-function on L_E,

(90) Theorem. *The* $M = (E,F)$ *of* (88) *is a matroid, called the* sum *of the matroids* M_i.

In [5], and in [2], it is shown that

(91) Theorem. $J \subseteq E$ *satisfies* (89) *iff* J *can be partitioned into sets* J_i *such that* $J_i \in F_i$.

(92) An algorithm, *MPAR*, is given there for either finding such a partition of J or else finding an $A \subseteq J$ which violates (89). That is, for recognizing whether or not $J \in F$.

(93) The algorithm is a good one, assuming:

(94) that a good algorithm is available for recognizing, for any $K \subseteq E$ and for each $i \in I$, whether or not $K \in F_i$.

(95) The definition of a matroid $M = (E,F)$ is essentially that, modulo the ease of recognizing, for any $J \subseteq E$, whether or not $J \in F$, one has what is perhaps the easiest imaginable algorithm for finding, in any $A \subseteq E$, a maximum cardinality subset J of A such that $J \in F$.

(96) In particular, by virtue of (90), assuming (94), *MPAR* provides a good algorithm for finding a maximum cardinality set $J \subseteq E$ which is partitionable into sets $J_i \in F_i$.

(97) Assuming (94), *MPAR* combined with (19) is a good algorithm for, given numbers c_j, $j \in E$, finding a set J which is partionable into sets $J_i \in F_i$ and such that $\sum_{j \in J} c_j$ is maximum.

Where r is the rank function of matroid $M = (E,F)$, let

(98) $r^*(A) = |A| + r(E-A) - r(E)$ for every $A \subseteq E$.

Substituting $r(E) = |E| - r^*(E)$, and A for $E-A$, in (98), yields

(99) $r(A) = |A| + r^*(E-A) - r^*(E)$.

(100) It is easy to verify that r^* is the rank function of a matroid $M^* = (E,F^*)$, e.g., that r^* satisfies (9). M^* is called *the dual* of M. By (99), $M^{**} = M$.

(101) By (98), $|J| = r^*(J)$ iff $r(E-J) = r(E)$. Therefore, $J \in F^*$ iff $E-J$ contains an M-basis of E, i.e., a *basis* of M. Thus, it can be determined whether or not $J \in F^*$ by obtaining an M-basis of $E-J$ and observing whether or not its cardinality equals $r(E)$.

Where r is the rank function of a matroid $M = (E,F)$, and where n is a non-negative integer, let

(102) $r^{(n)}(A) = \min[n, r(A)]$ for every $A \subseteq E$.

(103) Clearly, $r^{(n)}$ is the rank function of a matroid $M^{(n)} = (E,F^{(n)})$, called the *n-truncation* of M, such that $J \in F^{(n)}$ iff $J \in F$ and $|J| \leq n$.

(104) For matroids $M_1 = (E,F_1)$ and $M_2 = (E,F_2)$, and any integer $n \leq r_2(E)$, by (103) and (101), there is a set $J \in F_1 \cap F_2$ such that $|J| = n$ iff E can be partitioned into a set $J \in F_1$ and a set $J_2 \in F_2^{(n)*}$. Theorem (91) says this is possible iff $|A| \leq r_1(A) + r_2^{(n)*}(A)$ for every $A \subseteq E$. Using (102) and (98), this implies (69).

(105) Using *MPAR*, a maximum cardinality $J \in F_1 \cap F_2$ can be found as follows: Find a maximum cardinality set $H = J_1 \cup J_2$ such that $J_1 \in F_1$ and $J_2 \in F_2^*$. Extend J_2 to B, an M_2^*-basis of H. Clearly, B is an M_2^*-basis of E, and so $H - B \in F_1 \cap F_2$. It is easy to verify that $|H-B| = \max |J|$, $J \in F_1 \cap F_2$.

(106) It is more practical to go in the other direction, obtaining for a given family of matroids $M_i = (E, F_i)$, $i \in I$, an "optimum" family of mutually disjoint sets $J_i \in F_i$, by using the "matroid intersection algorithm" of (48) on the following two matroids $M_1 = (E_I, F_1)$ and $M_2 = (E_I, F_2)$. Let E_I consist of all pairs (j, i), $j \in E$ and $i \in I$. There is a 1-1 correspondence between sets $J \in E_I$ and families $\{J_i\}$, $i \in I$, of sets $J_i \subseteq E$, where J corresponds to the family $\{J_i\}$ such that $j \in J_i \Longleftrightarrow (j, i) \in J$. Let $M_1 = (E_I, F_1)$ be the matroid such that $J \subseteq E_I$ is a member of F_1 iff the corresponding sets J_i are mutually disjoint -- that is, if and only if the j's of the members of J are distinct. Let $M_2 = (E_I, F_2)$ be the matroid such that $J \subseteq E_I$ is a member of F_2 iff the corresponding sets J_i are such that $J_i \in F_i$.

(Nash-Williams has developed the present subject in another interesting way [13].)

VIII

(107) If $f(a)$ is a β-function on L and k is a not-too-large constant, then $f(a) - k$ is a β_0-function on L. It is useful to apply (15) to, non-β, β_0-functions.

(108) For example, let G be a graph having edge-set $E = E(G)$ and node-set $V = V(G)$. For each $j \in E$, let Q_j be the set of nodes which j meets. For every $A \subseteq E$, let $f(A) = |u(A)| - 1$. Then, by (70), $f(A)$ is a β_0-function on L_E.

(109) Applying (15) to this f yields a matroid, $M(G) = (E, F(G))$.

(110) The minimal dependent sets of a matroid $M = (E, F)$, i.e., the minimal subsets of E which are not members of F, are called the *circuits* of M.

(111) The circuits of $M(G)$ are the minimal non-empty sets $A \subseteq E$ such that $|A| = |u(A)|$.

(112) A set $J \subseteq E$ is a member of $F(G)$ iff J together with the set $u(J)$ of nodes is a forest in G.

IX

(113) Let G be a directed graph. For any $R \subseteq V(G)$, a *branching* B of G *rooted* at R, is a forest of G such that, for every $v \in V(G)$, there is a

unique directed path in B (possibly having zero edges) from some node in R to v.

(114) The following problem is **solved using matroid-intersection**

(115) Given any directed graph G, given a numerical weight c_j for each $j \in E = E(G)$, and given sets $R_i \subseteq V(G)$, $i \in I$, find edge-disjoint branchings B_i, $i \in I$, rooted respectively at R_i, which minimize $z = \sum_j c_j$, $j \in \bigcup_{i \in I} B_i$.

(116) The problem easily reduces to the case where each R_i consists of the same single node, $v_0 \in V(G)$. That is, find $n = |I|$ edge-disjoint branchings B_i, each rooted at node v_0, which minimize z.

(117) Where $F(G)$ is as defined in (109), let $J \subseteq E$ be a member of F_1 iff it is the union of n members of $F(G)$. By (91), $M_1 = (E, F_1)$ is a matroid.

(118) Let $J \subseteq E$ be a member of F_2 iff no more than n edges of J are directed toward the same node in $V(G)$ and no edge of J is directed toward v_0. Clearly, $M_2 = (E, F_2)$ is a matroid.

(119) Theorem. *A set $J \subseteq E$ is the edge-set of n edge-disjoint branchings of G, rooted at node $v_0 \in V(G)$, iff $|J| = n(|V(G)| - 1)$ and $J \in F_1 \cap F_2$.*

This is a consequence of the following.

(120) Theorem. *The maximum number of edge-disjoint branchings of G, rooted at v_0, equals the minimum over all C, $v_0 \in C \subset V(G)$, of the number of edges having their tails in C and their heads not in C.*

(121) **There is an** algorithm for finding such a family of branchings in G, and in particular for partitioning a set J as described in (119) into branchings as described in (119).

(122) Let P_1 and P_2 be the polyhedra of matroids M_1 and M_2 respectively. Let $H = \{x: x(E) = n(|V(G)| - 1)\}$.

It follows from (45) that:

(123) A vector $x \in R_E$ is a vertex of $P_1 \cap P_2 \cap H$ iff it is the incidence vector of a set J as described in (119).

(124) A varient of the matroid-intersection algorithm will find such an x which minimizes $c \cdot x$. The case $n = 1$ is treated in [4].

X

(125) Let each L_i be a commutative semigroup. We say $a \leqslant b$, for $\{a,b\} \subseteq L_i$, iff $a + d = b$ for some $d \in L_i$.

(126) A function f from L_0 into L_1 is called a *ψ-function* iff

(127) for every $\{a,d\} \subseteq L_0$, $f(a) \leqslant f(a+d)$; and

(128) for every $\{a,b,c\} \subseteq L_0$, $f(a+b+c) + f(c) \leqslant f(a+c) + f(b+c)$.

(129) L_i is called a *ψ-semigroup* iff, for $\{a,b,c\} \subseteq L_i$,

$$a + c + c = b + c + c \Rightarrow a + c = b + c.$$

For example, L_i is a ψ-semigroup if it is cancellative or if it is idempotent.

(130) Theorem. *If $f(\cdot)$ is a ψ-function from L_0 into L_1, $g(\cdot)$ is a ψ-function from L_1 into L_2, and L_1 is a ψ-semigroup, then $g(f(\cdot))$ is a ψ-function from L_0 into L_2.*

(131) Theorem. *A function f from a lattice, L_0, into the non-negative reals, L_1, satisfies* (128), *where "+" in L_0 means "$\vee$" and "+" in L_1 means ordinary addition, iff f is non-decreasing, i.e., satisfies* (127), *and f is submodular.*

(132) Thus, β-functions can be obtained by composing ψ-functions.

(133) Theorem. *A function f from the non-negative reals into the non-negative reals is a ψ-function, relative to addition in the image and pre-image, iff it is non-decreasing and concave.*

(134) Theorem. *A function f from a lattice, L_0, into a lattice, L_1, is a ψ-function, relative to joins "$\vee$" in each, iff it is a join-homomorphism, i.e., for every $\{a,b\} \subseteq L_0$, $f(a \vee b) = f(a) \vee f(b)$.*

Let $h(S)$ be any real (integer)-valued function of the elements $S \in L$ of a finite lattice L. In principle, an (integral) non-decreasing submodular function f on L can be obtained recursively from h as follows:

(135) Theorem. *For each $S \in L$, let $g(S) = \min[h(S), g(A) + g(B) - g(A \wedge B)]$ where $A < S$, $B < S$, and $A \vee B = S$. Then g is submodular. For each $S \in L$, let $f(S) = \min g(A)$ where $S \leqslant A \in L$. Then f is, submodular and non-decreasing. If h is submodular then $g = h$. If h is submodular and non-decreasing then $f = h$.*

(A similar construction was communicated to me by D. A. Higgs.)

(136) The β-functions on a finite lattice L correspond to the members of a polyhedral cone $\beta(L)$ in the space of vectors $y = [y_A]$, $A \in L - \{\phi\}$. Where $y_\phi = 0$, $\beta(L)$ is the set of solutions to the system:

(137) $y_A + y_B - y_{A \vee B} - y_{A \wedge B} \geq 0$ and $y_{A \vee B} - y_A \geq 0$ for every $A \in L$ and $B \in L$.

(138) Characterizing the extreme rays of $\beta(L)$, in particular for $L = \{A\colon A \subseteq E\}$, appears to be difficult.

References

1. Dilworth, R. P., Dependence Relations in a Semimodular Lattice, *Duke Math. J.*, 11 (1944), 575-587.

2. Edmonds, J. and Fulkerson, D. R., Transversals and Matroid Partition, *J. Res. Nat. Bur. Standards*, 69B (1965), 147-153.

3. Edmonds, J., Systems of Distinct Representatives and Linear Algebra, *J. Res. Nat. Bur. Standards*, 71B (1967), 241-245.

4. Edmonds, J., Optimum Branchings, *J. Res. Nat. Bur. Standards*, 71B (1967), 233-240, reprinted with [5], 346-361.

5. Edmonds, J., Matroid Partition, *Math. of the Decision Sciences*, *Amer. Math. Soc. Lectures in Appl. Math.*, 11 (1968), 335-345.

6. Gale, D., Optimal assignments in an ordered set: an application of matroid theory, *J. Combin. Theory*, 4 (1968) 176-180.

7. Hoffman, A. J., Some Recent Applications of the Theory of Linear Inequalities to Extremal Combinatorial Analysis, *Proc. Amer. Math. Soc. Symp. on Appl. Math.*, 10 (1960), 113-127.

8. Ingleton, A. W., A Note on Independence Functions and Rank, *J. London Math. Soc.*, 34 (1959), 49-56.

9. Kruskal, J. B., On the shortest spanning subtree of a graph, *Proc. Amer. Math. Soc.*, 7 (1956), 48-50.

10. Kuhn, H. W. and Tucker, A. W., eds., *Linear inequalities and related systems*, *Annals of Math. Studies*, no. 38, Princeton Univ. Press, 1956.

11. Lehman, A., A Solution of the Shannon Switching Game, *J. Soc. Indust. Appl. Math.*, 12 (1964) 687-725.

12. Mirsky, L. and Perfect, H., Applications of the Notion of Independence to Problems in Combinatorial Analysis, *J. Combin. Theory*, 2 (1967), 327-357.

13. Nash-Williams, C. St. J. A., An application of matroids to graph theory, *Proc. Int'l. Symposium on the Theory of Graphs*, Rome 1966, Dunod.

14. Rado, R., A theorem on Independence Relations, *Quart. J. Math.* 13 (1942), 83-89.

15. Rado, R., A Note on Independence Functions, *Proc. London Math. Soc.*, 7 (1957), 300-320.

16. Tutte, W. T., Menger's Theorem for Matroids, *J. Res. Nat. Bur. Standards*, 69B (1965), 49-53.

MATCHING: A WELL-SOLVED CLASS OF INTEGER LINEAR PROGRAMS

Jack Edmonds
National Bureau of Standards, Washington, D.C., U.S.A.
and
Ellis L. Johnson
I.B.M. Research Center, Yorktown Heights, NY, U.S.A.

A main purpose of this work is to give a good algorithm for a certain well-described class of integer linear programming problems, called *matching problems* (or the *matching problem*). Methods developed for simple matching [2, 3], a special case to which these problems can be reduced [4], are applied directly to the larger class. In the process, we derive a description of a system of linear inequalities whose polyhedron is the convex hull of the admissible solution vectors to the given matching problem. At the same time, various combinatorial results about matchings are derived and discussed in terms of graphs.

The general *integer linear programming problem* can be stated as:

(1) Minimize $z = \Sigma_{j \varepsilon E} c_j x_j$, where c_j is a given real number, subject to

(2) x_j an integer for each $j \varepsilon E$;

(3) $0 \leqslant x_j \leqslant \alpha_j$, $j \varepsilon E$, where α_j is a given positive integer or $+\infty$;

(4) $\Sigma_{j \varepsilon E} a_{ij} x_j = b_i$, $i \varepsilon V$, where a_{ij} and b_i are given integers;
V and E are index sets having cardinalities $|V|$ and $|E|$.

(5) The integer program (1) is called a *matching problem* whenever

$$\Sigma_{i \varepsilon V} |a_{ij}| \leqslant 2$$

holds for all $j \varepsilon E$.

(6) A *solution* to the integer program (1) is a vector $[x_j]$, $j \varepsilon E$, satisfying (2), (3), and (4), and an *optimum solution* is a solution which minimizes z among all solutions. When the integer program is a matching problem, a solution is called a *matching* and an optimum solution is an *optimum matching*.

If the integer restriction (2) is omitted, the problem becomes a linear program. An optimum solution to that linear program will typically have fractional values. There is an important class of linear programs, called transportation or network flow problems, which have the property that for any integer right-hand side b_i, $i \in V$, and any cost vector c_j, $j \in E$, there is an optimum solution which has all integer x_j, $j \in E$. The class of matching problems includes that class of linear programs, but, in addition, includes problems for which omitting the integer restriction (2) results in a linear program with no optimum solution which is all integer.

Many interesting and practical combinatorial problems can be formulated as integer linear programs. However, limitations in the known methods for treating general integer linear programs have made such formulations of limited value. By contrast with general integer linear programming, the matching problem is *well-solved*.

(7) Theorem. *There is an algorithm for the general matching problem such that an upper bound on the amount of work which it requires for any input is on the order of the product of* (8), (9), (10), *and* (11). *An upper bound on the memory required is on the order of* (8) *times* (11) *plus* (9) *times* (11).

(8) $|V|^2$, the number of nodes squared;

(9) $|E|$, the number of edges;

(10) $\Sigma_{i\varepsilon V}|b_i| + 2\Sigma_{\alpha_j<\infty} \alpha_j$;

(11) $\log(|V| \max |b_i| + |E| \max_{\alpha_j<\infty} \alpha_j) + \log(\Sigma_{j\varepsilon E} |c_j|)$.

(12) Theorem. *For any matching problem,* (1), *the convex hull* P *of the matchings, i.e., of the solutions to* [(2), (3), *and* (4)], *is the polyhedron of solutions to the linear constraints* (3) *and* (4) *together with additional inequalities:*

(13) $\Sigma_{j\varepsilon W} x_j - \Sigma_{j\varepsilon U} x_j \geqslant 1 - \Sigma_{j\varepsilon U} \alpha_j$.

There is an inequality (13) *for every pair* (T,U) *where* T *is a subset of* V *and* U *is a subset of* E *such that*

(14) $\Sigma_{i\varepsilon T} |a_{ij}| = 1$ for each $j \in U$;

(15) $\Sigma_{i\varepsilon T} b_i + \Sigma_{j\varepsilon U} \alpha_j$ is an odd integer.

The W in (13) *is given by*

(16) $W = \{j \varepsilon E : \Sigma_{i \varepsilon T} |a_{ij}| = 1\} - U.$

(17) Let Q denote the set of pairs (T,U). The inequalties (13), one for each $(T,U) \varepsilon Q$, are called the *blossem inequalities* of the matching problem (1).

By Theorem (12), the matching problem is the linear program:

(18) Minimize $z = \Sigma_{j \varepsilon E}\, c_j x_j$ subject to (3), (4), and (13).

(19) Theorem. *If c_j is an integral multiple of $\Sigma_{i \varepsilon V} |a_{ij}|$, and if the l.p. dual of* (18) *has an optimum solution, then it has an optimum solution which is integer-valued.*

Using *l.p.* duality, theorems (12) and (19) yield a variety of combinatorial existence and optimality theorems.

To treat matching more graphically, we use what we call *bidirected* graphs. All of our graphs are bidirected, so *graph* is used to mean bidirected graph.

(20) A *graph* G consists of a set $V = V(G)$ of *nodes* and a set $E = E(G)$ of *edges*. Each edge has one or two *ends* and each end *meets* one node. Each end of an edge is either a *head* or a *tail*.

(21) If an edge has two ends which meet the same node it is called a *loop*. If it has two ends which meet different nodes, it is called a *link*. If it has only one end it is called a *lobe*. An edge is called *directed* if it has one head and one tail. Otherwise it is called *undirected, all-head* or *all-tail* accordingly.

(22) The node-edge incidence matrix of a graph is a matrix $A = [a_{ij}]$ with a row for each node $i \varepsilon V$ and a column for each edge $j \varepsilon E$, such that $a_{ij} = +2, +1, 0, -1,$ or -2, according to whether edge j has two tails, one tail, no end, one head, or two heads meeting node i. (Directed loops are not needed for the matching problem).

(23) If we interpret the capacity α_j to mean that α_j copies of edge j are present in graph G^{α}, then x_j copies of j for each $j \varepsilon E$, where $x = [x_j]$ is a solution of [(2), (3), (4)], gives a subgraph G^x of G^{α}. The *degree* of node i in G^x is b_i, the number tails of G^x which meet i minus the number of heads of G^x which meet i. Thus, where x is an optimum matching, G^x can be regarded as an "optimum degree-constrained subgraph" of G^{α}, where the b_i's are the degree constraints.

(24) A Fortran code of the algorithm is available from either author. It was written in large part by Scott C. Lockhart, who also wrote many comments interspersed through the deck to make it understandable. Several random problem generators are included.

It has been run on a variety of problems on a Univac 1108, IBM 7094, and IBM 360. On the latter, problems of 300 nodes, 1500 edges, $b = 1$ or 2, $\alpha = 1$, and random c_j's from 1 to 10, take about 30 seconds. Running times fit rather closely a formula which is an order of magnitude better than our theoretical upper bound.

References

1. Berge, C., *Théorie des graphes et ses applications*, Dunod, Paris, 1958.
2. Edmonds, J., Paths, trees, and flowers, *Canad. J. Math.*, 17 (1965), 449-467.
3. Edmonds, J., Maximum matching and a polyhedron with 0,1-vertices, *J. Res. Nat. Bur. Standards* 69B (1965), 125-130.
4. Edmonds, J., An introduction to matching, preprinted lectures, Univ. of Mich. Summer Engineering Conf. 1967.
5. Johnson, E.L., Programming in networks and graphs, Operation Research Center Report 65-1, Etchavary Hall, Univ. of Calif., Berkeley.
6. Tutte, W.T., The factorization of linear graphs, *J. London Math. Soc.* 22 (1947), 107-111.
7. Tutte, W.T., The factors of graphs, *Canad. J. Math.*, 4 (1952), 314-328.
8. Witzgall, C. and Zahn, C.T. Jr., Modification of Edmonds' matching algorithm, *J. Res. Nat. Bur. Standards*, 69B (1965), 91-98.
9. White, L.J., A parametric study of matchings, Ph.D. Thesis, Dept. of Elec. Engineering, Univ. of Mich., 1967.
10. Balinski, M., A labelling method for matching, Combinatorics conference, Univ. of North Carolina, 1967.
11. Balinski, M., Establishing the matching polytope, preprint, City Univ. of New York, 1969.

THEORETICAL IMPROVEMENTS IN ALGORITHMIC EFFICIENCY FOR NETWORK FLOW PROBLEMS

Jack Edmonds
National Bureau of Standards, Washington, D.C., U.S.A.
and
Richard M. Karp
University of California, Berkeley, CA, U.S.A.
(Formerly at I.B.M. Thomas J. Watson Research Center)

This paper presents new algorithms for the maximum flow problem, the Hitchcock transportation problem and the general minimum-cost flow problem. Upper bounds on the numbers of steps in these algorithms are derived, and are shown to improve on the upper bounds of earlier algorithms.

<u>The Maximum Flow Problem</u>. A network N is a directed graph together with an assignment of nonnegative *capacity* $c(u,v)$ to each arc (u,v). A *flow* is an assignment of a real number $f(u,v)$ to each arc so that

(i) $0 \leq f(u,v) \leq c(u,v)$;

(ii) for any fixed u, $\sum_v f(u,v) = \sum_v f(v,u)$.

One arc (t,s) is distinguished, and a flow which maximizes $f(t,s)$ is called *maximum*. Let f^* denote the maximum value of $f(t,s)$.

Ford and Fulkerson [1] have given a labelling algorithm to compute a maximum flow by repeated flow changes along "flow-augmenting paths". They do not specify which flow-augmenting path to choose.

In Fig. 1, let M be any positive integer.

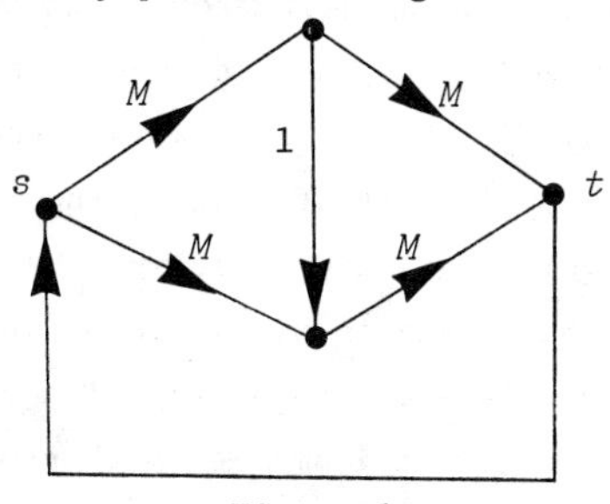

Figure 1.

Then, if a flow-augmenting path of length 3 is selected at each step, $2M$ augmentations will be needed to determine that $f^* = 2M$.

Let n denote the number of nodes of network N.

Theorem 1. *If each flow augmentation is made along an augmenting path having a minimum number of arcs, then a maximum flow will be obtained after no more than* $\frac{n^3-n}{4}$ *augmentations.*

Let $\bar{c}$ be the average capacity of an arc (excluding the distinguished arc (t,s)).

Theorem 2. *If each flow augmentation is chosen to produce a maximum increase in* $f(t,s)$, *then a maximum flow will be obtained after no more than*

$$1 + \frac{n^2}{4}\left(1 + \frac{2}{n^2-2}\right)(2 \ln n + \ln \bar{c})$$

augmentations.

The Minimum-Cost Flow Problem. Assign to each arc (u,v) a nonnegative *cost* $d(u,v)$. Define the cost of a flow f as $\sum d(u,v)\, f(u,v)$. We seek a flow f of minimum cost subject to the constraint that $f(t,s) = f^*$. Define the cost of a flow-augmenting path P as

$$\sum_{\substack{(u,v) \text{ a forward} \\ \text{arc in } P.}} d(u,v) \quad - \sum_{\substack{(u,v) \text{ a reverse} \\ \text{arc in } P.}} d(u,v)$$

The following algorithm solves the minimum-cost flow problem: start with the zero flow, and repeatedly augment along a minimum-cost flow-augmenting path. Stop when a maximum flow is obtained.

In a direct implementation of this algorithm, the selection of each flow-augmenting path requires the calculation of a minimum-cost path through a network which includes arcs of negative cost (namely, the reverse arcs of flow-augmenting paths). We show, however, that the algorithm can be modified so that, in each minimum-cost path calculation, all arcs have nonnegative cost. This is done by replacing $d(u,v)$ at the kth augmentation by $d(u,v) + \pi^k(u) - \pi^k(v)$, where the "node potentials $\pi^k(u)$ are derived as a by-product of finding the $(k-1)$th augmenting path. This refinement is significant, since minimum-cost path problems with all costs nonnegative can be solved in $0(n^2)$ steps (where a step is an operation on scalar quantities, such as addition or comparison), whereas existing methods for general minimum-cost path problems require $0(n^3)$ steps. This refinement reduces the number of steps in the assignment problem, for example, from $O(n^4)$ to $O(n^3)$.

A Scaling Method for the Hitchcock Problem*

An instance of the Hitchcock transportation problem is specified as follows:

$$\text{minimize} \sum_{i=1}^{m} \sum_{j=1}^{n} c_{ij} x_{ij}$$

subject to

$$\sum_{i} x_{ij} = b_j \qquad j=1, 2, \ldots, n$$

$$\sum_{j} x_{ij} = a_i \qquad i=1, 2, \ldots, m$$

$$x_{ij} \geq 0.$$

Here the "supplies" a_i and "demands" b_j are nonnegative integers such that $\sum_{i=1}^{m} a_i = \sum_{j=1}^{n} b_j = B$.

A Hitchcock problem can be exprssed as a minimum-cost flow problem on a network with $m + n + 2$ nodes, and can be solved within B flow augmentations. The scaling method reduces this bound by applying the technique of flow augmentations to a sequence of approximate problems. In the pth approximate problem the ith supply is $\left\lceil \frac{a_i}{2^p} \right\rceil$, and the jth demand is $\left\lceil \frac{b_j}{2^p} \right\rceil$. A fictitious supply or demand is added in a standard way to establish a balance of supplies and demands. The original problem is the zeroth approximate problem. Approximate problems are solved successively, using the flow-augmenting path technique. A saving is effected by using, as a starting solution for approximate problem $p-1$, twice the optimum solution for problem p. The over-all process is shown to require only $n\log_2(1 + \frac{B}{n})$ flow augmentations.

---

*The material of this section was presented by the present authors under the title "A Technique for Accelerating the Solution of Transportation Problems" at the Second Annual Princeton Conference on Information Sciences and Systems, March, 1968.

References

1. Ford, L.R. and Fulkerson, D.R., Flows in Networks, Princeton Universit
Press, 1962.

2. Busacker, R.G. and Saaty, T.L., Finite Graphs and Networks,
McGraw-Hill, 1965.

PROBLEMS IN COMBINATORIAL SET THEORY

Paul Erdös

Hungarian Academy of Sciences, Budapest, Hungary

In this lecture I shall mention a number of results and problems from combinatorial set theory.

The starting point for all these investigations is the well known theorem of F.P. Ramsey (1930) which states that

$$\aleph_0 \to (\aleph_0, \aleph_0)^r \tag{1}$$

holds for finite r. Here we are using the partition symbol

$$a \to (b_0, b_1)^r \tag{2}$$

which means that the following is true: *If* $|S| = a$ *and* $[S]^r = \{X \subset S: |X| = r\} = K_0 \cup K_1$, *then there are* $T \subset S$ *and* $i < 2$ *such that* $|T| = b_i$ *and* $[T]^r \subset K_i$. The symbol (2) was first used in a paper by Rado and myself (1952) but a number of papers dealing with generalizations of Ramsey's theorem appeared before then. The first of these was a paper by W. Sierpinski who proved that $2^{\aleph_0} \nrightarrow (\aleph_1, \aleph_1)^2$. References to the earlier literature will be found in [1] and [2]. Finally in [2] the truth value of (2) is determined for arbitrary cardinals a, b_0, b_1 provided the generalized continuum hypothesis is assured and that a is not larger than the first inaccessible cardinal. Here I want to state just one problem. Without assuming the continuum hypothesis, is it possible to split the pairs of real numbers into three classes so that in any uncountable subset of the reals there is a pair in each of the three classes? With the continuum hypothesis this and much more is true (see [2]).

The symbol (2) defined for cardinal numbers above, has an obvious interpretation when the symbols are replaced by ordinal numbers or order types. Rado and I conjectured that $\omega^2 \to (3,\omega^2)^2$ and more generally that

$$\omega^2 \to (m, \omega^2)^2 \qquad (m < \omega).$$

This conjecture was settled by E. Specker [3] and he also showed, much to our surprise, that $\omega^k \nrightarrow (3,\omega^k)^2$ $(3 \leq k < \omega)$. This suggested the problem whether

$$(?) \quad \omega^\omega \to (3,\omega^\omega)^2.$$

This seems to be very difficult and I offered $250 for either a proof or disproof. It is now rumoured that C.C. Chang has settled the problem affirmatively, but his method does not seem to give

$$(?) \quad \omega^\omega \to (4,\omega^\omega)^2.$$

Other results of this kind have been found. E.C. Milner [4] proved $\omega^4 \to (3,\omega^3)^2$. I proved $\omega^{2n+1} \to (4,\omega^{n+1})^2$ and Milner [4] showed $\omega^{3k+1} \nrightarrow (3,\omega^{2k+1})^2$. Independently of each other A. Hajnal and F. Galvin (unpublished) have reduced the question of deciding the truth status of

$$\omega^l \to (m,\omega^n)^2 \qquad (l,m,n < \omega)$$

to a finite but difficult combinatorial problem. In particular their method gives $\omega^4 \to (4,\omega^3)^2$, $\omega^4 \nrightarrow (s,\omega^3)^2$. Also, for every integer n, there is an integer $f(n)$ such that $\omega^n \nrightarrow (f(n), \omega^3)^2$.

If λ is the order type of the real numbers (or more generally the order type of any uncountable subset of the reals), Rado and I [1] proved $\lambda \to (\omega+n,\omega+n)^2$ $(n<\omega)$. In fact, if the pairs of reals are split into three classes, we have $\lambda \to (\omega+n, \omega+n, \omega+n)^2$. We also showed that $\omega_1 \to (\omega+n,\omega+n)$ Hajnal [5] proved $\lambda \to (\omega n,\alpha)^2$ for $\alpha < \omega_1$. We conjectured that $\lambda \to (\alpha,\alpha)^2$ holds for every $\alpha < \omega$, and I offered $100 for a proof of this. Recently F. Galvin deservedly collected the prize although his method does not see to give our further conjectures

$$(?) \quad \lambda \to (\alpha, \alpha,\ldots,\alpha)^2,$$
$$(?) \quad \omega_1 \to (\alpha, \alpha,\ldots,\alpha)^2$$

when the splitting involves an arbitrary finite number of classes. The simplest unsolved problems in this connection are

$$(?) \quad \omega_1 \to (\omega 3,\omega 3)^2$$

(Hajnal has shown $\omega_1 \to (\omega 2+n, \omega 2+n)^2$),

$$(?) \quad \omega_1 \to (\omega+n, \omega+n, \omega+n)^2,$$

$$(?) \quad \lambda \to (\omega+n,\ \omega+n,\ \omega+n,\ \omega+n)^2.$$

Not much work has been done about partitioning of triples rather than pairs. Rado and I [1] proved that $\lambda \to (\omega+n,\ 4)^3$ and $\lambda \to (\omega+2, \omega)^3$. But the relations

$$(?) \quad \lambda \to (\omega+n, 5)^3\ , \quad \lambda \to (\omega.2, 5)^3\ , \quad \lambda \to (\omega+1, \omega)^3$$

are all undecided.

Rado and I [1] proved that for given m,n there is an integer $f(m,n)$ such that

$$\omega_\alpha f(m,n) \to (m,\ \omega_\alpha n)^2$$

holds for every α. This left open the question whether $\omega_1\omega \to (3, \omega_1\omega)^2$. Hajnal and I showed that this was false and more generally $\omega_1\alpha \nrightarrow (3, \omega_1\omega)^2$ for every $\alpha < \omega_1$. We also proved that $\omega_1{}^2 \to (3,\ \omega_1\omega)^2$ but we do not know if

$$(?) \quad \omega_1{}^2 \to (3,\ \omega_1{}^2)^2.$$

A set mapping is a function from a set S into the subsets of S such that $x \notin f(x)$. Specker and I [6] showed that if $\overline{S} = \omega_1$ and f is a set mapping of order $\alpha(<\omega_1)$ on S (i.e. $\overline{f(n)} < \alpha$ for all $x \in S$), then there is a free set S_1 of type ω_1, i.e. $S_1 \cap f(S_1) = \emptyset$. Recently, Hajnal, Milner and I have extended this and shown that if $\overline{S} = \omega_1^{\rho+1}$ $(\rho \leq \omega)$ and f is a set mapping of type $\alpha < \omega$, then there is a free set of type $\omega_1^{\rho+1}$. The corresponding result is false if $\overline{S} = \omega_1{}^{\omega}$ unless $\alpha \leq \omega$. If $\overline{S} = \omega_1^{\rho} \geq \omega_1{}^{\omega+2}$ the result is false even for $\alpha = \omega$, i.e. we can construct a finite set mapping on $\omega_1{}^{\rho}$ $(\geq \omega_1{}^{\omega+2})$ so that there is no free set of type $\omega_1{}^{\rho}$. Using these results we were able to show that if $\gamma = \omega\beta < \omega_1{}^{\omega+2}$, then

$$\gamma \to (\gamma,\ \text{infinite path})^2. \qquad (3)$$

This means that if G is any graph on a set of type γ, then either there is an independent set of type γ or there is an infinite path in the graph. Our method using the set mapping result completely breaks down if $\gamma \geq \omega_1{}^{\omega+2}$ but (3) may very well hold for every $\gamma = \omega\beta$. We were able to prove that

$$\gamma \to (\omega, \square)^2 \qquad (4)$$

holds for $\gamma = \omega\beta$, i.e. either there is an independent set of type γ or there is a circuit in the graph of length 4. We conjecture that (4) holds for any order type γ which has no fixed point (x is a fixed point of the ordered set S if $\overline{S - \{x\}} \nleq \overline{S}$).

Finally I mention some problems concerning the order type η_α of all 0,1 sequences of length ω_α having a final 1 and ordered lexicographically. η_0 is the order type of the rationals and it is an easy result of Rado and myself that $\eta_0 \to (\eta_0, \aleph_0)^2$. Milner, Rado and I generalized this and we showed $\eta_\alpha \to (\eta_\alpha, \aleph_\beta)^2$ holds provided $\aleph_\alpha$ is regular and $\aleph_\gamma^k < \aleph_\alpha$ ($\gamma < \alpha$; $k < \aleph_\beta$). With the generalized continuum hypothesis Hajnal, Milner and I showed that $\eta_\alpha \to (\eta_\alpha, [m, \eta_\alpha])^2$ if $m^+ < \aleph_\alpha = \aleph_{cf\alpha}$. Here $[m, \eta_\alpha]$ indicates a complete bipartite graph on sets A, B with $|A| = m$ and $\overline{B} = \eta_\alpha$. We cannot prove any of the following

$$(?) \quad \eta_\omega \to (\eta_\omega, 3)^2, \qquad (?) \quad \eta_\omega \to (\eta_\omega, \aleph_0)^2$$

$$(?) \quad \eta_\omega \to (\eta_\omega, \text{infinite path})^2.$$

References

1. Erdös, P. and Rado, R., A partition calculus in set theory, *Bull.Amer. Math. Soc.*, 62, (1956), 427-489.

2. Erdös, P., Hajnal, A. and Rado, R., Partition relations for cardinal numbers, *Acta. Math. Acad. Sci. Hung.*, 16 (1965), 93-196.

3. Specker, E., Teilmengen von Mengen mit Relationen, *Comment. Math. Helv.*, 31 (1957), 302-314.

4. Milner, E.C., Partition relations for ordinal numbers, *Canadian J. Math.*, 21 (1969), 317-334.

5. Hajnal, A., Some results and problems on set theory, *Acta. Acad. Sci. Hung.* 11 (1960), 277-298.

6. Erdös, P. and Specker, E., On a theorem in the theory of relations and a solution of a problem of Knaster, *Coll. Math.* 8 (1961), 19-21.

7. Erdös, P. and Rado, R., Partition relations and transitivity domains of binary relations, *J. Lond. Math. Soc.* 42 (1967), 624-633.

GROUP TRANSITIVITY AND A MULTIPLICATIVE FUNCTION OF A PARTITION

H. O. Foulkes

University College, Swansea, Great Britain

A numerical function introduced by Tsuzuku (Nagoya Math. J. 18 (1961), 93-109) in connection with multiple transitivity is here regarded as a function $\tau(\lambda)$ of a partition $(\lambda) = (\kappa_1^{r_{\kappa_1}} \kappa_2^{r_{\kappa_2}} \dots \kappa_x^{r_{\kappa_x}})$ where no two of $\kappa_1, \kappa_2, \dots, \kappa_x$ are equal. It is found that

(i) $\tau(\kappa^r) = \sum_{t=1}^{r} \frac{S(r,t)}{\kappa^t}$

where $S(r,t)$ is a Stirling number of the second kind,

(ii) $\tau(a) = \tau(b)\tau(c)$, where a is a product of the positive integers b, c,

(iii) $e^{(e^z-1)/\kappa} = \sum_{r=0}^{\infty} \tau(\kappa^r) \frac{z^r}{r!}$,

(iv) $\tau(\kappa_1^{r_{\kappa_1}} \kappa_2^{r_{\kappa_2}} \dots \kappa_x^{r_{\kappa_x}}) = \tau(\kappa_1^{r_{\kappa_1}})\tau(\kappa_2^{r_{\kappa_2}}) \dots \tau(\kappa_x^{r_{\kappa_x}})$.

A full account is to appear in Journ. Comb. Theory.

EXTREMUM PROBLEMS CONCERNING GRAPHS AND THEIR GROUPS

Allan Gewirtz
Brooklyn College CUNY, Brooklyn, N.Y. U.S.A.

Anthony Hill
Chelsea School of Art, London, England

Louis V. Quintas
Pace College, New York, N.Y., U.S.A.

I.

By a *graph* we mean a finite undirected graph (as defined in [1, p.2]) without loops and without multiple edges. The *automorphism group* of a graph consists of those permutations of the vertex set of the graph which preserve adjacency relations. The pioneer work concerning graphs and their groups was done by Frucht [2] when in 1938 he showed that for any finite group G there is a graph whose automorphism group is isomorphic to G. This problem was first posed by König [4] in 1936. In 1949 Frucht [3] showed that for any finite group G there exists a regular graph of degree 3 with automorphism group isomorphic to G. In 1957 Sabidussi [5] proved that for any finite group G of order greater than one and an integer J, $1 \leq J \leq 4$, there exists infinitely many non-homeomorphic connected fixed-point-free graphs X (i.e. no vertex is fixed under all automorphisms of the graph) such that

1) The automorphism group of X is isomorphic to G

and

2) X has property P_J

where

P_1 is the property that the connectivity of X is n, $n \geq 1$
P_2 is the property that the chromatic number of X is n, $n > 1$
P_3 is the property that X is regular of degree n, $n > 2$
P_4 is the property that X is spanned by a graph $\tilde{Y} \simeq Y$,

where Y is a given connected graph.

Then Izbicki in 1957 [7]) and in 1960 [8] showed (among other things) that for graphs of degree 3, 4 or 5 that more than one of properties P_1 - P_4 could hold simultaneously.

The next stage in this development was the consideration of extremum problems, e.g., given a finite group G what is the least number of vertices or edges that a graph can have and have automorphism group isomorphic to G? Certain upper bounds were established by Frucht [3] when for a group G with a finite number $r > 1$ of generators and of order n he produced a graph on $2rn$ vertices which had automorphism group isomorphic to G. For cyclic groups ($r = 1$) when $n = 3$ he produced a graph on 10 vertices and for all $n > 3$ a graph on $3n$ vertices. In 1963 Meriwether (cf. [9]) completely answered the vertex extremum problem for cyclic groups of all finite orders n as follows. Let $\alpha(n)$ be the least number of vertices for which a group has automorphism group isomorphic to C_n, the cyclic group of order n. Let $n = p_1^{e_1} p_2^{e_2} \ldots p_r^{e_r}$ be the prime decomposition of n and if $n = p_1^{e_1}$ write $n = p^e$. Then,

	n	$\alpha(n)$
1	2	2
2	2^e, $e > 1$	$2^e + 6$
3	3^e	$3^e + 6$
4	5^e	$5^e + 10$
5	p^e, $p \geq 7$	$p^e + p$
6	Let n be such that $60 \mid n$, $8 \nmid n$ and $25 \nmid n$	$\left[\sum_1^r \alpha(p_i^{e_i})\right] - 4$
7	Let n be such that $15 \mid n$, $25 \nmid n$ and if $4 \mid n$ this implies $8 \mid n$	$\left[\sum_1^r \alpha(p_i^{e_i})\right] - 3$

8	Let n be such that $(12\|n$ and $8 \nmid n$ or $24\|n$ and $9 \nmid n)$ and if $5\|n$ this implies $25\|n$	$\left[\sum_1^r \alpha(p_i^{e_i})\right] - 1$
9	For all other n	$\sum_1^r \alpha(p_i^{e_i})$

Meriwether also proved that once $\alpha(G)$ the minimum number of vertices for a graph to have G as its group is established then graphs on any larger number of vertices can be constructed that realize G.

Harary and Palmer, [10] in 1966 established that the minimum number of edges in a graph with automorphism group isomorphic to C_3 and C_5 is 15 and 25 respectively. Also in this paper the authors present a graph on 12 vertices, having 20 edges and automorphism group isomorphic to C_4. The graph of Meriwether, in which he establishes that the minimum number of vertices for a graph with group C_4 is 10, also has 20 edges. In unpublished private correspondence (1969) to Quintas, Frucht asserts the existence of a graph, with group C_4, on 10 vertices and having 18 edges. For $p \geq 7$, Sabidussi [8] in 1959 established an upper bound of $4\,p^e$ for the least number of edges a graph could have and have automorphism group isomorphic to C_n where $n = p^e$.

Another approach to extremum problems is the following. Let $e(G,v)$ and $E(G,v)$ represent respectively the least and greatest number of edges a graph can have if it has v vertices and automorphism group isomorphic to G. When the graph is asymmetric (i.e. G is the identity) the complete answer was given in 1967 by Quintas [12] as follows.

Let a_n denote the number of asymmetric trees having n vertices. The numbers a_n were determined for all n by Harary and Prins [11]. For every integer $v \geq 8$ let N be defined by

$$\sum_{n=1}^{N} a_n\, n \leq v < \sum_{n=1}^{N+1} a_n\, n$$

and w by

$$v = \sum_{n=1}^{N} a_n\, n + w(N+1) + r, \quad (0 \leq w < a_{N+1};\ 0 \leq r < N+1).$$

Then $e(id, v)$ is undefined for $2 \leqslant v \leqslant 5$ and

$$\text{(i)}\quad e(id, v) = \begin{cases} 0 & \text{if } v = 1 \\ 6 & \text{if } v = 6,7 \\ v - \sum_{n=1}^{N} a_n - w & \text{if } v \geqslant 8 \end{cases}$$

$$\text{(ii)}\quad E(id, v) = \begin{cases} 0 & \text{if } v = 1 \\ 9 & \text{if } v = 6 \\ 15 & \text{if } v = 7 \\ \frac{v(v-3)}{2} + \sum_{n=1}^{N} a_n + w & \text{if } v \geqslant 8 \end{cases}$$

If the graph is connected (ii) holds as above, (i) holds if $v = 1,6$ and $e(id, v) = v-1$ for $v \geqslant 7$. Quintas also establishes these numbers for topological and connected topological graphs and for each of these types of graphs having asymmetry equal to 1 (part of a more general problem on the asymmetry of asymmetric graphs as defined by Erdös and Rényi [16]) in 1963.

For $G \simeq S_m$ the symmetric group of degree m, $e(S_m, v)$ and $E(S_m, v)$ were determined for all graphs, in 1968, by Quintas [13] and for connected graphs in 1969 by Gewirtz and Quintas [14]. These results are given as,

If X is a graph on v vertices with automorphism group isomorphic to S_m, $m \geqslant 2$, then,

$e(S_m, v)$ is undefined for $v < m$

$e(S_2, v) = v-2 \qquad v = 2,\ldots,8$

$$\text{if } m \geqslant 3,\ e(S_m, v) = \begin{cases} 0 & v = m \\ m & v = m+1,\ m+2 \\ m+2 & v = m+3,\ m+4 \\ m+3 & v = m+5 \\ 6 & v = m+6 \end{cases}$$

and

if $m \geqslant 2$, $e(S_m, v) = e(id, v-m+1) \qquad v = m+1,\ldots$.

also $\quad E(S_m, v) = \binom{v}{2} - e(S_m, v)$.

If X is connected then

$$e(S_m, v) \text{ is undefined for } v < m$$

$$e(S_2, v) = v-1,\ v = 2,3,\ldots$$

$$\text{if } m \geq 3,\ e(S_m, v) = \begin{cases} \binom{m}{2} & v = m \\ m & v = m+1 \\ \binom{m+1}{2}+1 & v = m+2 \\ v-1 & v = m+3,\ldots\ . \end{cases}$$

Except for $v = m+1$, $m \geq 3$ $E(S_m, v)$ is as above and for $m \geq 3$ $E(S_m, m+1) = m = e(S_m, m+1)$.

Now let a graph be asymmetric and regular of degree d. Let $V(d)$ denote the least number of vertices for which such a graph exists. Then the following theorem is true.

Theorem. (i) *$V(0) = 1$.*

(ii) *There are no graphs of degree 1 or 2 that are asymmetric.*

(iii) *$V(3) = 12$.*

(iv) *$V(4) = V(5) = 10$.*

(v) *$V(6) = 11$.*

(vi) *$V(2n+i) = 2(n+2+i)$, $2n+i \geq 7$, $i = 0,1$.*

Sketch of proof

(i) The singleton graph yields $v(0) = 1$.

(ii) Regular graphs of degree 1 have as components complete 2 points and regular graphs of degree 2 have as components simple cycles neither of which are asymmetric.

(iii) (cf. [17]).

(iv) (cf. [15]).

(v) (cf. [15]).

(vi) (cf. [15]).

The arguments in (iv), (v) and (vi) are rather long and make use of the notions of the type of a vertex as defined by Frucht in [3] and a variation of the heirarchy of a graph.

II. Unsolved Problems.

In [17] it was verified that there are exactly two non-isomorphic planar graphs with $V(3) = 12$ vertices. It is hoped that we can shortly announce the number of non-planar graphs with $V(3) = 12$ vertices and answer the question as to the uniqueness of the graphs with $V(4) = V(5) = 10$ and $V(6) = 11$ vertices.

Many unanswered questions remain, in particular the determination of $e(G, v)$ and $E(G, v)$ for exotic G.

References

1. Harary, F., Graph Theory, Reading, Mass., Addison-Wesley, 1969.
2. Frucht, R., Herstellung von Graphen mit vorgegebener abstrakten Gruppe, *Compositio Math.* 6 (1938), 239-250.
3. Frucht, R., Graphs of degree 3 with given abstract group, *Canad. J. Math.* 1 (1949), 365-378.
4. König, D., Theorie der endlichen un unendlichen Graphen, Leipzig, 1936.
5. Sabidussi, G., Graphs with given group and given graph-theoretical properties, *Canad. J. Math.* 9 (1957), 515-525.
6. Sabidussi, G., On the minimum order of graphs with given automorphis group, *Monatsh. Math.* 63 (1959), 124-127. [For corrections see *Math Reviews*, 33 (1967), #2563.]
7. Izbicki, H., Reguläre Graphen 3, 4 und 5 grades mit vorgegbenen abstrakten Automorphismengruppen, Farbenzahlen und Zusammenhängen, *Monatsh. Math.* 61 (1957), 42-50.
8. Izbicki, H., Reguläre Graphen beliebigen Grades mit vorgegebenan Eigenschaften, *Monatsh. Math.* 64 (1960), 15-21.
9. Meriwether, R.L., Smallest graphs with a given cyclic group, (1963) unpublished, but see *Math Reviews*, 33 (1967), #2563.
10. Harary, F., and Palmer, E.M., The smallest graph whose group is cyclic, *Czech. Math. J.* 16 (1966), 70-71.
11. Harary, F., and Prins, G., The number of homeomorphically irreducib trees and other species, *Acta Math.* 101 (1959), 141-162.

12. Quintas, L.V., Extrema concerning asymmetric graphs, *J. Comb. Theory* 3 (1967), 57-82.

13. Quintas, L.V., The least number of edges for graphs having symmetric automorphism group, *J. Comb. Theory* 5(1968), 115-125.

14. Gewirtz, A., and Quintas, L.V., Connected extremal edge graphs having symmetric automorphism group, Recent Progress in Combinatorics, Academic Press, (in press).

15. Gewirtz, A., Hill, A., and Quintas, L.V., El numero minimo de puntos para grafos regulares Y asimetricos, *Scientia* (in press).

16. Erdös, P., and Rényi, A., Asymmetric Graphs, *Acta Math. Acad. Sci. Hungar.* 14 (1963), 295-315.

17. Balaban, A.T., Davies, R.O., Harary, F., Hill, A., and Westwick, Cubic Identity graphs and Planar Graphs derived from Trees, *Australian Math. Soc.* (in press).

[illegible] Some [illegible] concerning [illegible] graphs, [illegible]
[illegible]

[illegible] The [illegible] number of edges [illegible] graphs having [illegible]
[illegible]

12. [illegible] Connected extremal [illegible]
[illegible] Proc. [illegible] Combinatorial [illegible]
[illegible]

13. [illegible] minimal [illegible]
[illegible] graphs [illegible] in press).

14. [illegible] asymmetric graphs, [illegible]
[illegible]

15. [illegible] and [illegible]
[illegible] Graphs and [illegible] derived from [illegible]
[illegible]

QUASISYMMETRIC BLOCK DESIGNS

J.M. Goethals
M.B.L.E. Research Laboratory, Brussels, Belgium

and

J.J. Seidel
Technological University, Eindhoven, Netherlands

0. Abstract

Block designs are called quasisymmetric if the number of points in the intersection of any pair of blocks attains only two values. The relations between these designs and the strongly regular graphs formed by their blocks are discussed. This leads to the construction of new strongly regular graphs. The present paper reports on joint work [2] by the authors.

1. Steiner triple systems.

A *block design* is defined in terms of its point-block incidence matrix N, of size $v \times b$, as follows:

$$NN^T = (r-\lambda)I + \lambda J, \quad NJ = rJ, \quad JN = kJ,$$

where the parameters v, b, k, r, λ satisfy

$$bk = vr, \quad r(k-1) = \lambda(v-1), \; 0 < \lambda, \quad 0 < k < v.$$

Steiner triple systems on n symbols, denoted by $S(n)$, are block designs which have the parameters $v = n$, $b = \frac{1}{6}\, n(n-1)$, $k = 3$, $r = \frac{1}{2}(n-1)$, $\lambda = 1$, $n > 3$. They exist if and only if $n \equiv 1, 3 \pmod 6$.

$S(7)$ is the projective geometry $PG(2,2)$ of dimension 2 over the Galois field $GF(2)$. It is a symmetric block design and any two blocks (= lines) intersect in one point. $S(g)$ is the affine geometry $AG(2,3)$ of dimension 2 over $GF(3)$. For the intersection of any two lines we have two possibilities: either they intersect in one point or they are parallel.

There are 2 nonisomorphic $S(13)$. Each has the following property. Given any point P and line l, nonincident, then there are 3 lines through P which intersect l, and 3 lines through P which do not intersect l. Therefore, these $S(13)$ might be called finite Bolyai - Lobatchevsky geometries. However, one of the $S(13)$ has a better claim on this name than the other. Indeed, the collineation group of one $S(13)$ acts transitively on the points (but not on the lines), whereas the collineation group of the other $S(13)$ is not transitive.

There are 80 nonisomorphic $S(15)$. Any $S(n)$, $n \geq 9$, and more generally any nonsymmetric block design with $\lambda = 1$, has the property that the intersection of any pair of blocks consists of 1 or 0 points. It is this property that we shall investigate in a more general setting.

2. Quasisymmetric block designs.

A block design with point-block incidence matrix N and parameters v, b, k, r, λ is called *quasisymmetric* if there exist integers $x+y$, $x-y$, and a symmetric matrix A of size $b \times b$, having elements 0 on the diagonal and ± 1 elsewhere, such that

$$N^TN = kI+x(J-I) - yA, \quad 0 < y \leq x < k.$$

This condition says that any two distinct blocks have either $x+y$ or $x-y$ points of intersection. The notion of quasisymmetric block design essentially goes back to Shrikhande, who investigated the dual of block designs with $\lambda = 1$; cf. also Shrikhande and Bhagwandas, [5], concluding remarks. Stanton and Kalbfleisch introduced the name quasisymmetric for a more restricted class of such block designs.

Since NN^T and N^TN have the same nonzero eigenvalues with the same multiplicities the b eigenvalues of the matrix A are readily calculated. Indeed, from the following table of eigenvalues:

Multiplicity	NN^T	N^TN	kI	$x(J-I)$	A
1	rk	rk	k	$x(b-1)$	ρ_0
$v-1$	$r-\lambda$	$r-\lambda$	k	$-x$	ρ_2
$b-v$		0	k	$-x$	ρ_1

it follows that these eigenvalues are

$$\rho_1 = \frac{1}{y}(k-x)$$

$$\rho_2 = \frac{1}{y}(k-x-r+\lambda),$$

$$\rho_0 = \frac{1}{y}(xb-x-rk+k),$$

with multiplicity $b-v$, $v-1$, 1, respectively. The eigenvector belonging to ρ_0 is the all-one vector j. If ρ_0 coincides with ρ_1 or with ρ_2, then we have

$$(A-\rho_1 I)\ (A-\rho_2 I) = 0.$$

If $(\rho_0-\rho_1)(\rho_0-\rho_1) \neq 0$, then the eigenspaces of ρ_1 and of ρ_2 are perpendicular to j. From these observations we have the following theorem.

Theorem. *The matrix A of a quasisymmetric block design satisfies*

$$AJ = \rho_0 J,\ b(A-\rho_1 I)(A-\rho_2 I) = (\rho_0-\rho_1)(\rho_0-\rho_2)J,$$

where ρ_0, ρ_1, ρ_2 *are the eigenvectors of A and* $\rho_1 > \rho_2$.

An alternative formulation of this theorem is:

Theorem. *The matrix of a quasisymmetric block design is the adjacency matrix of a strongly regular graph.*

In order to explain this, we first agree to describe ordinary graphs by their (-1,1,0) adjacency matrix which has elements 0 on the diagonal and -1 or +1 elsewhere, according as the corresponding vertices are adjacent or nonadjacent, respectively. A *strongly regular graph* on b vertices is defined in terms of its adjacency matrix A by the relations

$$AJ = \rho_0 J,\ b(A-\rho_1 I)(A-\rho_2 I) = (\rho_0-\rho_1)(\rho_0-\rho_2)J$$

for some real numbers ρ_0, ρ_1, ρ_2, where $\rho_1 > \rho_2$. These numbers, which appear to be the only eigenvalues of A, do not assume arbitrary values. It is known that ρ_0 is an integer and that ρ_1 and ρ_2 are odd integers unless $\rho_1 + \rho_2 = 0$. For details and for an alternative graph-theoretic definition for strongly regular graphs we refer to [3], [4].

Any two graphs on the same number of vertices are called *equivalent*, if their adjacency matrices A_1 and A_2 are related by

$$A_1 = DA_2 D$$

for some diagonal matrix D with diagonal elements +1 or -1.

3. Examples.

We calculate the eigenvalues of the (adjacency matrix of the) strongly regular graphs belonging to the Steiner triple systems introduced in 1, by use of the relations obtained in 2. The matrix of $S(13)$ satisfies

$$AJ = -5J, \quad A^2 = 25I.$$

Thus, two orthogonal matrices with zero diagonal of order 26 are obtained. The corresponding two strongly regular graphs have the same spectrum. We remark that, apart from these graphs, there exist two other strongly regular graphs with this spectrum; these are derived from the two Latin squares of order 5. These four strongly regular graphs are pairwise nonisomorphic and even nonequivalent.

The matrix of $S(15)$ satisfies

$$AJ = -2J, \quad (A-5I)(A+7I) = -J,$$

and has order 35. By adding a row and a column of ones and an element zero, we obtain the matrix

$$B = \begin{bmatrix} 0 & j^T \\ j & A \end{bmatrix}$$

of order 36, which satisfies

$$(B-5I)(B+7I) = 0, \text{ whence } (B+I)^2 = 36I.$$

Thus, 80 symmetric Hadamard matrices with a constant diagonal of order 36 are obtained. The row sums of these matrices are not constant. It has been observed that, by suitable multiplication of corresponding rows and columns of B by -1, all 80 matrices can be made regular. Thus 80 strongly regular graphs of order 36 are obtained. We remark that, apart from these graphs, there exist 12 other strongly regular graphs with the same spectrum, derived from the 12 Latin squares of order 6. Among this total of 92 strongly regular graphs only 2, originating from Latin squares, are equivalent.

4. The Steiner System (24,8,5).

Witt proved the existence and the uniqueness of the Steiner system (24,8,5) which has the Mathieu group M_{24} as its automorphism group. The Steiner system (24,8,5) is a set of 24 symbols and a collection of 8-symbols, called blocks, such that every 5-subset is contained in exactly one block. For $i = 0,1,2,3,4,5$, the number b_i of blocks containing any i-subset is independent of the choice of that i-subset, and satisfies $b_i(8-i) = b_{i+1}(24-i)$, $b_5 = 1$. Therefore, the Steiner system is a block design with $b_2 = \lambda$, $b_1 = r$, $b_0 = b$. By some counting it is observed that any pair of blocks intersects in 4, 2 or 0 points. Hence this Steiner system is not quasisymmetric. However, from this system other block designs may be derived, which have the property that the number s of points contained in any two blocks assumes only two values. These so called "derived" and "residual" designs have parameters as follows:

No.	v	k	b_0	b_1	b_2	b_3	b_4	b_5	s
1	24	8	759	253	77	21	5	1	4,2,0
2	23	7	253	77	21	5	1		3,1
3	22	6	77	21	5	1			2,0
4	21	5	21	5	1				1
5	23	8	506	176	56	16	4		4,2,0
6	22	7	176	56	16	4			3,1
7	21	6	56	16	4				2,0
8	22	8	330	120	40	12			4,2,0
9	21	7	120	40	12				3,1
10	21	8	210	80	28				4,2,0

By application of the theorem of 2 the following strongly regular graphs are obtained:

No.	b		
2.	253	$(A-5I)(A+51I) = -3J$,	$AJ = -28J$,
3.	77	$(A-5I)(A+11I) = 21J$,	$AJ = -44J$,
6.	176	$(A-5I)(A+35I) = 0$,	$AJ = -35J$,
7.	56	$(A-5I)(A+7I) = 20J$,	$AJ = -35J$,
9.	120	$(A-5I)(A+23I) = 4J$,	$AJ = -35J$,
11.	100	$(A-5I)(A+15I) = 24J$,	$AJ = -55J$,

The graph No. 11 is an extension of the graph No. 3. The existence and the uniqueness of the graph No. 7 was first proved by Gewirtz [1], who discovered its complement as a graph of diameter 2, girth 4, valency

10, order 56. Gewirtz also proved the uniqueness of the graph of order 100 which, together with its subgraph of order 77, was first constructed by Higman and Sims, while discovering the simple group which carries their names. The present strongly regular graphs, indicated by the number of their vertices, are partially ordered by containment of one graph by another as a subgraph according to the following diagram.

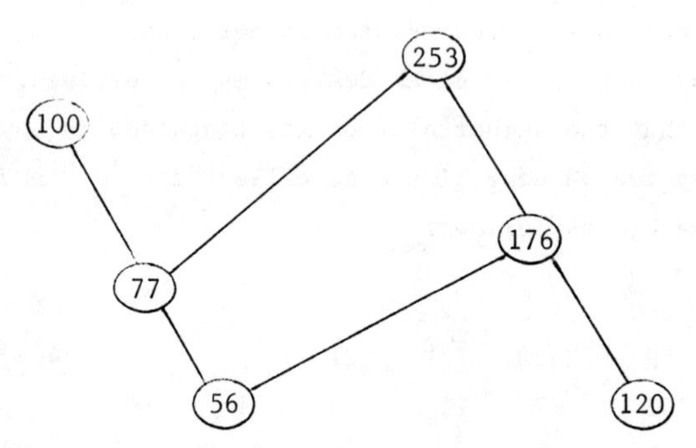

For further details we refer to [2].

References

1. Gewirtz A., The uniqueness of g(2,2,10,56), *Transact. New York Acad. Sci.*, 1969.

2. Goethals, J.M. and J.J. Seidel, Strongly regular graphs derived from combinatorial designs,

3. Seidel, J.J., Strongly regular graphs, *Proc. 3rd Waterloo Conf. Combin.*, (to appear),

4. Seidel, J.J., Strongly regular graphs with (-1,1,0) adjacency matrix having eigenvalue 3, *Lin. Alg. and Appl.* 1 (1968), 281-298,

5. Shrinkhande, S.S., and Bhagwandas, Duals of incomplete block designs, *Journal Indian Statist. Assoc.* 3 (1965), 30-37.

SOME POLYHEDRA RELATED TO COMBINATORIAL PROBLEMS

R. E. Gomory

I.B.M. Watson Research Center, Yorktown, NY, U.S.A.

In the study of integer programming, the convex hull of the integer points within the polyhedron defined by the corresponding linear programming problem plays a key role. Unfortunately, in most cases, information about this integer polyhedron is very hard to come by. In this paper, we replace the study of the convex hull of the integer points by the study of the convex hull of the integer points lying in a polyhedral cone formed by dropping all the linear programming constraints except those which provide a single vertex of the polyhedron. More precisely, if B is the basis corresponding to a particular vertex of the linear programming problem, the nonnegativity constraints on the basic variables are dropped. The properties of this new object, called the corner polyhedron, depend on a group equation depending on the group G, which is the factor group of all integer vectors modulo the lattice generated by the columns of B and a few other factors. The corner polyhedron can be shown to be one cross section of a higher dimensional polyhedron which depends only on G and the automorphism class of G_0, where G_0 is the group element into which the right-hand side of the underlying system of inequalities is mapped by the transformation sending the integer vectors onto G.

Various properties of these higher dimensional polyhedra are then discussed. In particular, it is shown that the faces of these polyhedra can be obtained as vertices of another polyhedron whose faces can be written down explicitly. The integer polyhedra corresponding to all groups of order 11 or less have been explicitly computed.

ON COMBINATORIAL SPHERES*

Branko Grünbaum

University of Washington, Seattle, WA, U.S.A.

A *combinatorial n-sphere* is a simplicial n-complex obtainable from the boundary of the $(n+1)$-simplex by a finite sequence of transformations each of which is either an elementary subdivision or the inverse of an elementary subdivision (for details compare Alexander [1]).

It is well known that the boundary complex of every simplicial $(n+1)$-polytope is a combinatorial n-sphere, and that for $n = 2$ the converse is also true. However, for $n \geq 3$ there are combinatorial n-spheres which are not isomorphic to the boundary complex of any $(n+1)$-polytope.

The problem of enumerating the different isomorphism-types of simplicial $(n+1)$-polytopes (or of all $(n+1)$-polytopes) with v vertices has attracted considerable interest; Grünbaum [4] contains a survey of the known results, together with references to the literature, as well as detailed proofs of the results discussed in the present paper. The analogous question of enumerating the different isomorphism-types of combinatorial n-spheres ($n \geq 3$) with v vertices appears not to have been considered in the literature. A few simple results are described here.

It is well known (see [3] for details of proofs, and for references) that there are $[\frac{1}{2}(n+1)]$ different types of simplicial $(n+1)$-polytopes with $n+3$ vertices. It is not hard to establish

Theorem 1. *Every combinatorial n-sphere with $n+3$ vertices is isomorphic to the boundary complex of an $(n+1)$-polytope.*

*Research supported in part by the Office of Naval Research under Grant N00014-67-A-0103-0003.

The $(n+1)$-polytopes with $n+4$ vertices may be enumerated using the so-called *Gale diagrams* (see Chapters 5 and 6 of [3]). Concerning combinatorial spheres we have only:

Conjecture 1. Every combinatorial n-sphere with $n+4$ vertices is isomorphic to the boundary complex of an $(n+1)$-polytope.

This is a contrast to:

Theorem 2. *There exist 2 different combinatorial 3-spheres with 8 vertices which are not isomorphic to the boundary complex of any 4-polytope.*

One of those 3-spheres (the "Brückner 3-sphere") was described in Grünbaum-Sreedharan [5] (it was there called "Complex M"), the other (the "Barnette 3-sphere") in Barnette [2]. In Table 1 we present the two spheres by listing their 3-simplices, each 3-simplex being designated by its four vertices.

Bruckner 3-sphere		Barnette 3-sphere	
1234	2345	1234	2345
1237	2358	1237	2358
1248	2367	1247	2378
1267	2368	1345	2456
1268	2458	1358	2468
1347	3456	1367	2478
1478	3467	1368	2568
1567	3568	1456	3678
1568	4567	1467	4678
1578	4578	1568	

Table 1. Two combinatorial 3-spheres not isomorphic to the boundary complex of any 4-polytope.

It may be conjectured that there are no other combinatorial 3-spheres with 8 vertices except those of Table 1 and the 37 which correspond to simplicial 4-polytopes. Moreover, we make the following:

Conjecture 2. For every $n \geq 3$ there exist combinatorial n-spheres with $n+5$ vertices which are not isomorphic to the boundary complex of any $(n+1)$-polytope.

Centrally symmetric simplicial $(n+1)$-polytopes have as boundaries *symmetric n-spheres* (that is, combinatorial n-spheres which have an involution ϕ such that $\phi(F) \cap F = \emptyset$ for every face F of the sphere). Clearly, every symmetric n-sphere has at least $2n + 2$ vertices. We have:

Theorem 3. *For every even $v \geq 10$ there exist symmetric 3-spheres with v vertices which are not isomorphic to the boundary complex of any centrally symmetric 4-polytope.*

For v = 10 such a 3-sphere (probably the only one) is given in Table 2 (where ±1, ±2, ±3, ±4, ±5 are the vertices).

In Table 3 we similarly describe such a sphere with v = 12. That sphere is of special interest since it contains an edge for every pair of non-antipodal vertices - a situation which is not possible for a centrally symmetric 4-polytope with 12 or more vertices (see [3, Chapter 6] or [6]).

-1	2	3	4	1	-2	-3	-4
-1	2	-3	4	1	-2	3	-4
1	-2	-3	4	-1	2	3	-4
1	-2	3	5	-1	2	-3	-5
-1	-2	3	5	1	2	-3	-5
1	2	-3	5	-1	-2	3	-5
1	-2	4	5	-1	2	-4	-5
-1	-2	4	5	1	2	-4	-5
-1	3	4	5	1	-3	-4	-5
2	3	4	5	-2	-3	-4	-5
1	-3	4	5	-1	3	-4	-5
2	-3	4	5	-2	3	-4	-5
1	2	-4	5	-1	-2	4	-5
1	3	-4	5	-1	-3	4	-5
2	3	-4	5	-2	-3	4	-5

Table 2. A symmetric 3-sphere with 10 vertices.

-1	2	3	4	1	-2	-3	-4
1	-2	-3	4	-1	2	3	-4
-1	-2	3	5	1	2	-3	-5
-1	3	4	5	1	-3	-4	-5
2	3	4	5	-2	-3	-4	-5
2	-3	4	5	-2	3	-4	-5
1	3	-4	5	-1	-3	4	-5
2	3	-4	5	-2	-3	4	-5
-1	-2	3	6	1	2	-3	-6
-1	2	4	6	1	-2	-4	-6
-1	-3	4	6	1	3	-4	-6
2	-3	4	6	-2	3	-4	-6
-1	2	-4	6	1	-2	4	-6
-1	3	-4	6	1	-3	4	-6
1	2	5	6	-1	-2	-5	-6
1	-4	5	6	-1	4	-5	-6
2	-4	5	6	-2	4	-5	-6
1	2	-5	6	-1	-2	5	-6
-1	-2	-5	6	1	2	5	-6
-2	3	-5	6	2	-3	5	-6
-1	-3	-5	6	1	3	5	-6
2	-3	-5	6	-2	3	5	-6
1	-4	-5	6	-1	4	5	-6
3	-4	-5	6	-3	4	5	-6

Table 3. A 2-neighbourly symmetric 3-sphere with 12 vertices.

Concerning other unsolved problems on spheres, we venture the following guesses.

Conjecture 3. For each $n \geq 3$ there exist symmetric n-spheres with $2n + 4$ vertices which are not isomorphic to the boundary complex of any centrally symmetric $(n+1)$-polytope.

Conjecture 4. For all n, v, and k with $2k \leq n + 1$ and even $v \geq 2n$ there exist k-neighbourly symmetric n-spheres with v vertices.

(A symmetric n-sphere S is *k-neighbourly* if every k vertices of S, no two of which are antipodal, determine a $(k-1)$-simplex of S.)

Conjecture 5. The enumeration problem of combinatorial (or of symmetric) spheres is not algorithmically solvable. More precisely, there exists no algorithm that would enumerate, for each fixed n, first of all the n-spheres with $n + 2$ vertices, then those with $n + 3$, $n + 4$, etc.

It should be noted that polytopes are known (see [3], [6]) to behave in a manner contrasting conjectures 4 and 5.

References

1. Alexander, J. W., The combinatorial theory of complexes. *Ann. Math.*, 31 (1930), 292-320.

2. Barnette, D. W., Diagrams and Schlegel diagrams. (to appear)

3. Grünbaum, B., Convex polytopes. Wiley, New York, 1967.

4. Grünbaum, B., On the enumeration of convex polytopes and combinatorial spheres. *ONR Technical Report, University of Washington,* May, 1969.

5. Grünbaum, B. and Sreedharan, V. P., An enumeration of simplicial 4-polytopes with 8 vertices. *J. Combinat. Theory,* 2 (1967), 437-465.

6. McMullen, P. and Shephard, G. C., Diagrams of centrally symmetric polytopes. *Mathematika,* 15 (1968), 123-138.

A COLOUR PROBLEM FOR INFINITE GRAPHS

R. Halin

Köln-Lindenthal, Germany

Erdös and Hajnal [1] introduced the *colouring number* $col(G)$ of a graph[1] G in the following way. Consider all cardinals k for which there exists a well ordering $<$ of the vertex set of G such that every vertex v of G is adjacent to less than k vertices $< v$. Then $col(G)$ is defined to be the minimum of all these cardinals k.

There is a certain relationship between $col(G)$ and the chromatic number $\chi(G)$ of G. It is important that for every G holds $col(G) \geq \chi(G)$.

Erdös and Hajnal proved: If $col(G) \geq \aleph_1$, then, for every natural number n, G contains a complete bipartite $n,\aleph_1$-graph $K_{n,\aleph_1}$. The analogous question for the existence of complete graphs of order ≥ 3 in a graph G with large colouring number can easily be answered negatively: Obviously the complete bipartite graph $K_{a,a}$ does not even contain a triangle, but $col(K_{a,a}) = a + 1$ may be arbitrarily large.

In what follows, our aim will be to show that, on the other hand, graphs with a sufficiently large colouring number must contain subdivisions[2] of complete graphs of arbitrarily large cardinality.

First we have:

(1) $col(G) = 1 + \max_{H \subseteq G} \min_{v \in H} \rho_H(v)$ *for every finite graph* G,

---

1 All graphs considered in this paper are undirected and do not contain loops or multiple edges.

2 A subdivision (or refinement) of a graph G is a graph which arises from G by inserting new vertices (of valency 2) on some of the edges of G. A subdivision of G will be denoted by $U(G)$.

where $\rho_H(v)$ denotes the valency of the vertex v of H relative H.

From a theorem by W. Mader [4], Satz 2 we conclude that for every finite n there exists a least finite cardinal $g(n)$ such that every finite G with $col(G) \geq g(n)$ contains a subdivision $U(K_n)$ of the complete n-graph and moreover,

(2) $g(n) \leq \frac{4}{3} 2^{\binom{n}{2} - 2(n-1)} + 1.$

From Theorem 9.1 by Erdös-Hajnal [1] one concludes that the analogue is also correct for infinite graphs G (but finite n) with some $g^*(n)$ (instead of $g(n)$) where $g^*(n) \leq 2\, g(n) - 2$.

Now let us consider the graphs with infinite colouring number. Our aim is to show the following:

Theorem. *If $col(G) \geq \aleph_0$, then G contains a subdivision of the complete a-graph for every cardinal $a < col(G)$.*

I shall now sketch the ideas of the proof. First some notations and lemmas will be stated.

Definition. Let G be a graph, let σ be an ordinal number > 0, and let, for every ordinal $\lambda < \sigma$, G_λ be a section graph[3] of G. Then this family G_λ $(\lambda<\sigma)$ is called a *simplicial decomposition* of G, if the following conditions are fulfilled:

i) $G = \bigcup_{\lambda<\sigma} G_\lambda$;

ii) For every τ, $0 < \tau < \sigma$, $(\bigcup_{\lambda<\tau} G_\lambda) \cap G_\tau = S_\tau$ is a complete graph;

iii) S_τ is properly contained in $\bigcup_{\lambda<\tau} G_\lambda$ and in G_τ for every τ, $0 < \tau < \sigma$.

The concept of simplicial decomposition has a certain relation to that of colouring number, as is shown by the following lemma:

(3) *Let G_λ $(\lambda<\sigma)$ be a simplicial decomposition of G. Let b be an infinite cardinal such that the cardinality of every G_λ is less than or equal to b. Then it follows: $G \supseteq K_b$, or $col(G) \leq b$.*

In order to prove (3), we choose a well ordering $<_\lambda$ of the vertices of $G_\lambda - S_\lambda$ (for every $\lambda, \lambda < \sigma$; S_0 is defined to be the empty graph) such that the corresponding ordinal number is less than or equal to the initial ordinal number of cardinality b. For any two vertices $v \neq w$ of G we set

---

3 H is called a section graph of G, if the following two conditions hold: 1. H is a subgraph of G; 2. if two vertices of H are adjacent in G, then they are also adjacent in H.

$v < w$ if either $v \in G_\lambda - S_\lambda$, $w \in G_\mu - S_\mu$ with $\lambda < \mu$ or if v,w are elements of the same $G_\lambda - S_\lambda$ and $v <_\lambda w$. Then $<$ is a well ordering of the vertex set of G. If $G \nsupseteq K_b$, then in particular every S_λ has less than b vertices. It follows that every $v \in G$ has less than b neighbours which preceed v (with respect to $<$). Hence $col(G) \leq b$ by definition of the colouring number.

The section graphs G_λ which appear in some simplicial decomposition of G may be considered as something like the direct summands of G. It is possible to characterize these special subgraphs without using the notation of simplicial decomposition.

Definition. Let H be a section graph of G. We call H a *separation-invariant* subgraph of G if the following condition is fulfilled: If a,b are any vertices of H separated in H by a subgraph T of H, then a,b are also separated in G by T.

Then we have

(4) *Let H be a section graph of G, and suppose H is not complete. Then there exists a simplicial decomposition of G containing H as one of its members if and only if H is separation-invariant in G.*

(See [2], p. 93 for the proof of (4).)

It is not difficult to show that the intersection of any family of separation-invariant subgraphs of a graph G is again separation-invariant in G (see [2], p. 94). Since of course G itself is separation-invariant in G, it follows that for every subgraph H of G there exists the *separation-invariant closure* in G; we denote it by $P_G(H)$.

If $a \neq b$ are two non-adjacent vertices of a graph G, the minimum number of vertices which are necessary to separate a from b in G will be denoted as the *Menger number* $\mu_G(a,b)$ with respect to G. Menger's graph theorem says that $\mu_G(a,b)$ is also equal to the maximum number of disjoint a,b-paths contained in G. (This result also holds in the infinite case.) Furthermore let us define the *Menger number* $\mu(G)$ of the graph G as the least upper bound of the set of the cardinals $\mu_G(a,b)$ where a,b run through all pairs of non-adjacent vertices of G.

The proof of our theorem depends essentially on the fact that it is possible to estimate the cardinality of the separation-invariant closure of some subgraph of G by means of the Menger number $\mu(G)$. We have the following result (where $\alpha(G)$ denotes the cardinality of some graph G):

(5) *If* $H \subseteq G$, *then* $\alpha(P_G(H)) \leq \mathrm{Max}(\alpha(H), \mu(G), \aleph_0)$.

In order to prove (5) we construct a chain of subgraphs $G_0 \subseteq G_1 \subseteq G_2 \subseteq \ldots$ of G in the following way: Let G_0 be the graph H. Now suppose that, for some finite $n \geq 0$, G_n is already constructed. For every pair of non-adjacent vertices a,b of G_n choose a maximal system $S_{a,b}$ of disjoint a,b -paths $\subseteq G$. Then let G_{n+1} be the union of G_n and all these $S_{a,b}$. It follows that $\bigcup_{n=0}^{\infty} G_n$ is a separation-invariant section graph of G which contains H and has cardinality less than or equal to the asserted number.

Now it can be shown:

(6) *If* $\mu(G) \leq b$, $b \geq \aleph_0$ *and* G *does not contain a* K_{b^+} (b^+ *denotes the smallest cardinal greater than* b), *then* G *has a simplicial decompositio* G_λ $(\lambda<\sigma)$ *such that* $\alpha(G_\lambda) \leq b$ *holds for every* $\lambda < \sigma$.

Outline of the proof: Let τ be an ordinal number, and suppose G_λ to be defined already for every $\lambda < \tau$ in such a way that every G_λ is a separation-invariant section graph of G, has cardinality $\leq b$, and that the family G_λ $(\lambda<\tau)$ forms a simplicial decomposition of the graph $G' = \bigcup_{\lambda<\tau} G_\lambda$ which is assumed to be separation invariant in G. If G' is a proper section graph of G, we choose a vertex v of G not contained G'. Let T denote the set of the terminal vertices of all paths of G starting at v, ending in G', and not having an inner vertex in common with G'. It follows from the separation-invariance of G' that the section graph of G generated by T is a complete graph S_τ. Now we choose G_τ as the section graph $P_G(S_\tau \cup \{v\})$. From $\alpha(S_\tau) \leq b$ and $\mu(G) \leq b$ we conclude by (5) that $\alpha(G_\tau) \leq b$ also holds. Now we see that the simplicial decomposition G_λ $(\lambda<\tau$ can be continued by G_τ. In this way we finally obtain a simplicial decomposition of the whole graph G, which has the desired properties.

From (6) and (3) we conclude immediately:

(7) *If* $col(G) > b \geq \aleph_0$ and $\mu(G) \leq b$, *then* $G \supseteq K_b$.

Otherwise G has a simplicial decomposition in which every member has cardinality $\leq b$, and by (3) one finds $col(G) \leq b$.

Now we are ready to complete the proof of the above mentioned theorem. Let G be a graph with $col(G) \geq b^+$, $b \geq \aleph_0$.[4] For any two vertices x,y of (with $\mu_G(x,y) > b$ we add a new edge to G, joining x,y. Let the graph obtai

---

[4] The case $col(G) = \aleph_0$ of our theorem is obviously clear by the the existence of $g^*(n)$ which was mentioned above.

in this way be denoted by $\hat{G}$. It can be shown that $\mu(\hat{G}) \leq b$. By (7) we find $\hat{G} \supseteq K_b$. Possibly, some edges of this K_b do not lie in G; but each edge of $\hat{G}$ not in G can be replaced by a path in G connecting its end vertices, and because of the condition on the Menger numbers which was imposed when we constructed $\hat{G}$, this replacing can be done in such a way that these paths, together with the vertices and edges of K_b which are contained in G, form a $U(K_b)$ in G. The theorem follows.

As a concluding remark I mention that the theorem is the best possible, as is shown by the complete bipartite a,a^+-graph ($a \geq \aleph_0$) which has colouring number a^+ but of course does not contain a $U(K_{a^+})$. Since $col(G) \geq \chi(G)$, every G with $\chi(G) \geq a^+$ must contain a $U(K_a)$. However, the question whether $\chi(G) \geq a^+$ implies $G \supseteqq U(K_{a^+})$ is an unsolved problem. This question may be considered as an infinite analogue of Hadwiger's conjecture on n-chromatic finite graphs. Of course, for limit cardinals a (instead of the cardinals a^+) our result is the best possible also for the case of the chromatic number.

References

1. Erdös, P. and Hajnal, A., On chromatic number of graphs and set systems, *Acta Math. Acad. Sci. Hungar.*, 17 (1966), 61-99.
2. Halin, R., Ein Zerlegungssatz für unendliche Graphen und seine Anwendung auf Homomorphiebasen, *Math. Nachr.*, 33 (1967), 91-105.
3. Halin, R., Unterteilungen vollständiger Graphen in Graphen mit unendlicher chromatischer Zahl, *Abh. Math. Sem. Univ. Hamburg*, 31 (1967), 156-165.
4. Mader, W., Homomorphieeigenschaften und mittlere Kantendichte von Graphen, *Math. Ann.*, 174 (1967), 265-268.

COMBINATORIAL DESIGNS*

Haim Hanani

Technion, I.I.T., Haifa, Israel

1. Finite Planes. Finite affine and projective planes as combinatorial designs. Construction of lines (blocks) by powers of primitive marks of the respective Galois fields.

2. Steiner Triple Systems. A simple proof by induction that a necessary and sufficient condition for the existence of a Steiner triple system with v elements is that $v \equiv 1$ or $3 \pmod 6$. A direct proof of the same theorem due to Skolem.

3. Balanced Incomplete Block Designs. BIBD's and their generalizations: pairwise balanced designs and group divisible designs. A proof that if $k = 3$ a necessary and sufficient condition for the existence of a BIBD is that $\lambda(v - 1) \equiv 0 \pmod{k-1}$ and $\lambda v(v - 1) \equiv 0 \pmod{k(k-1)}$. Incidence matrix. Proof of Fisher's theorem that $b \geq k$ and Chowla and Ryser's theorem that in a symmetric BIBD (i.e. BIBD with $b = v$) if v is even then $k - \lambda$ must be a square.

4. Latin Squares. Mutually orthogonal Latin squares. Equivalence of existence of a group divisible design with m groups and blocks having m elements each with the existence of $m - 2$ m.o.l.s.. MacNeish theorem that if $n = \Pi p_i^{\alpha_i}$ (p_i distinct primes, α_i positive integers), then the number of m.o.l.s. of order n, $N(n) \geq \min p_i^{\alpha_i} - 1$. Euler's conjecture that there are no orthogonal Latin squares of order n if $n \equiv 2 \pmod 4$ and its disproval by Bose, Parker and Shrikhande.

---

*Synopsis of Instructional Series of Lectures.

5. Resolvable Designs (with $k = 3$). Kirkman's 15 school-girls problem and its generalization for every $v \equiv 3 \pmod 6$. The solution for $v = 3q$ where $q \equiv 1 \pmod 6$ is a prime power. Harrison's theorem that if resolvable designs exist for $v = 3m$ and $v = 3n$, then such design exists also for $v = 3mn$. Ray-Chaudhuri and Wilson's complete solution for the existence of resolvable designs with $k = 3$.

THE RECONSTRUCTION CONJECTURE FOR LABELED GRAPHS*

Frank Harary and Bennetoe Manvel

University of Michigan, Ann Arbor, MI, U.S.A.

Since the Reconstruction Conjecture (R.C.) appears to be intractable for graphs, we propose the corresponding problem for labeled and partially labeled graphs. We first summarize those special classes of graphs for which the R.C. has been proved. Then we formulate conjectures for the reconstruction number $r(p,n)$ of one-point deleted subgraphs required to reconstruct an arbitrary p-point graph having n unlabeled points. A simple example yields lower bounds for $r(p,n)$ and theorems are derived which demonstrate certain upper bounds. Combining these, we obtain exact values of $r(p,n)$ for all $p \leqslant 7$, after also invoking a computer search. We conclude by proposing further unsolved problems concerning the reconstruction of graphs.

Motivation. In his stimulating book [17], Ulam states a problem which in graph theoretic terms proposes the following:

Conjecture. If G and H are two graphs with $p \geqslant 3$ points $u_1, u_2, \ldots, u_p$ and $v_1, v_2, \ldots, v_p$, and $G - u_i \simeq H - v_i$ for all i, then $G \simeq H$.

Kelly [12] showed that this is true when G and H are trees, and stated that it holds for disconnected (and complement-disconnected) graphs as well. It occurred to one of us [6] that it might be better to view this problem as one of reconstructing a graph G from its maximal subgraphs $G_i = G - v_i$. Obviously this can be done uniquely only if the original conjecture is valid. Thus we reformulated the problem as follows.

Reconstruction Conjecture (R.C.). Any graph G with $p \geqslant 3$ points can be reconstructed uniquely from its collection of maximal subgraphs $G_i = G - v_i$.

* Research supported in part by a grant from the U.S. Air Force Office of Scientific Research.

The progress to date on this question can be summarized very briefly. Trees have received the most attention, with Harary and Palmer [10], Bondy [2], and Manvel [14] using various proper subcollections of maximal subgraphs to reconstruct them. Other classes of graphs with connectivity one have been reconstructed: those with no endpoints by Bondy [3], unicyclic graphs by Manvel [13], cacti by Geller and Manvel [4], Greenwell and Hemminger [5]. Many invariants of G have been deduced from the subgraphs G_i (see [6], [7], and [15]), and certain special graphs such as cycles and other regular graphs are instantly reconstructed. For digraphs, Harary and Palmer [7] have reconstructed non-strong tournaments with at least five points, but little else has been done except for the discovery by Beineke and Parker [1] of two small counterexamples to the R.C. for strong tournaments with $p = 5$ and 6 points. The R.C. appears to hold for all tournaments with $p \geq 7$ points.

Since the general problem appears to be very difficult, it seems desirable to modify it in some way to make it less intractable. One obvious ploy is to strengthen it to the point where counterexamples can be found. This has been attempted in at least four different ways by one of us [15], but unfortunately all of the stronger versions are holding up as well as the original R.C. The other possibility is to try to prove some weakened, but still general, form. The most obvious way to do this is to consider graphs with distinct labels on some of the points, so-called *partially labeled graphs*. If all of the points have distinct labels, we have a *labeled graph*. For labeled graphs reconstruction is very easy, as we observe in Theorem 1. This is an unusual case because exactly three subgraphs G_i and no fewer are required to reconstruct any labeled graph. If we look at partially labeled graphs, we can usually find pairs of p-point graphs which do not share k subgraphs G_i, and other pairs which do have k common subgraphs. We therefore naturally define $r(p,n)$, the number of subgraphs G_i required to distinguish p-point graphs with n points unlabeled, to be just one greater than the maximum number of subgraphs shared by any two such non-isomorphic graphs. Since $r(p,0) = 3$, and the R.C. asserts that $r(p,p)$ is at most p, our natural inclination, when we began to investigate the R.C. for labeled graphs, was to form the following table, anchored in those values.

Table 1. Conjectured values of $r(p,n)$

n	0	1	2	...	$p-1$	p
$r(p,n)$	3	4	5	...	$p-1$	p

The obvious question in the values listed in Table 1 deals with the apparent nonlinear nature of $r(p,n)$. There appears to be a "slippage" of 3 units somewhere and one wonders where. To investigate this issue, we concentrate in the next two sections on lower bounds for $r(p,n)$. We then derive corresponding upper bounds and thus establish exact values of $r(p,n)$ for all $p \leq 7$ and also for $n \leq 4$. We conclude with a list of related unsolved problems.

Lower bounds on $r(p,n)$. Early in our investigation we found a simple class of graphs which give remarkably good lower bounds for $r(p,n)$. Each of the two nontrivial graphs in Figure 1 consists of one star together with several isolated points. They originated in Manvel and Stockmeyer [16], where the reconstruction of matrices from principal submatrices was studied.

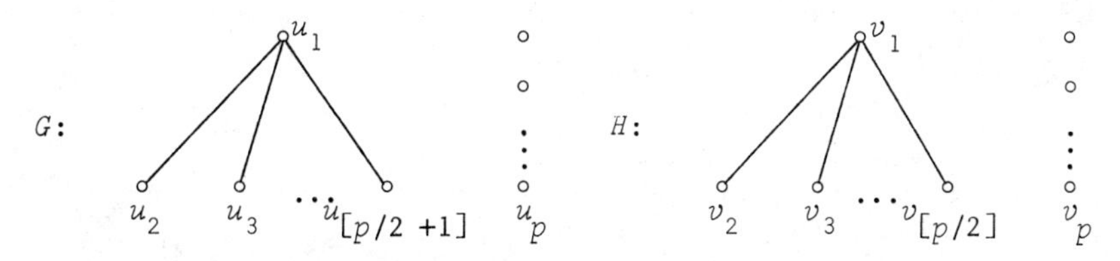

Figure 1. Two p-point graphs.

They have $[p/2] + 1$ isomorphic maximal subgraphs as follows:

$$G_1 \simeq H_1 \simeq \overline{K}_{p-1}$$

and $$G_{i+1} \simeq H_{p-i+1}, \qquad i = 1, \ldots, [p/2].$$

This gives us the bound for unlabeled graphs,

$$r(p,p) \geq [p/2] + 2. \tag{1}$$

We utilize the full power of G and H only when we label the points u_1 and v_1. Then we have two rooted (1-point labeled) graphs, G' and H', which have the same isomorphic subgraphs as G and H, and thus rooted graphs have the same bound

$$r(p,p-1) \geq [p/2] + 2. \tag{2}$$

To obtain bounds for more labeled points, we need only note that when we add k labeled, isolated points to G' and H', we obtain non-isomorphic $(p+k)$-point graphs, with $p-1$ unlabeled points, having $[p/2] + 1$ common subgraphs. Hence we see that

$$r(p+k,p-1) \geq [p/2] + 2,$$

or, more conveniently,

$$r(p,n) \geq [(n+1)/2] + 2, \qquad p > n > 0. \tag{3}$$

For $n = 0$, we have a labeled graph and, as we observed in the preceding section and will later prove, such a graph is reconstructable from exactly 3 subgraphs:

$$r(p,0) = 3. \tag{4}$$

The simplicity of these examples might be expected to limit their usefulness but, in fact, the bounds (1) to (4) are very often best possible. The known cases in which they can be improved are discussed in the next section.

<u>Computer Search</u>. The bound of (1) is obviously not always best possible since its value $r(5,5) \geq 4$ may be improved to 5 by the pair of five-point graphs shown in Figure 2. There are three other such pairs of five-point graphs.

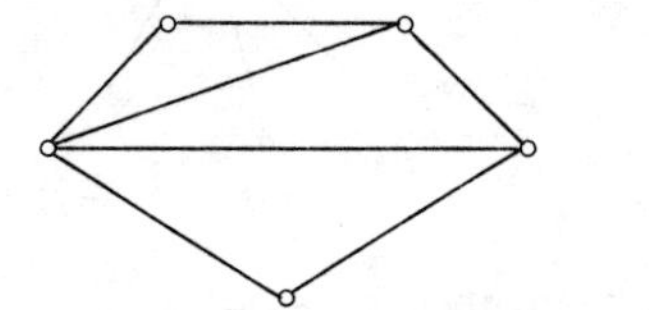
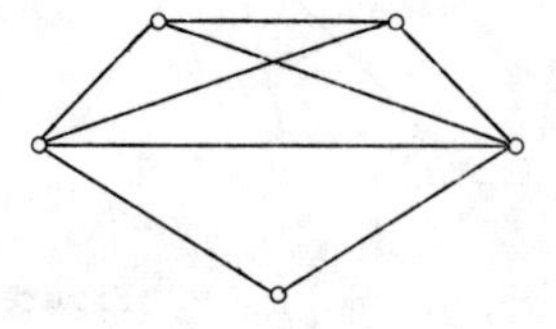

Figure 2. Graphs with four common maximal subgraphs.

For six- and seven-point graphs, the job of finding pairs sharing many subgraphs becomes tedious. Using a program written by D. P. Geller, an IBM 1800 found the unique pair of six-point graphs with five common subgraphs shown in Figure 3. This example shows that $r(6,6) \geq 6$. Obviously the two graphs of Figure 3 are complements of each other. Incidentally, if these two graphs were not complementary, they could not constitute a unique pair as their complements would also have five common subgraphs.

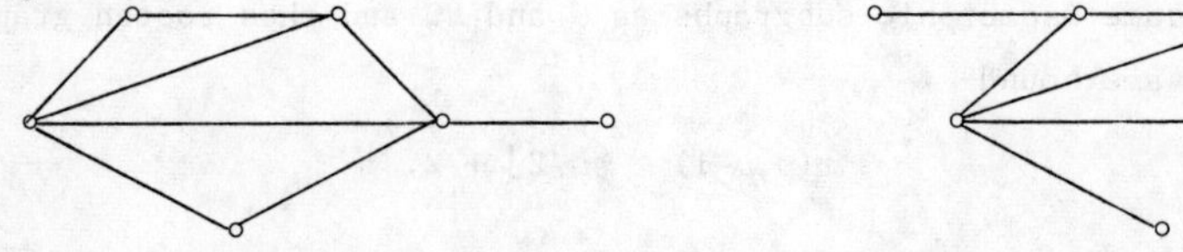

Figure 3. The pair of six-point graphs with five common subgraphs.

A computer search of the seven-point graphs utilizing the same program produced several pairs of graphs with five common subgraphs, including the one

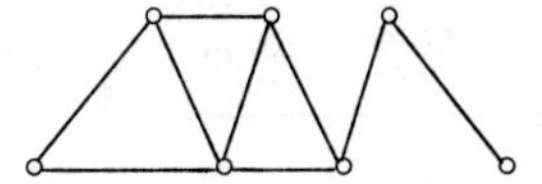
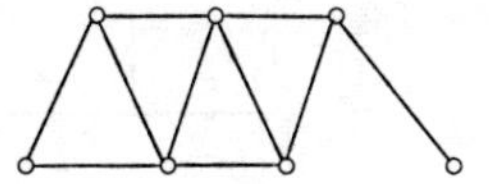

Figure 4. Seven-point graphs with five common subgraphs.

shown in Figure 4. Thus we see that $r(7,7)$ is at least 6. If a labeled isolated point is added to each graph of Figure 3, we obtain an example which shows that $r(7,6) \geq 6$.

These special examples do not seem to generalize in any way. Since a search of 8-point pairs is out of the question at present, we must be satisfied with the lower bounds found so far; see Table 2. The entries which were improved by the special examples in Figures 2-4 are circled.

Table 2. Lower bounds for $r(p,n)$

p \ n	0	1	2	3	4	5	6	7	8	9
3	3	3	3	3						
4	3	3	3	4	4					
5	3	3	3	4	4	(5)				
6	3	3	3	4	4	5	(6)			
7	3	3	3	4	4	5	(6)	(6)		
8	3	3	3	4	4	5	5	6	6	
9	3	3	3	4	4	5	5	6	6	7
	.	.	.	.	.	.	.	.	.	.
	.	.	.	.	.	.	.	.	.	.
	.	.	.	.	.	.	.	.	.	.

Upper bounds. Our computer search confirmed earlier rumors by Kelly [12] and Harary [6] that the R.C. is true through seven-point graphs, so that

$$r(p,p) \leq p, \qquad 3 \leq p \leq 7. \tag{5}$$

Furthermore, since no point of the two graphs of Figure 3 can be labeled without reducing the number of isomorphic subgraphs below five, we see that $r(6,5) \leq 5$. For seven-point graphs, the computer found no pair with six common subgraphs, so $r(7,7) \leq 6$. This result seems rather surprising until we note the relatively small number of graphs for which all maximal subgraphs are needed. The data in Table 3, which compares g_p, the number of p-point graphs, with r_p, the number of these requiring all p maximal subgraphs for reconstruction, tend to indicate that the R.C. is not sharp for $p \geq 7$.

Table 3. The sharpness of the R.C.

p	3	4	5	6	7
g_p	4	11	34	156	1044
r_p	4	9	8	1	0

Since $r(7,6)$ is at most $r(7,7)$ we see that $r(7,6) \leq 6$. Finally, by examining all labelings of two points of the graphs obtained from Figure 3 by adding an isolate to each, we find that $r(7,5) \leq 5$.

All other cases for $p \leq 7$ involve at most four unlabeled points, and are settled by the following theorems. In order to simplify our somewhat lengthy arguments, we first introduce a bit of special notation.

Let u, v, w be unlabeled points of a given graph G. Let U and L be the subgraphs of G induced by its unlabeled and labeled points respectively. We denote the set of labeled points adjacent to an unlabeled point u by $N(u)$. The points v_1, v_2, ... are deleted to form subgraphs G_1, G_2, ... The following simple lemma is used repeatedly in the proof of Theorems 1 and 2.

Lemma 1. If, in reconstructing a graph G, three labeled points v_1, v_2, and v_3 are among the deleted points and v_1 is adjacent to all (or none) of the points of U, then G is reconstructable from G_1, G_2, and G_3.

Proof. From G_2 and G_3 we can see which labeled points are adjacent to v_1. Furthermore, G_2 shows us that v_1 is adjacent to all (or no) unlabeled points. Thus G_1 may be augmented to G by inserting v_1 adjacent (not adjacent) to the points of U and the appropriate labeled points.

Theorem 1. *A graph G with at most two unlabeled points is reconstructable from any three of its subgraphs G_i, that is, $r(p,0) \leq 3$, $r(p,1) \leq 3$, and $r(p,2) \leq 3$.*

Proof. Suppose, first, that all points of G are labeled, and we have deleted v_1, v_2, and v_3. From G_1 we can read off all lines not on v_1, from G_2 those not on v_2, and from G_3, those not on v_3. Since no line is on v_1, v_2, and v_3, we have all lines and so have G.

If just one point is unlabeled, that point is thereby distinguished, and we are thus dealing again with a graph which is labeled for all practical purposes.

Finally, if there are two unlabeled points, u and v, we must distinguish three cases. If $v_1 = u$ and $v_2 = v$ are both deleted, we have L from G_1, $N(u)$ from G_2, and $N(v)$ from G_1. Since G_3 shows whether or not u and v are adjacent we have G. If $v_1 = u$, but v_2, $v_3 \neq v$, then we find L and $N(v)$ from G_1, and determine $N(u)$ and the presence or absence of the line uv from G_2 and G_3. If neither u nor v is deleted, then we have U and, by the first part of this theorem, L. If any one of the three deleted labeled points v_1, v_2, or v_3 is adjacent to neither or both of u and v, we are done by Lemma 1. Hence assume that each of v_1, v_2, v_3 is adjacent to exactly one unlabeled point. But then, in any case, u and v can be distinguished in every G_i, so we are again done by the first part of this theorem.

The following lemma is useful in proving Theorem 2.

Lemma 2. Any graph G with 3 or 4 unlabeled points can be reconstructed from the subgraphs G_i resulting from the deletion of those points only.

Proof. From these subgraphs G_i, we can easily find the subgraph L and also the labeled neighborhoods of all the unlabeled points. In order to reconstruct G, we must find the subgraph U with the neighborhoods assigned to the proper points. Thus the proof reduces to showing that a graph on 3 or 4 points, perhaps with a partial labeling, can be reconstructed from its subgraphs G_i, so as to display the labeling. But this is a straightforward exercise since there are only 14 graphs with 3 or 4 points.

We have already encountered, in the proofs of Theorem 1 and Lemma 2, situations in which an unlabeled point can be distinguished from other points in the subgraphs G_i because it is adjacent to certain labeled points. Let us agree to say, rather loosely, that a point is *essentially labeled* in G_i if it can be recognized in G_i. More generally, if a set of unlabeled points can be distinguished from all other unlabeled points but not necessarily from each other, we will call that set of points *essentially labeled*. Such points and sets of points are used repeatedly in the proof of Theorem 2.

Theorem 2. *Any graph G with p points, at most four of which are unlabeled, is reconstructable from any four of its subgraphs G_i, that is, $r(p,3) \leq 4$ and $r(p,4) \leq 4$.*

Proof. If two or fewer points are unlabeled, any three G_i will do, as shown in Theorem 1. Thus we proceed to deal with graphs with three or four unlabeled points. In almost every case, we can easily reconstruct the labeled subgraph L and find the labeled neighborhoods of the unlabeled points. The difficult step is to reconstruct U and distinguish which point receives which neighborhood.

Consider a graph with three unlabeled points u, v, and w, and delete the points v_1, v_2, v_3, and v_4. There are four cases to consider, which correspond to one, two, three, or four labeled points being deleted.

Case 1. $v_1 = u$, $v_2 = v$, $v_3 = w$. That is, all unlabeled points are among the deleted points. Then Lemma 2 shows that G can be reconstructed.

Case 2. $v_1 = u$, $v_2 = v$, but w is not deleted. From G_1 and G_2 we can find L and the sets $N(u)$, $N(v)$, and $N(w)$ of labeled points adjacent to u, v, and w, respectively. We can also decide whether both (or neither) of u and v are adjacent to w. If so, we can reconstruct G, since G_3 gives us the number of lines in U, and hence shows whether or not u and v themselves are adjacent. If exactly one of u and v is adjacent to w (say u) and there are two lines in U, then G_1 can be completed to G by inserting u adjacent to both unlabeled points and the appropriate labeled ones. But if there is only one line in U, then G_2 can be similarly completed to G by inserting v adjacent only to the labeled points in $N(v)$.

Case 3. $v_1 = u$, but v and w are not deleted. By Lemma 1, if v_2, v_3, or v_4 is adjacent to all or none of u, v, and w, we are done. So suppose each of v_2, v_3, and v_4 is adjacent to either one or two of u, v, and w. If v_2 is adjacent to v, but not to u or w, then v is essentially labeled in G_1, G_3, and G_4, so we can use Theorem 1. This is also true if v_2 is adjacent to u and w but not to v. Since similar arguments work for w, we may assume that v_2 is adjacent (non-adjacent) to v and w if and only if it is non-adjacent (adjacent) to u, and similarly for v_3 and v_4. Since we know which of v_2, v_3, and v_4 are adjacent to v and w, we can immediatel reconstruct G from G_2, since we obtain L from G_1 and v and w are distinguished from u in G_2 by their relationship to v_3.

Case 4. u, v, and w are not deleted. Again, by Lemma 2, we may assume that each of v_1, v_2, v_3, and v_4 is adjacent to just one or two of u, v, and w. If v_1 is adjacent to u and not to v or w, then u is essentially labeled in G_2, G_3, and G_4, so we can apply Theorem 1. On the other hand, if v_1 is adjacent to u and v and not to w, then w is essentially labeled in G_2, G_3, and G_4, so Theorem 1 works in this case also.

In order to handle graphs with four unlabeled points, t, u, v, and w, we must consider five cases. These are called Cases 0 - 4 corresponding to the number of labeled points deleted to form maximal subgraphs G_i.

<u>Case 0</u>. $v_1 = t$, $v_2 = u$, $v_3 = v$, $v_4 = w$ are deleted. This case is covered by Lemma 1.

<u>Case 1</u>. $v_1 = t$, $v_2 = u$, $v_3 = v$, but $w \neq v_4$. We can find the labeled neighborhoods of t, u, v, and w from G_1, G_2, and G_3. If $N(t) = N(u) = N(v) = N(w)$, we are done, since L and U are found from G_1 and G_4 respectively, and G is reconstructed. On the other hand, if the labeled neighborhoods of the unlabeled points are all different, then those neighborhoods serve to label the points in G_1, G_2, and G_3, so we can apply Theorem 1. If two of the neighborhoods match, but the others are different, we can still use Theorem 1 since we then have only two essentially unlabeled points. Thus the only remaining possibility is that among $N(t)$, $N(u)$, $N(v)$, and $N(w)$ there are exactly two different sets. We have three subcases.

<u>Case 1.1</u>. $N(t) = N(u) = N(v) \neq N(w)$. Here we have a graph with three unlabeled points (as w is essentially labeled in G_1, G_2, and G_3 because it has a different labeled neighborhood), which is reconstructable by Lemma 2.

<u>Case 1.2</u>. $N(t) = N(u) = N(w) \neq N(v)$. Then from G_3, we know both L and the subgraph induced by t, u, and w. If G_1 and G_2 both display two lines from unlabeled points to the essentially labeled point v, then t, u, and w are all adjacent to v, and we have reconstructed G. If they display one and two such lines, respectively, then t and w are adjacent to v, but u is not. Similarly, if they have one and zero such lines, respectively, then u is adjacent to v, but t and w are not. Finally, if each of G_1 and G_2 show one such line to v, then either t and u are adjacent to v and w is not, or w is and t and u are not. But since we know the number of lines in U from the subgraph G_4, we can distinguish between these possibilities. Hence in every case we can find just how t, u, and w are joined to v, and complete our picture of U and hence of G.

<u>Case 1.3</u>. $N(t) = N(u) \neq N(v) = N(w)$. Here the points of U are essentially labeled in pairs in G_1, G_2, and G_3, and a point of each pair is deleted, so we know whether or not the lines tu and vw are in G. Furthermore, from G_4 we know how many lines are in U, so we know how many lines join $\{t,u\}$ with $\{v,w\}$. If this number is 0, 1, 3, or 4, we have a complete

picture of U, which is sufficient, with our knowledge of L and the labeled neighborhoods, to reconstruct G. If there are two lines joining the pairs, we can tell from G_2 whether they are both on t (or u), and from G_3 whether they are both on v (or w). Thus we can reconstruct U and hence G.

Case 2. $v_1 = t$, $v_2 = u$, v and w are not deleted. Here again, in G_1 and G_2, we immediately can find L, and using all four G_i, we can determine $N(t)$, $N(v)$, and $N(w)$, so we need only find U with the appropriate neighborhood attached to each point. If the neighborhoods are all equal, we are done as in Case 1. If they are all different, then G_1 and G_2 give us a picture of all of G except for determining whether line tu is present, since all points are essentially labeled in them. But since G_3 shows the number of lines in U, we can tell whether tu is in G. Thus there remain just four subcases involving two or three different N-neighborhoods, which we treat very briefly.

Case 2.1. If there are only two different N-neighborhoods, we use G_3 or G_4 to distinguish the points of U as follows: For $N(t) = N(u) = N(v) \neq N(w)$, w is clearly indicated by G_3 (or G_4) and if $N(t) \neq N(u) = N(v) = N(w)$, then the same is true of t. When, on the other hand, $N(t) = N(u) \neq N(v) = N(w)$, or $N(t) = N(v) \neq N(u) = N(w)$, the pairs are distinguished in either G_3 or G_4, which is all we need to find G.

Case 2.2. $N(t) = N(u) \neq N(v) \neq N(w)$ and $N(t) \neq N(w)$. This is easy, since G_1 and G_2 are essentially labeled, and G_3 tells us whether or not t and u are adjacent by revealing the number of lines in U.

Case 2.3. $N(t) \neq N(u) = N(v) \neq N(w)$ and $N(t) \neq N(w)$. Here G_1 tells us whether line uv is present, and G_2 gives this information about lines tv and tw. Since G_1 also shows the number of lines in $U - t$, and we know the number of lines in U from G_3, we know whether t and u are adjacent. If neither or both of the lines uw and vw are present, we are done. If there is one such line and only one line from t to u or v, then, ignoring tw and uv, if they are present, U is connected or not according as the line from w is incident with the same point as the single line on t. If t is adjacent to neither or both of u and v, the position of the line on w is irrelevant.

Case 2.4. $N(t) \neq N(u) \neq N(v) = N(w)$ and $N(t) \neq N(v)$. From G_1 we can see whether v and w are adjacent, find the number of lines in $U - t$, and discern L. From G_2 we can find the number of lines in $U - u$. If u (or t) is adjacent to both v and w or neither, that information is sufficient to reconstruct G. So suppose exactly one of the lines uv, uw and one of tv, tw are present. Then uv and tv (or uw and tw) are present if U is either $P_3 \cup K_1$, $K_3 \cup K_1$, $\overline{P_3 \cup K_1}$, or $\overline{K_3 \cup K_1}$, and otherwise we have the lines uv and tw or tv and uw. Since we know the number of lines in U, we know whether t and u are adjacent, and can reconstruct G.

Case 3. $v_1 = t$, but u, v, and w are not deleted. If any one of the points v_2, v_3, and v_4 is adjacent to all or none of the points t, u, v, and w, we are done by Lemma 1, so assume otherwise. From G_1 we have $N(u)$, $N(v)$, $N(w)$, and hence from G_2 and G_3, we can find $N(t)$ also. Also from G_1 we know $U - t$. If $U - t$ is point-symmetric, the number of lines in U, formed from G_2, suffices to complete our picture of G. If $U - t$ is $K_2 \cup K_1$, and there is a line from t to just one of the points u, v, and w, then there is a problem if that line goes to a point, say u, on the unique line uv of $U - t$. But then the line can be recognized as going to u instead of v in at least one of G_2, G_3, and G_4 if, in fact, u and v are not similar. Since all other possibilities are exactly analogous, we have G in any case.

Case 4. None of t, u, v, and w are deleted. By Theorem 1 we know L, so we are seeking U and the connections between L and U. Once again, Lemma 1 tells us that we may as well assume that each of v_1, v_2, v_3, and v_4 is adjacent to one, two, or three unlabeled points. In fact, adjacency to three points and adjacency to one point are such analogous situations that we may as well assume that each labeled point is adjacent to one or two unlabeled points. For all points $x \in U$, let $N^*(x) = N(x) - \{v_1, v_2, v_3, v_4\}$. Note we have the sets N^* from any one of the subgraphs G_i. If $N^*(t) \neq N^*(x)$ for $x = u, v, w$, then t is essentially labeled in G_1, G_2, G_3, and G_4 and so we can refer to the first part of this proof, which reconstructs graphs with three unlabeled points. Thus we have two subcases, depending on whether or not there are two sets of N^* neighborhoods.

Case 4.1. $N^*(t) = N^*(u) \neq N^*(v) = N^*(w)$. If v_1 is adjacent to t and u, then we can recognize that fact from G_2 and reconstruct G using G_1. If v_1 is adjacent to t and v and v_2 goes to one of t and u and one of v and w, then this can be seen in G_3, and using that information, G can be reconstructed from G_1. If v_1 is adjacent to t and v, and each of v_2, v_3, and v_4 is adjacent to exactly one of the four unlabeled points, then by using those three lines, we can use G_1 to find G. Finally, suppose each of v_1 through v_4 is adjacent to exactly one unlabeled point. If each is adjacent to a different point, any G_i gives G. But if v_1 and v_2 are adjacent to the same point, then G is reconstructable from G_1 (or G_2).

Case 4.2. $N^*(t) = N^*(u) = N^*(v) = N^*(w)$. Each deleted labeled point is adjacent to one or two unlabeled points. Suppose first that every deleted labeled point is adjacent to exactly two unlabeled points. Then two of them must be adjacent to the same pair or two must be adjacent to complementary pairs. In either case, it is easy to find G.

Suppose, then, that v_1 is adjacent to only one unlabeled point, t. From G_2 and G_3 we can find $N(t)$ and we have G from G_1 unless $N(t) - v_1 = N(u)$, say. Since no labeled point is adjacent or non-adjacent to all of the unlabeled ones, we cannot have $N(t) - v_1 = N(u) = N(v) = N(w)$. There are only two other possibilities.

If $N(t) - v_1 = N(u) = N(v) \neq N(w)$, then since every labeled point is adjacent to some unlabeled one, w must be adjacent to some deleted labeled point, say v_4, which is non-adjacent to all of u, v, and w. Thus t and w are both essentially labeled in G_2 and G_3 and, with the other information we have, we can easily find G.

If $N(u)$, $N(v)$, and $N(w)$ are unequal in paris, then we claim that in G_2, G_3, or G_4, u, v, and w can be distinguished. For if u and v are confused in G_2 then $N(u) = N(v) + v_2$ (or $N(v) - v_2$) so they cannot be confused in G_3 or G_4. Thus if we have u and v confused in G_2, and v and w confused in G_3, the only possible pair for confusion in G_4 is u and w. But $N(u)$ and $N(w)$ differ only in v_2 and v_3, so there is no confusion in G_4. Thus in every case G can be reconstructed.

Summary. The known values of $r(p,n)$, derived in the last three sections, are collected in Table 4. These data, together with Table 3, certainly lend additional plausability to the R.C!

Table 4. Known values of $r(p,n)$

p \ n	0	1	2	3	4	5	6	7
3	3	3	3	3				
4	3	3	3	4	4			
5	3	3	3	4	4	5		
6	3	3	3	4	4	5	6	
7	3	3	3	4	4	5	6	6
⋮	↓	↓	↓	↓	↓			

In the light of this table, we return to Table 1 for which we have now proved the values of $r(p,n)$ for $n < 5$. We may therefore write a revised table in which it is assumed that p is sufficiently large that $r(p,n)$ no longer varies with p.

Table 1'. Values of $r(p,n)$

	KNOWN					CONJECTURED			
n	0	1	2	3	4	5	6	7 ...	p
$r(p,n)$	3	3	3	4	4	5	6	7	p

It is appropriate to close with some related problems. We have treated the R.C. for graphs with small numbers of unlabeled points. A more interesting but also more difficult problem is the reconstruction of graphs with only a few labeled points. The reconstruction of arbitrary graphs with one or two labeled points would certainly be a major breakthrough.

In speaking about labeled and partially labeled graphs, we have been careful to assume that the labels are distinct. If this assumption is dropped, we have, instead of a labeling, a *coloring* of the points. One should not confuse this with a *proper coloring* in which adjacent points must receive different colors.

Coloring Reconstruction Conjecture. Every graph G with $p \geq 3$ points in which the points have been colored using $n \leq p$ colors can be reconstructed from its subgraphs G_i.

For $n = p$, this reduces to the case of labeled graphs which is settled, and clearly for $n = p - 1$, $p - 2$, and so on, G should be easy to reconstruct. But for $n = 1$, we again have the R.C. which appears hopeless. Similar, presumably easier, problems arise if one assumes proper colorings of the points.

Further, we must raise the basic question of whether partial labeling of the points actually might make reconstruction more difficult. Suppose, for example, a graph G has point v labeled. The presence of v in the subgraphs G_i will undoubtedly assist us in reconstructing G, but if we are required to reconstruct G with v labeled we may have trouble since, having reconstructed G, we must put v in its proper place. We will always be able to do this, unambiguously, if and only if the following conjecture is true. This asserts that if two points look the same in corresponding subgraphs G_i, they must be similar in G.

Labeled Point Conjecture. Suppose that for a graph G with points v_1 and v_2, $G_1 \simeq G_2$ and there exists a one-to-one correspondence $\tau:\{2,3,4,\ldots,p\} \to \{1,3,4,\ldots,p\}$ and isomorphisms θ_i between G_i and $G_{\tau i}$ such that for all i, $\theta_i(v_1) = v_2$. Then there must exist an automorphism of G sending v_1 to v_2.

The much simpler assumption $G_1 \simeq G_2$ is not sufficient to show that v_1 is similar to v_2, as shown by Harary and Palmer [8,9]. Anyone who feels that this conjecture is obviously true should notice that a proof would show that a graph consisting of a homeomorphically irreducible block together with a single protruding endline can be reconstructed. Such graphs have been excluded from consideration by various investigators (Bondy [3] and Manvel [15]) as being especially difficult to reconstruct. The R.C. does not imply the Labeled Point Conjecture in any instant fashion.

We conclude with two conjectures dealing with $r(p,p)$, due to D. Kleitman

Conjecture 1. Whenever $p \geq n \geq 4$, $r(p,n) = r(n,n)$.

Note that a proof of this conjecture would enable us to extend the last three columns of Table 4 down indefinitely.

Conjecture 2. The exact values of $r(p,n)$ may be expressed in terms of $r(n,n)$:

$$r(p,n) = \begin{cases} r(n,n) & n \text{ even} \\ \max\{r(n,n), \frac{n+5}{2}\} & n \text{ odd}, p \geq 5. \end{cases}$$

It is already known by (3) that this is a lower bound for $r(p,n)$.

References

1. Beineke, L. W. and Parker, E. M., A six point counterexample to the reconstruction of strongly connected tournaments. *J. Combinatorial Theory*, (to appear).

2. Bondy, J. A., On Kelly's congruence theorem for trees. *Proc. Cambridge Philos. Soc.* 65 (1969) 387-397.

3. Bondy, J. A., On Ulam's Conjecture for separable graphs. *Pacific J. Math.* (to appear).

4. Geller, D. P. and Manvel, B., Reconstruction of cacti. *Canad. J. Math.* (to appear).

5. Greenwell, D. L. and Hemminger, R. L. Reconstructing graphs. *The Many Facets of Graph Theory.* (G. T. Chartrand and S. F. Kapoor, eds.) Springer-Verlag, New York, 1969.

6. Harary, F., On the reconstruction of a graph from a collection of subgraphs. Theory of Graphs and its Applications (M. Fiedler, ed.) Prague, 1964, 47-52; reprinted, Academic Press, New York, 1964.

7. Harary, F. and Palmer, E. M., On the problem of reconstructing a tournament from subtournaments, *Monat. für Mathematik*, 71 (1967) 14-23.

8. Harary, F. and Palmer, E. M., A note on similar points and similar lines of a graph. *Rev. Roum. Math. Pures et Appl.* 10 (1965) 1489-1492.

9. Harary, F. and Palmer, E. M., On similar points of a graph. *J. Math. Mech.* 15 (1966) 623-630.

10. Harary, F. and Palmer, E. M., The reconstruction of a tree from its maximal proper subtrees. *Canad. J. Math.* 18 (1966) 803-810.

11. Hemminger, R. L., On reconstructing a graph. *Proc. Amer. Math. Soc.* (to appear).

12. Kelly, P. J., A congruence theorem for trees. *Pacific J. Math.* 7 (1957) 961-968.

13. Manvel, B., Reconstruction of unicyclic graphs. Proof Techniques in Graph Theory (F. Harary, ed.) Academic Press, New York, 1969.

14. Manvel, B., Reconstruction of trees. *Canad. J. Math.* (to appear).

15. Manvel, B., On reconstruction of graphs. The Many Facets of Graph Theory (C. T. Chartrand and S. F. Kapoor, eds.) Springer-Verlag, New York, 1969.

16. Manvel, B. and Stockmeyer, P. K., On reconstruction of matrices, *SIAM Rev.* (submitted).

17. Ulam, S. M., A Collection of Mathematical Problems. Wiley (Interscience), New York, 1960, p. 29.

HOW CUTTING IS A CUT POINT?

Frank Harary

University of Michigan, Ann Arbor, MI, U.S.A.

and

Phillip A. Ostrand

University of California, Santa Barbara, CA, U.S.A.

The *cutting number* $c(v)$ of a point v of a connected graph G is the number of pairs of points $\{u,w\}$ of G such that $u, w \neq v$ and every $u - w$ path contains v. Obviously $c(v) > 0$ if and only if v is a cut point. For a graph G of connectivity 1 (see [1] for definitions), let $c = c(G)$ be the maximum cutting number of a point. Then the *cutting center* of G is the set of all points v such that $c(v) = c(G)$.

It is easy to see that every tree T with $p \geq 3$ points has a cutting center which induces a connected subgraph, i.e., a subtree.

Let n be the number of points in the cutting center of T and call $c = c(T)$ the *cutting number* of T.

Theorem 1. *For any tree T with $p \geq 3$ points, the cutting center of T is a path.*

In order to prove this assertion, it is sufficient to verify numerically that no tree T contains a subtree of the form $K_{1,3}$ in which all 4 points u, v_1, v_2, v_3 have the same cutting number. For this purpose, it is convenient to include a modest lemma.

Lemma. Let u, v and w be three points of T with equal cutting numbers and uv, vw lines of T such that in $T - uv - vw$, the components containing u, v, w have a, b, c points respectively. Then $2b \leq a$ and $2b \leq c$.

Proof of Lemma. Let a^* be the cutting number of u in its component, and let b^*, c^* be defined similarly.

Note that $0 \leq a^* \leq \binom{a-1}{2}$ so that $2(a^*+a) < (a+1)^2$. Without loss of generality, suppose $c \leq a$. Obviously in T we find the cutting numbers

$$c(u) = a^* + (a-1)(b+c) \tag{1}$$

$$c(v) = b^* + (b-1)(a+c) + ac \tag{2}$$

$$c(w) = c^* + (c-1)(a+b). \tag{3}$$

By hypothesis $c(u) = c(v)$ and $c(v) = c(w)$, so that

$$a^* + a = b^* + (c+1)b \tag{4}$$

$$c^* + c = b^* + (a+1)b. \tag{5}$$

Since $b^* \geq 0$, we see by (4) that

$$2(c+1)b < (a+1)^2 \tag{6}$$

and similarly by (5) that

$$2(a+1)b < (c+1)^2, \tag{7}$$

so that

$$4b^2 < (a+1)(c+1). \tag{8}$$

Thus our supposition that $c \leq a$ yields

$$4b^2 < (a+1)^2 \tag{9}$$

and we have

$$2b \leq a. \tag{10}$$

By (7) and (10) we obtain at once

$$4b^2 < 2b(a+1) < (c+1)^2, \text{ whence } 2b \leq c.$$

Proof of theorem. Consider a tree T containing a subtree $K_{1,3}$ (see Figure 1) in which the four points have equal cutting numbers in T.

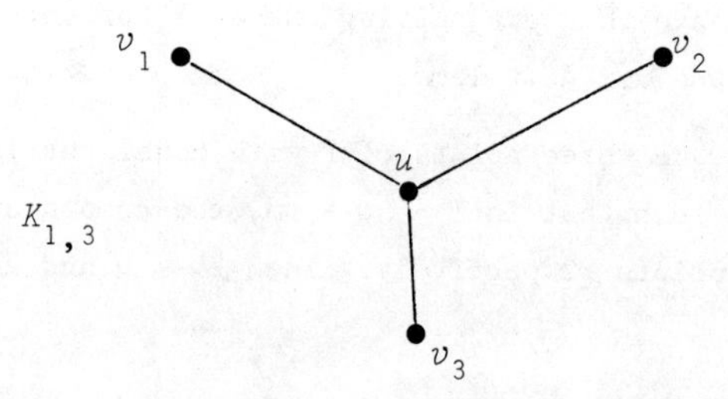

Figure 1. An impossible cutting center

In the graph obtained from T by removing the lines of $K_{1,3}$ each of the four points v_1, v_2, v_3, u lies in a different component; let these contain a, b, c, d points respectively.

Again without loss of generality, let $a \leq b \leq c$. Upon applying the lemma to T and regarding v_1, u, v_2 as u, v, w, we have $2(c+d) \leq b \leq c$. But this contradicts $c > 0$ and completes the proof of the theorem.

The strongest possible assertion subject to the restriction imposed by Theorem 1 can now be made!

Theorem 2. *For every positive integer n, there exists a tree T with a cutting center consisting of n points on a path.*

The proof is too long for inclusion here, and will eventually appear elsewhere.

The trees which we believe to have the smallest number p of points with cutting center containing $n \leq 5$ points are shown in Figure 2, while Table 1 contains the information p, n, and $c(T)$ for each tree in the figure. In general, it is an unsolved problem to determine this minimum p, given n.

Table 1

n	p	c
1	3	1
2	4	2
3	7	9
4	10	20
5	50	670

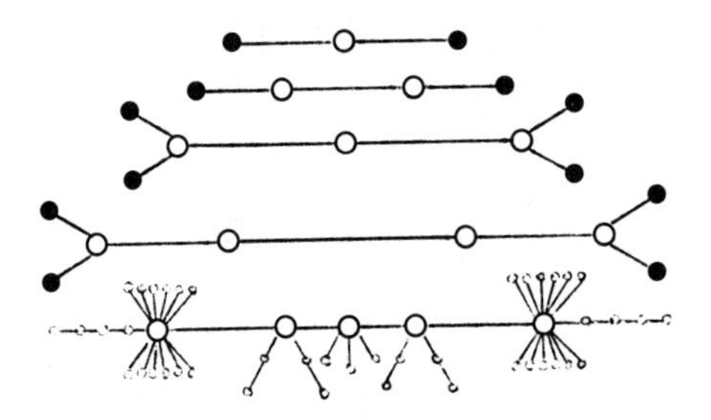

Figure 2

Reference

1. Harary, F., Graph Theory, Addison-Wesley, Reading, 1969.

NUMBER OF INTERSECTIONS OF DIAGONALS IN REGULAR n-GONS

Heiko Harborth

Technische Universität Braunschweig, Germany

If a regular n-gon with all its diagonals is drawn in the plane, then the questions arise, what are the numbers of originating regions $F(n)$, edges $K(n)$, and vertices $E(n)$. It can be shown, that:

$$F(n) = \binom{n-1}{2} + \sum_{\nu=2}^{[n/2]} (\nu-1)E_\nu(n); \quad K(n) = \binom{n}{2} + \sum_{\nu=2}^{[n/2]} \nu E_\nu(n);$$

$$E(n) = n + \sum_{\nu=2}^{[n/2]} E_\nu(n),$$

where $E_\nu(n)$ denotes the number of vertices being intersections of exactly ν diagonals and $[x]$ means the greatest integer not exceeding x.

For odd n it was conjectured [1], that there are no concurrences of more than two diagonals (excluded the corners of the n-gon). This has been proved in [2] (for odd primes) and in [3]. Thus for $n \equiv 1 \pmod 2$ the numbers are

$$F(n) = \binom{n-1}{2} + \binom{n}{4}; \quad K(n) = \binom{n}{2} + 2\binom{n}{4}; \quad E(n) = n + \binom{n}{4}.$$

If n is even then $\frac{n}{2}$ diameters cross each other in the center, so that $E_2(n) \lesseqgtr \binom{n}{4} - \frac{n(n-2)}{8}$. If in addition n is not divisible by 3, in [4] and [5] it was independently shown, that no concurrences of more than three diagonals occur (except center and corners). In [5] the numbers in question are determined; here they follow from the above formulas with:

$$E_3(n) = E_3^{(1)}(n) + E_3^{(2)}(n) = \begin{cases} \dfrac{n(n-2)(n-6)}{16} + \dfrac{n(n-2)(n-10)}{24} \\[2ex] \dfrac{n(n-4)^2}{16} + \dfrac{n(n-4)(n-8)}{24} \end{cases}$$

according as $n \equiv \pm 2$ or $\pm 4 \pmod{12}$. From these triple intersections $E_3^{(1)}(n)$ are located on the diameters and the remaining $E_3^{(2)}(n)$ are distributed on circular lines with a corner of the n-gon as center and the distances to the other corners as radii. In the case $n \equiv \pm 2 \pmod 6$ $E(n)$ is so diminished, that $E(n-3) < E(n) < E(n-1)$ holds for $n \geq 20$.

For $n \equiv 0 \pmod 6$ the numbers become again smaller, but are not known till now. If the corners of the regular n-gon are labelled with natural numbers, and (x,y) denotes the diagonal joining the corners x and y, then $(a+c, 2b+c)$, $(2a+c, b+c)$, and $\left(c, a+b+c+\frac{n}{2}\right)$ cross each other in one point for all even n. However

$$\left(a+b, -a+b-\frac{n}{6}\right), \left(-a+b, b+\frac{n}{6}\right), \left(b, -2a+b+\frac{n}{2}\right) ;$$

$$\left(-a+b+\frac{n}{6}, -\frac{a}{4}+b-\frac{n}{24}\right), \left(-\frac{a}{2}+b+\frac{n}{12}, -\frac{a}{4}+b-\frac{5n}{24}\right), \left(b, -\frac{a}{2}+b+\frac{n}{4}\right) ;$$

$$\left(a, a+\frac{2n}{5}\right), \left(a+\frac{n}{10}, a-\frac{n}{3}\right), \left(a+\frac{n}{3}, a-\frac{n}{30}\right) ;$$

$$\left(a, a+\frac{13n}{42}\right), \left(a+\frac{n}{6}, a-\frac{n}{7}\right), \left(a+\frac{3n}{14}, a-\frac{n}{14}\right)$$

are only 4 examples of existing further triples of diagonals having a point in common. These triples may concur, so that intersections of more than three diagonals occur. This can be seen for example from the following numbers of vertices being exact: $E(6) = 13$, $E(12) = 301$, $E(18) = 1837$, $E(24) = 7321$, and $E(30) = 18151$, where $E_3(30) = 1590$, $E_4(30) = 420$, $E_5(30) = 180$, $E_6(30) = 120$, and $E_7(30) = 30$.

References

1. Steinhaus, H.: Problem 225. *Colloq. Math.* 5(1958).

2. Croft, H.T. and Fowler, M.: On a Problem of Steinhaus About Polygons. *Proc. Cambridge philos. Soc.* 57, 686-688 (1961).

3. Heineken, H.: Regelmäbige Vielecke und ihre Diagonalen. *Enseignement math.*, II. sér. 8, 275-278 (1962).

4. Heineken, H.: Regelmäbige Vielecke und ihre Diagonalen II. *Rendiconti Sem. U. Padova* 41, 332-344 (1968).

5. Harborth, H.: Diagonalen im regulären *n*-Eck. *El. Math.* (to appear).

THE NUMBER OF SOLUTIONS OF $\alpha^{j+p} = \alpha^j$ IN SYMMETRIC SEMIGROUPS

Bernard Harris
Mathematics Research Center, University of Wisconsin, Madison, WI, U.S.A.

Lowel Schoenfeld
State University of New York, Buffalo, NY, U.S.A.

1. Introduction. Let $X_n = (x_1, x_2, \ldots, x_n)$ be a set of n distinct elements and let T_n be the set of all mappings of X_n into X_n. T_n is a semigroup with identity under composition. This semigroup is called the symmetric semigroup on n elements. Since any two finite sets X, Y with n elements have isomorphic symmetric semigroups, there is no loss of generality in choosing $X_n = \{1,2,\ldots,n\}$.

It is convenient to identify $\alpha \in T_n$ with the functional digraph constructed in the following manner. For every $x \in X_n$, draw the directed arc from x to $\alpha(x)$. It is easily seen that there is a one-to-one correspondence between the functional digraphs obtained in this manner and the elements of T_n.

In this paper we will give a combinatorial characterization of the set:

(1) $$B_{jpn} = \{\alpha \in T_n : \alpha^{j+p} = \alpha^j\},$$

where p is a prime and $j \geq 0$ is an integer. We also give an enumerating formula for B_{jpn} and the exponential generating function of U_{jpn} which is the number of elements in B_{jpn}.

The space limitations imposed in this volume prevent our providing details of the proofs of these results and necessitate the omission of the many references that we would wish to cite. The details of proofs will be produced in a forthcoming paper, which will contain the results

---

Sponsored by the Mathematics Research Center, United States Army, Madison, Wisconsin, under Contract No. DA-31-124-ARO-D-462.

stated here along with other closely related results.

2. The Enumeration of B_{jpn}. It is desirable to state some related results first. Let S_n be the symmetric group on n elements and let ε be the identity element of this group. Then the number of solutions of the equation $\beta^p = \varepsilon$, $\beta \in S_n$ is known to be given by

$$(2) \qquad W_{pn} = \sum_{i=0}^{\left[\frac{n}{p}\right]} \frac{n!}{i!\, p^i (n - pi)!}$$

and the corresponding exponential generating function is

$$(3) \qquad \psi_{0,p}(z) = \sum_{n=0}^{\infty} W_{pn}\, z^n/n! = e^{z+z^p/p} \quad .$$

Now let U_{jn} be the number of solutions of $\alpha^{j+1} = \alpha^j$, $\alpha \in T_n$. In a forthcoming paper, the authors show that the generating function $\phi_j(z)$ satisfies

$$(4) \qquad \phi_j(z) = \sum_{n=0}^{\infty} U_{jn}\, z^n/n! = e^{z\phi_{j-1}(z)}, \quad j \geq 1,$$

and

$$(5) \qquad \phi_0(z) = e^z.$$

In terms of functional digraphs, the solution set of the equation $\alpha^{j+1} = \alpha^j$, $\alpha \in T_n$, coincides with the set of all functional digraphs on n vertices, all of whose cycles are loops and such that the longest directed path from every acyclic vertex to the cycle has length not exceeding j. Interpreting the loops as roots, we can see that the enumeration of this set is precisely the enumeration of forests of rooted labeled trees on n vertices of height not exceeding j.

In a similar manner, it can be shown that B_{jpn} is equivalent to the set of functional digraphs on n vertices, all of whose cycles are of length either p or 1, and such that the longest directed path from any acyclic vertex to the cycle has length not exceeding j. These observations facilitate the enumeration of B_{jpn}. In particular, define $U_{jpo} = 1$; then

$$(6) \quad U_{jpn} = \sum_{k_i \geq 0, \Sigma k_i = n} \frac{n!}{k_0! k_1! \ldots k_j!} \sum_{i=0}^{\left[\frac{k_0}{p}\right]} \frac{k_0!}{i!\, p^i (k_0 - pi)!}\, k_0^{\,k_1} k_1^{\,k_2} \ldots k_{j-1}^{\,k_j}.$$

For $j \geq 1$, let $\psi_{jp}(z) = \sum_{n=0}^{\infty} U_{jpn} z^n/n!$, let

$$\tau_{jp}(z) = e^{[z\phi_{j-1}(z)]^p/p}, \tag{7}$$

and let

$$\tau_{0p}(z) = e^{z^p/p} .$$

Then the following can be established:

$$\psi_{jp}(z) = \phi_j(z)\, \tau_{jp}(z). \tag{8}$$

SOME RESULTS ON COMBINATORIAL SET THEORY

E. Harzheim

Köln, Germany

A well known theorem of Kuratowski/Birkhoff states: If S is an arbitrary set, $R(S)$ the set of all subsets of S, then $R(S)$ taken with the inclusion order $\subset$ is $|S|$-universally ordered: Every partially ordered set M of cardinality $|M| \leqq |S|$ is order-isomorphic to a subset of $R(S)$.

In this connection the following combinatoric question arises: If $R(S)$ is divided into (not too many) classes, does one of these still have the property of being an $|S|$-universally ordered set?

I proved the following.

Theorem 1. *If $\aleph_\alpha$ is a regular cardinal, $|S| = R_\alpha$ $(= \Sigma 2^{\aleph_\nu} | \nu < \alpha$, which equals $\aleph_\alpha$ when the GCH is adopted) and if $R(S)$ is divided into $\aleph_\alpha$ many classes, at least one of these is $\aleph_\alpha$-universal.*

This theorem can be regarded as sharp, for if the number of classes is $\aleph_{\alpha+1}$ and if we adopt the GCH, then we have $|R(S)|$ many classes.

And even if we take only two classes there doesn't follow essentially more because of

Theorem 2. *(with GCH): $R(S)$ can be divided into two classes, A, B such that neither A nor B has a subset of the same order-type as $R(S)$.*

For the case where $\aleph_\alpha$ is not restricted to be a regular cardinal, I proved:

Theorem 3. *If $|S| = R_\alpha$ and if $R(S)$ is divided into $\aleph_\alpha$ many classes, then there is at least one class K with the following property: For every ordinal number $\tau < \omega_{\alpha+1}$ there is a subset of K of order-type $\tau^* + \tau$ (τ^* is the inverse of τ).*

The proof of Theorem 3 makes use of several notions: If we have a (possibly transfinite) sequence s of numbers 0 and 1 (a so called dyadic sequence) we define the change number of s ($ch(s)$) as the ordinal number corresponding to the set of maximal blocks of zeros and ones in s. For instance the change number of 0011001100...111 is $\omega_0 + 1$, and the change number of 11100111 is 3.

Further $2((\omega_\alpha))$ shall denote the set of all dyadic sequences of length ω_α. It is totally ordered according to the principle of first differences.

The subset of $2((\omega_\alpha))$ which consists of all those sequences which have a last digit 1 is denoted by R_α. (For $\alpha = 0$ R_α is similar to the set of rational numbers in their usual order).

A splitting system of a totally ordered set M is a subset S of $R(M)$ obtained in the following way: We split M into a nonvoid initial segment $S_0 \neq M$ and its complementary final segment $S_1 = M - S_0$. Analogously we split S_0 into segments S_{00}, S_{01} and S_1 into segments S_{10}, S_{11} and so on (as far as possible). Each (transfinite) dyadic sequence $\alpha_0, \alpha_1, \ldots,$ of maximal length for which $S_{\alpha 0}$, $S_{\alpha 0 \alpha 1}$, ... are defined, is called a dyadic sequence occurring in the splitting system S. We say

$ch(M) \lesssim \xi$ iff there exists a splitting system S of M such that

$ch(s) \lesssim \xi$ for every dyadic sequence s occurring in S.

The key for the proof of Theorem 3 is the following.

Lemma 1. Let $M = \bigcup M_v | v < \omega_e$ be a totally ordered set and $ch(M_v) \lesssim \omega_e$ for all $v < \omega_e$. Then there holds also $ch(M) \lesssim \omega_e$.

Lemma 2. ω_α be a regular initial ordinal. Then in every splitting system S of R_α there occurs a dyadic sequence s with $ch(s) \gtrsim \omega_\alpha$.

Lemma 3. Let $M \supset N$ be totally ordered. If in every splitting system of N there occurs a dyadic sequence s with $ch(s) \gtrsim \omega_\alpha$ then the same is true for M.

The notion of change number is related to the notion of scattered sets. A totally ordered set M is called scattered, if it has no dense subset (the same: if it has no subset of type $\eta_0 = \text{tp } R_0$).

Then we can characterize the scattered sets by means of the change number as follows:

Theorem 4. *A totally ordered set M is scattered iff there exists a splitting system of M such that all occurring dyadic sequences s have finite change number $ch(s)$.*

Corollary. If M is scattered, it is embeddable into the lexicographic product $(\tau^* + \tau) \times (\tau^* + \tau) \times \ldots$ of ω_0 many factors, where τ is the least ordinal, such that neither τ nor τ^* is a subtype of tp (M).

In this connection the following change-number-hypothesis seems to be of interest:

(CNH): If M is totally ordered and if in every splitting of M there occurs a sequence s with $ch(s) \geqq \omega_\alpha$, then M has a subset of type $\eta_\alpha = \text{tp } R_\alpha$.

For $\alpha = 0$ this is true (as follows from theorem 4) and the converse statements of (CNH) is also true for regular ω_α (take Lemma 3 and 2).

If (CNH) were true, then Theorem 3 could be stated for every total order type τ with $|\tau| \leqq \aleph_\alpha$ (not only for ordinals τ).

Further (CNH) would give a positive answer to the following

Problem. (with GCH) If $\aleph_\alpha$ is regular, M a totally ordered set of cardinality $\aleph_{\alpha+1}$ which has no subsets of type $\omega_{\alpha+1}$, $\omega^*_{\alpha+1}$, then M has a subset of order type η_α?

If we adopt the GCH then this problem turns out to be equivalent to the following weakened form (CNH)' of (CNH):

(CNH)' : If M is totally ordered, $|M| = \aleph_{\alpha+1}$, $\aleph_\alpha$ regular, $\omega_{\alpha+1}$ and $\omega^*_{\alpha+1}$ no subtypes of tp (M), and if in every splitting system of M there occurs a dyadic sequence s with $ch(s) \geqq \omega_\alpha$, then M has a subset of type η_α.

A quite different result in combinatorial set theory (on which I referred already in July, 1968 in Oberwolfach) is the following.

Theorem. *If $(S, \leqq)$ is a partially order set, n a natural number and if each finite subset of S has the Dustmik-Miller-dimension $\leqq n$, then also S has dimension $\leqq n$.*

My proof used the theory of η_α-sets and some topology. Subsequently two of my colleagues gave other proofs, B. Koppelberg by using the compactness theorem and H.A. Jung by using a theorem of R. Rado.

DISCONNECTED-COLORINGS OF GRAPHS

Stephen Hedetniemi

University of Iowa, Iowa City, IA, U.S.A.

A *disconnected-coloring* (or *D-coloring*) of a graph $G = (V,E)$ is a partition $\pi = \{V_1, V_2, \ldots, V_n\}$ of V such that for every i, the subgraph induced by the subset V_i is disconnected. The *D-chromatic number* $\chi_d(G)$ of G is the smallest number of color classes (subsets) in any D-coloring of G. In spite of the unnaturalness of the concept of disconnected colorings, more than two dozen results have been constructed for D-colorings and χ_d which are virtually identical to corresponding results that have been established for the traditional colorings and chromatic number $\chi(G)$ of graphs. These new results indicate that the established results reveal much less about properties of colorings than they do about concepts which are much more general; they also reveal that most of the established results on coloring can be proved using little more than purely set theoretic arguments. A conclusion that one can easily draw from this is that we really have not obtained many results of any significance about the concept of coloring a graph.

The following list contains a series of results about colorings of graphs, on the left, and for each an analogous result for D-colorings, on the right. For completeness we also include a brief summary of definitions.

$G = (V,E)$,	a graph G with point set V and line set E.
p,	number of points of G, $p = \lvert V \rvert$.
α_0,	point covering number; minimum number of points needed to cover all the lines of G.
β_0,	point independence number; maximum number of points in a set, no two of which are adjacent.

α_d,	minimum number of points in a set S such that $G - S$ is disconnected.
κ,	point connectivity of G, $\kappa = \alpha_d$.
β_d,	maximum number of points in a set S such that the subgraph induced by S is disconnected.
χ,	chromatic number of G.
$\overline{\chi}$,	chromatic number of the complement $\overline{G}$ of G.
ψ,	achromatic number; largest order of any complete coloring of G.
ψ_d,	largest order of any complete D-coloring of G.
K_n,	complete graph on n points.
$G + H$,	the join of G and H, $V(G+H) = V(G) \cup V(H)$, $E(G+H) = E(G) \cup E(H) \cup \{V(G) \times V(H)\}$.
n-critical;	$\chi(G) = n$ and for any line $uv \in E$ $\chi(G-uv) = n-1$.
uniquely n-colorable;	$\chi(G) = n$ and every n-coloring of G produces the same partition $V_1, V_2, \ldots, V_n$ of V into color classes.

	Totally disconnected, colorings, and chromatic number	Disconnected, D-colorings, and D-chromatic number
(Gallai)[4]	$\alpha_0 + \beta_0 = p$	$\alpha_d + \beta_d = p = \kappa + \beta_d$.
(HHP)	$\chi \leq p - \beta_0 + 1$	$\chi_d \leq p - \beta_d + 1 = \kappa + 1$.
(HH)	$\psi \leq p - \beta_0 + 1$	$\psi_d \leq p - \beta_0 + 1$.
		$\psi_d \nleq p - \beta_d + 1$.

Theorem. (Gupta) *For any integer m, if $V_1, V_2, \ldots, V_m$ is any complete partition of $V(G)$ of order m, then $m \leq p - \beta_0 + 1$.*

(Ore, Berge)	$\frac{p}{\beta_0} \leq \chi$	$\frac{p}{\beta_d} \leq \chi_d$.
(Folklore)	$\chi \leq \max \deg G + 1$	$\chi_d \leq \min \deg G + 1$.
(Brooks)	If $\max \deg G = n > 2$ and no component of G is K_n+1, then $\chi \leq n$.	If $\min \deg G = n \geq 2$ and K_n is not a subgraph of G, then $\chi_d \leq n$.
(Four Color Conjecture)	If G is planar, then $\chi \leq 4$.	Theorem. *If G is planar then $\chi_d(G) \leq 4$.*
(Grötzsch)	If G is planar and contains no triangles, then $x \leq 3$.	Theorem. *If G is planar and contains no triangl then $x_d \leq 2$.*

(NG)	$\chi + \overline{\chi} \leq p + 1$	$\chi_d + \overline{\chi}_d \leq \beta_0 + \overline{\beta}_0 \leq \chi + \overline{\chi} \leq p + 1$.
(HH,Gupta)	$\chi + \overline{\psi} \leq p + 1$	$\chi_d + \overline{\psi}_d \leq p + 1$.
(Konig)	For any graph G, $\chi = 2$ iff G contains no odd cycles.	If G connected and contains no triangles, then $\chi_d = 2$.
(Folklore)	$\overline{\beta}_0 \leq \chi$	$\chi_d \leq \beta_0 \leq \chi \leq \max \deg G + 1 \leq \overline{\beta}_d$.
(Folklore)	$\chi(G+H) = \chi(G) + \chi(H)$	$\chi_d(G+H) = \chi_d(G) + \chi_d(H)$.
(H)	$\psi(G+H) = \psi(G) + \psi(H)$	$\psi_d(G+H) = \psi_d(G) + \psi_d(H)$.
(Ore)	If G is n-critical, then min deg $G \geq n - 1$.	If G is n-D-critical, then min deg $G \geq n - 1$.
(Gallai)[5]	$G + H$ is n-critical if and only if G is n_1-critical, H is n_2-critical and $n_1+n_2=n$.	$G + H$ is n-D-critical if and only if G is n_1-D-critical, H is n_2-D-critical and $n_1+n_2=n$.
(Ore)	If G is n-critical, then G has no cutpoint.	If G is n-D-critical, $n > 2$, then G has no cutpoint.
(Dirac)	If G is n-critical then in every $(n-1)$-coloring of $G - uv$, points u and v receive the same color.	If G is n-D-critical, then in every $(n-1)$-D-coloring of $G - uv$, points u and v receive the same color.
(Folklore)	Every graph G contains a maximal independent set S such that $\chi(G-S) = \chi(G) - 1$.	Not every graph G contains a maximal D-set for which $\chi_d(G-S) = \chi_d(G) - 1$.
(CG)	For any n-coloring of a uniquely n-colorable graph G, the subgraph induced by the union of any two color classes is connected.	For any n-D-coloring of an n-D-chromatic graph the subgraph induced by the union of any two color classes is connected.

(CG)	If G is uniquely n-colorable then G is $(n-1)$-connected.	If G is uniquely n-D-colorable then G is $(n-1)$-connected.
(CG)	If G is uniquely n-colorable then min deg $G \geq n-1$.	If G is uniquely n-D-colorable then min deg $G \geq n-1$.
		If G is uniquely n-D-colorable and $G \neq K_1 + H$ then G is $2(n-1)$-connected, and min deg $G \geq 2(n-1)$.
		If G is a planar uniquely 4-D-colorable graph then $G = K_1 + H$.
(CG)	There does not exist a uniquely 5-colorable planar graph.	There exist only two uniquely 4-D-colorable planar graphs; they are K_4 and $\overline{K}_2 + K_3$.
(Vizing)	For any graph G, max deg $G \leq \chi(L(G)) \leq$ max deg G+1.	For any connected graph G, either $2 \leq \chi_d(L(G)) \leq 3$ or $G = K_1 + \overline{K}_n$ is a stargraph, in which case $\chi_d(L(G)) = \chi_d(K_n) = n$.

References

1. Berge, C., The Theory of Graphs and its Applications, Methuen, London, 1962.

2. Brooks, R.L., On Colouring the Nodes of a Network, *Proc. Cambridge Philos. Soc.*, 37 (1941), 194-197,

3. Chartrand, G. and Geller, D., On Uniquely Colorable Planar Graphs, *J. Combinatorial Theory* 6 (1969), 271-278. (CG)

4. Gallai, T., Uber extreme Punkt - und Kantenmengen. *Ann. Univ. Sci. Budapest,* Eotvos Sect. Math. 2 (1959), 133-138.

5. Gallai, T., Critical Graphs. I. *Magyar Tud. Akad. Mat. Kutato Int. Kozl.* 8 (1963), 165-192.

6. Gupta, R.P., Bounds on the chromatic and achromatic numbers of complimentary graphs., Proceedings of Third Waterloo Conference (1968), (to appear).

7. Harary, F. and Hedetniemi, S., The achromatic number of a graph. *J. Combinatorial Theory*, to appear. (HH)

8. Harary, F., Hedetniemi, S. and Prins, G., An interpolation theorem for graphical homomorphisms, Port. Math. (to appear). (HHP)

9. Hedetniemi, S., On partitioning planar graphs, *Canad. Math. Bull.* 11 (1968), 203-211. (H)

10. Hedetniemi, S., On hereditary properties of graphs, *J. Combinatorial Theory*, submitted.

11. Konig, D., Theorie der Graphen, Chelsea, New York, 1950.

12. Nordhaus, E. and Gaddum, J., On complementary graphs, *Amer. Math. Monthly* 63 (1956), 175-177. (NG)

13. Ore, O., Theory of Graphs. *Amer. Math. Soc. Colloq.* Publ. 38, Providence, 1962.

14. Vizing, V., On an estimate of the chromatic class of p-graph, *Disket. Analiz.* 3 (1964), 25-30.

Added in proof

Dirac, G.A., Note on the Coloring of Graphs, *Math. Z.* 54 (1951), 347-353.

Grötzsch, H., Ein Dreifarbensatz für dreilereisfreie Netz auf der Kugel., *Wiss. Z. Martin-Luther Univ. Halle-Wittenberg. Math. Naturwiss. Reihe* 8 (1958), 109-120.

RIGID AND INVERSE-RIGID GRAPHS

Pavol Hell and Jaroslav Nešetřil

McMaster University, Hamilton, Ont., Canada

Let $G = (X,R)$ mean undirected graph without multiple edges, X its vertex-set, R its edge-set. The semigroup of all compatible mappings G into itself (i.e. mapping $f:X \to X$ such that $(a,b) \in R$ implies $(f(a),f(b) \in R)$ will be denoted by $C(G)$. In the whole paper we exclude the trivial graphs $G = (X,R)$ with $|X| = 1$. Graphs satisfying $C(G) = \{id\}$ are called rigid. It is proved in [1] ([2] resp.) that there are no rigid graphs with less than 8 vertices (14 edges resp.) and that for any integer $n \geq 8$ ($m \geq 14$ resp.) or for any infinite cardinal n (m resp.) there is a rigid graph $G = (X,R)$ with $|X| = n$ ($|R| = m$ resp.) The last result for infinite edge-set is not mentioned in [2], but it follows easily from [1] and the fact, that for every infinite rigid graph $|R| = |X|$ (clearly $|R| \leq |X|$, and since degree of each vertex in a rigid graph is at least 2 also $|X| \leq |R|$.)

There is a "minimal" rigid graph, i.e. graph $G = (X,R)$ with $|X| = 8$, $|R| = 14$:

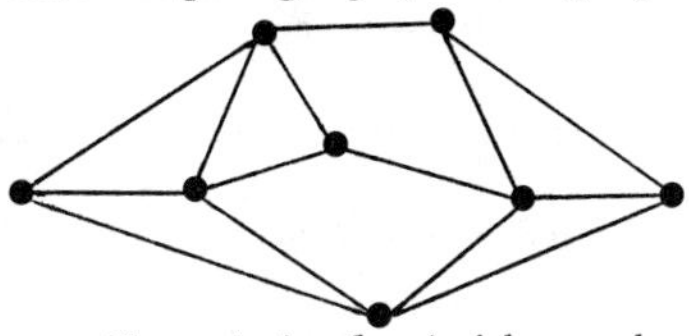

The minimal rigid graph

Let us denote by r_n (R_n resp.) the minimal (resp. maximal) number of edges of a rigid graph with n vertices (i.e. there is a rigid graph $G = (X,R)$ with $|X| = n$, $|R| = r_n$ resp. $|R| = R_n$ and for every rigid graph $G = (X,R)$ with $|X| = n$ holds $r_n \leq |R| \leq R_n$).

We shall find r_n and R_n for every integer n. By our previous remark $|X| = |R|$ for every rigid infinite graph $G = (X,R)$, thus the generalisation to n an infinite cardinal is trivial.

Let us look now at a different kind of mappings. Z. Hedrlin suggested to investigate the inverse-compatible mappings (IC-mappings):

Definition. Let $G = (X,R)$, $H > (Y,S)$ be graphs. A mapping $f: X \to Y$ is said to be an IC-mapping if $(a,b) \in S$, $a,b \in f(X)$ implies $f^{-1}(\{a,b\}) = \{x,y\}$ and $(x,y) \in R$ (with a,b and x,y resp. not necessarily distinct).

Let us denote $C_{-}(G)$ the semigroup of all IC-mappings $G \to G$, let inverse-rigid graph mean graph with $C_{-}(G) = \{id\}$.

Lemma 1. Graph $G = (X,R)$ is inverse-rigid if and only if it satisfies the following conditions:

(i) G is asymmetric (i.e. does not admit 1-1 compatible mappings onto - see [3])

(ii) G has all loops (i.e. $(x,x) \in R$ for every $x \in X$)

(iii) G has at most one isolated vertex y_0

(iv) $|R(A)| > |A|$ for every independent set $A \subset X - \{y_0\}$

$(R(A) = \{x \in X;$ there exists an $a \in A$, $a \neq x$, such that $(a,x) \in R\})$.

Theorem 1. *There is no inverse-rigid graph* $G = (X,R)$ *with* $|X| < 7$ $(|R| < 16$ *resp.) and for every integer* $n \geq 7$ $(m \geq 16$ *resp.) or any infinite cardinal* n $(m$ *resp.) there is an inverse-rigid graph* $G = (X,R)$ *with* $|X| = n$ $(|R| = m$ *resp.)*

The "minimal" inverse-rigid graph $G = (X,R)$ with $|X| = 7$ and $|R| = 16$ exists again:

The minimal inverse-rigid graph

We will denote the minimal (maximal resp.) number of edges of an inverse-rigid graph with n vertices by s_n (S_n resp.).

Lemma 2. $r_n = n + 2$ for $n \geq 17$, and $s_n = 2n + 1$ for $n \geq 10$. We can moreover construct inverse-rigid graphs $G = (X,R)$ with $|X| = n$, $|R| = s_n$ for $n \geq 10$ in such a way that they are without triangles.

Definition. Complement $G = coG = (X,coR)$ for $G = (X,R)$, where $(x,y) \notin R \iff (x,y) \in coR$. Note that if G has all loops, then coG is without loops.

Lemma 3. a) $C_{-}(G) \subset C(coG)$

b) If G has all the loops and no triangle, then $C_{-}(G) = C(coG$

Corollary. a) G rigid implies coG inverse-rigid

b) G inverse-rigid without triangles implies coG rigid

By corollary a) we have $R_n \leq \binom{n}{2} + n - s_n$, by corollary b) and remark about lemma 2, $R_n = \binom{n}{2} - n - 1$ for $n \geq 10$.

Theorem 2. *Let $G = (X,R)$ be a rigid graph, $|X| = n$. Then $n \geq 8$ and $r_n \leq |R| \leq R_n$ where*

n	8	9	10	11	12	13	14	15	16	17	k
r_n	14	16	14	14	15	15	17	17	19	19	$k+2$
R_n	17	25	34	43	53	64	76	89	103	118	$\binom{k}{2}-k-1$

and the bounds are best possible in the sense, that for every $n \geq 8$ there are graphs $G = (X,R)$ with $|X| = n$ and $|R| = r_n$ ($|R| = R_n$ resp.).

Theorem 3. *Let $G = (X,R)$ be an inverse-rigid graph, $|X| = n$. Then $n \geq 7$ and $s_n \leq |R| \leq S_n$, where $s_7 = 16$, $s_8 = 17$, $s_9 = 20$, $s_n = 2n + 1$ for $n > 9$*

$$S_n = \binom{n}{2} + n - m_n \quad (m_n \text{ described in [3, Theorem 1]})$$

and the bounds are best possible again in the obvious sense.

Proof of the theorem 3. In [3, Theorem 1] L. Quintas found the minimal number of edges m_n of an asymmetric graph with n vertices, and constructed for every $n \geq 7$ an asymmetric graph G_n without cycles (forest) with n vertices and m_n edges. By lemma 1 (i) obviously $S_n \leq \binom{n}{2} + n - m_n$. Graph coG_n satisfies clearly conditions (i), (ii), (iii) of lemma 1 ((iii) follows from (i)) and the condition (iv) is also satisfied since independent set in coG_n is a complete subgraph, thus an edge (a,b) in G_n and $|R_{coG_n}(\{a,b\})| = n$.

The proofs ommited here can be found in a paper of the same name, which is to appear.

References

1. Hedrlin, Z. and Pultr, A., On rigid undirected graphs, *Can. J. Math.* 18 (1966), 1237-1242.

2. Hell, P., Rigid undirected graphs with given number of edges, *CMUC* 9,1 (1968), 51-69.

3. Quintas, L. V., Extrema Concerning Asymmetric Graphs, *Journal of Comb. Th.* 3 (1967), 57-82.

$-1-\sqrt{2}$? *

A. J. Hoffman

IBM Research Center Yorktown Heights, NY, U.S.A.

1. Introduction

Let G be a graph (finite, undirected, with at most one edge joining a pair of vertices, and no edge joining a vertex to itself). Number the vertices of G arbitrarily, and define $A(G)$--the adjacency matrix of G--by the rule

$$A(G) = (a_{ij}) = \begin{cases} 1 & \text{if } i \text{ and } j \text{ are adjacent vertices} \\ 0 & \text{otherwise} \end{cases}$$

Since $A(G)$ is a real symmetric matrix, it has real eigenvalues. Denote the algebraically least of these by $\lambda(G)$.

It is known from earlier work (cf the surveys in [1], [2], [3]) and more recent (unpublished) research of Michael Doob, Charles Sims and the author that knowledge of a lower bound for $\lambda(G)$ is very useful in proving properties of G. These results suggest the following two problems.

Problem 1. Let $\mathcal{G}$ be an infinite class of graphs. It may or may not be true that there exists a λ such that

$$\lambda(G) \geq \lambda \text{ for all } G \in \mathcal{G}.$$

Characterize those $\mathcal{G}$ for which (1.1) holds.

Problem 2'. Let λ be given, define $\mathcal{G}(\lambda)$ to be the set of all graphs G such that $\lambda(G) \geq \lambda$, and characterize $\mathcal{G}(\lambda)$.

Precedent indicates that Problem 2' is too difficult to be attacked now, so we replace it by the more modest

---

*This research was supported in part by the Office of Naval Research under Contract No. Nonr 3775(00).

Problem 2. Let $d(G)$ be the minimum of the valences of the vertices of G, and define $\mathrm{G}(d)$ to be the set of all graphs G such that $d(G) \geq d$. Find a class of graphs G such that

$$\mathrm{G} \subset \mathrm{G}(\lambda),$$

and

$$\text{for some } d(\lambda), \quad \mathrm{G}(\lambda) \cap \mathrm{G}(d(\lambda)) \subset \mathrm{G}.$$

Put another way, Problem 2 seeks a characterization of $\mathrm{G}(\lambda)$ confined to graphs of sufficiently large minimum valence. Even simplified, Problem 2 is solved only for $\lambda > -1-\sqrt{2}$ (whence the title of this paper). The results given below include and supersede (except for the value of $d(\lambda)$) the relevant parts of [2] and [4]. For Problem 1, the results are more complete. We provide two characterizations, one "local" (in terms of excluded subgraphs), the other "global" (describing how each graph in G is, modulo small adjustments, composed by fitting cliques together).

2. Results for Problem 1

Theorem 1. *Let* G *be an infinite class of graphs. Then the following are equivalent:*

(2.1) *There exists a λ such that $\lambda(G)>\lambda$ for all $G\in$ G.*

(2.2) *There exists an integer $l > 0$ such that $K_{1,l} \not\subset G$, $K_{l+1}(K_l)K_{2l} \not\subset G$ for all $G \in$ G, where*
$K_{1,l}$ is a claw of order l, and
$K_{l+1}(K_l)K_{2l}$ is the graph on $2l+1$ vertices formed by the union of cliques K_{l+1} and K_{2l}, such that $K_{l+1} \cap K_{2l} = K_l$.

(2.3) *There exists an $L > 0$ such that, for each $G \in$ G, there exist graphs $\hat{G}$ and H with*
(2.3a) *$A(G) + A(\hat{G}) = A(H)$,*
every vertex of $\hat{G}$ has valence at most L, and H contains some family of cliques $K^1, K^2, \ldots$ with the properties.
(2.3b) *Every edge of H is in at least one K^j*
(2.3c) *Every vertex of H is in at most L K^j's*
(2.3d) *$|K^i \cap K^j| \leq L$ for $i \neq j$.*

(There exist examples to show that the theorem is false if no mention is made of $\hat{G}$.)

3. Results for Problem 2

Theorem 2. *Let* $\lambda \in [-1-\sqrt{2}, -1]$. *There exists a function* $d(\lambda)$ *such that*

(3.1) *If* $-1 \geq \lambda > -2$, $\lambda(G) \geq \lambda$, $d(G) \geq d(\lambda)$, *then* $\lambda(G) = -1$, *and* G *is the union of disconnected cliques. Conversely, if* G *is the union of disconnected cliques (and has at least one edge),* $\lambda(G) = -1$.

(3.2) *If* $-2 \geq \lambda > -1 - \sqrt{2}$, $\lambda(H) \geq \lambda$, $d(H) \geq d(\lambda)$, *then* $\lambda(H) \geq -2$, *and* $H = L(G_n; a_1, \ldots, a_n)$ *where* G_n *is a graph on* n *vertices,* $a_1, \ldots, a_n$ *are nonnegative integers, and* $L(G_n; a_1, \ldots, a_n)$ *is the following graph:*

The vertices of $L(G_n; a_1, \ldots, a_n)$ *are all (unordered) pairs* (i,j), *where* i *and* j *are adjacent vertices of* G_n, *and all ordered pairs of the form* $[i, \alpha']$ *or* $[i, \alpha'']$, *where* $1 \leq \alpha \leq a_i$. *(If* $a_i = 0$, *there are no vertices of the form* $[i,\alpha']$ *or* $[i,\alpha'']$.*) Every vertex of the form* (i,j) *is adjacent to every vertex of the form* (k,l), *where* $|\{k,l\} \cap \{i,j\}| = 1$, *and every vertex* $[i,\alpha']$, $[i,\alpha'']$, $[j,\alpha']$, $[j,\alpha'']$ *and to no other vertices. Additionally every vertex of the form* $[i,\alpha']$ *or* $[i,\alpha'']$ *is adjacent to every vertex of the form* $[i,\beta']$ *or* $[i,\beta'']$ *(same* i*), but* $[i,\alpha']$ *and* $[i,\alpha'']$ *are* <u>*not*</u> *adjacent, and to no other vertices.*

Conversely, if $H = L(G_n; a_1, \ldots, a_n)$, *then* $\lambda(H) \geq -2$.

Corollary. Define, for any graph G,

$$\mu(G) = \lim_{d \to \infty} \sup_{\substack{G \subset H \\ d(H) \geq d}} \lambda(H).$$

(Results on $\mu(G)$ were given in [3].) From Theorem 2, we can infer

$$\mu(G) > -1 - \sqrt{2} \text{ if and only if } G = L(G_n; a_1, \ldots, a_n)$$

for some graph G_n on n vertices and some nonnegative integers $a_1, \ldots, a_n$.

4. Sketch of Proofs

Full proofs will be published elsewhere. We confine ourselves here to a sketch of the proof of Theorem 1; the proof of Theorem 2 is similar in spirit, but pays greater attention to detail and uses the formula for $\mu(G)$ given in [3].

To prove (2.1) implies (2.2), we observe that $\lambda(K_{1,l}) \to -\infty$, $\lambda(K_{l+1}(K_l)\ K_{2l}) \to -\infty$, and invoke the principle that $H \subset G$ implies $\lambda(H) \geq \lambda(G)$ To prove (2.3) implies (2.2), we first consider the (0,1) matrix K describing incidence of vertices of H with cliques $K^1, K^2, \ldots$. The matrix $KK^T = A(H) + M$, where M is a nonnegative matrix with spectral radius bounded by a function of L (this comes from (2.3b, c, d)). Since KK^T is positive semi-definite, this implies $\lambda(H) \geq$ some function of L. By (2.3a), the eigenvalues of $A(G)$ differ respectively from the eigenvalues of $A(H)$ by at most L in absolute value. Hence, (2.1).

To show that (2.2) implies (2.3) is an entirely graph theoretic argument. One looks for "large" cliques in G, and says that two large cliques are equivalent if each vertex of one clique is adjacent to all but at most l-1 vertices of the other clique. This relation turns out to be symmetric and transitive, so the large cliques can be grouped into equivalence classes. The union of the vertices in each equivalence class is almost a clique; one can show that by adding to G a graph $\hat{G}$ of modest valence, we get a clique. These become the cliques $K^1, K^2, \ldots$ of H, along with any edges left over in G (each regarded as a clique with two vertices), and one verifies (2.3b, c, d).

References

1. A. J. Hoffman, "The Eigenvalues of the Adjacency Matrix of a Graph," to appear in Proceedings of a Symposium on Combinatorial Mathematics, University of North Carolina, Chapel Hill, N. C.

2. _______, "Some Recent Results on Spectral Properties of Graph," in Beiträge zur Graphentheorie, edited by H. Sachs, H. J. Voss and H. Walther, published by Teubner, Leipzig, 1968.

3. _______, "The Change in the Least Eigenvalue of the Adjacency Matrix o[f a] Graph under Imbedding," *Siam J. Appl. Math.*, 17 (1969), to appear.

4. D. K. Ray Chaudhuri, "Characterizations of Line Graphs," *Journal of Combinatorial Theory*, 3 (1967), 201-214.

PLANARITY PROPERTIES OF THE GOOD-DE BRUIJN GRAPHS

Diane M. Johnson and N.S. Mendelsohn

University of Manitoba, Winnipeg, Man., Canada

Abstract.

The graphs $G_t^{(r)}$ are generalizations of graphs introduced by N.G. de Bruijn [1] and I.J. Good [2] in 1946. In this paper it is shown that the only planar graphs of this class are the graphs $G_t^{(1)}$, $G_1^{(2)}$, $G_1^{(3)}$, $G_1^{(4)}$, $G_2^{(2)}$, $G_2^{(3)}$, $G_3^{(2)}$. The genus of $G_t^{(r)}$ in general is still an open question.

Definition of $G_t^{(r)}$.

The graph $G_t^{(r)}$ is defined as follows. It is a directed graph with r^t vertices where each vertex is a t-sequence $\alpha_1\alpha_2\alpha_3\ldots\alpha_t$ with each of the α_u taken from the r-set $\{0, 1, 2, \ldots, r-1\}$. A vertex $\alpha_1\alpha_2\alpha_3\ldots\alpha_t$ is joined to a vertex $\beta_1\beta_2\beta_3\ldots\beta_t$ by a directed edge, if and only if $\beta_1 = \alpha_2, \beta_2 = \alpha_3, \ldots, \beta_{t-1} = \alpha_t$. Such a graph has r^{t+1} edges and each vertex has both in-degree and out-degree equal to r. The graph is connected since between any two of its vertices there is a unique directed path of length r viz.,

$$\alpha_1\alpha_2\ldots\alpha_t \to \alpha_2\alpha_3\ldots\alpha_t\beta_1 \to \alpha_3\alpha_4\ldots\beta_1\beta_2 \to \ldots \to \beta_1\beta_2\ldots\beta_t.$$

The graph has r loop vertices (i.e. vertices joined to themselves by an edge), namely, the vertices $00\ldots0$, $11\ldots1$, $22\ldots2$, ..., $r-1\ r-1\ r-1\ldots r-1$. It also contains $\frac{1}{2}r(r-1)$ directed bi-gons since the vertices $ijij\ldots$, $jiji\ldots$, are the ends of such a bi-gon for any distinct pair of symbols i, j taken from $\{0, 1, 2, \ldots, r-1\}$.

In figure 1, are illustrated the graphs $G_1^{(3)}$, $G_3^{(2)}$.

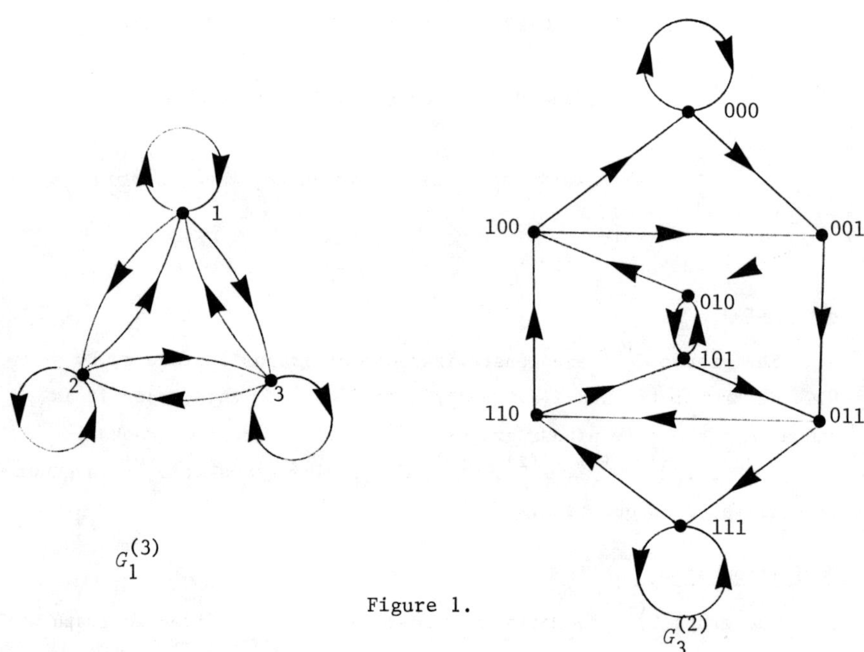

Figure 1.

<u>The Planarity Theorem.</u>

<u>Theorem</u>. *The only planar graphs among* $G_t^{(r)}$ *are* $G_t^{(1)}$, $G_1^{(2)}$, $G_1^{(3)}$, $G_1^{(4)}$, $G_2^{(2)}$, $G_2^{(3)}$, $G_3^{(2)}$.

<u>Proof</u>. The graph $G_t^{(1)}$ consists of a single point. The graphs $G_1^{(r)}$, $r = 1, 2, 3, 4, \ldots$ are complete directed K_r graphs and hence by Kuratowski's theorem are planar for $r = 1, 2, 3, 4$ and non-planar for $r \geq 5$. The graphs $G_1^{(2)}$, $G_2^{(3)}$, $G_3^{(2)}$ are all easily drawn in the plane. The graph $G_3^{(3)}$ is non-planar as it contains the topological $K_{3,3}$ as shown in Figure 2.

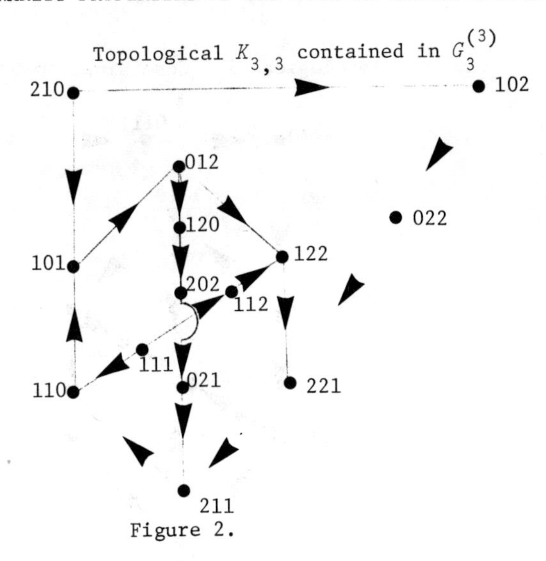

Figure 2.

Also, since $G_t^{(s)}$ is a subgraph of $G_t^{(r)}$ for $s < r$, it follows that $G_3^{(s)}$ is non-planar for $s \geqslant 3$.

We now consider $G_t^{(r)}$ with $t \geqslant 2$, $r \geqslant 3$. If we remove from $G_t^{(r)}$ its r loop-edges and one edge from each of the $\frac{1}{2}r(r - 1)$ bi-gons we have a graph with no bi-gons and each vertex of degree greater than or equal to 3. If V is the number of vertices and E the number of edges in such a graph then $V = r^t$, $E = r^{t+1} - r - \frac{1}{2}r(r - 1)$. A necessary condition for planarity is that $E \leqslant 3V - 6$. This becomes $r^{t+1} - 3r^t - \frac{1}{2}r^2 - \frac{1}{2}r + 6 \leqslant 0$. This is not satisfied if $r \geqslant 4$, $t \geqslant 2$. Hence, all other cases of non-planarity are decided if we can decide the cases $G_t^{(2)}$, $t = 4, 5, 6, \ldots$. These cases will now all be shown to have a topological $K_{3,3}$.

In Figure 3, a topological $K_{3,3}$ in $G_4^{(2)}$ is shown while the case of $G_5^{(2)}$ is depicted in Figure 4.

Topological $K_{3,3}$ contained in $G_4^{(2)}$.

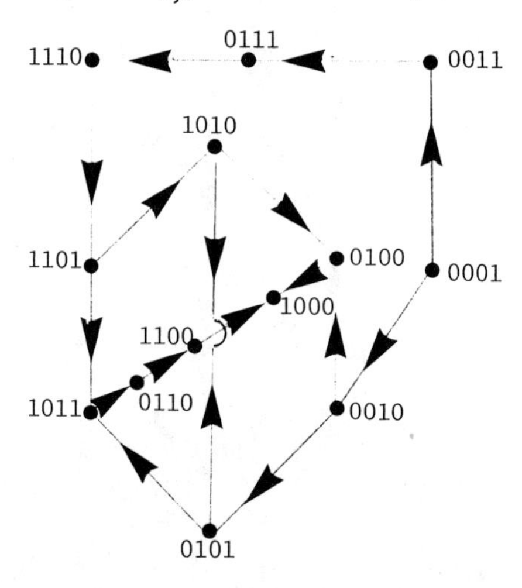

Figure 3.

Topological $K_{3,3}$ contained in $G_5^{(2)}$.

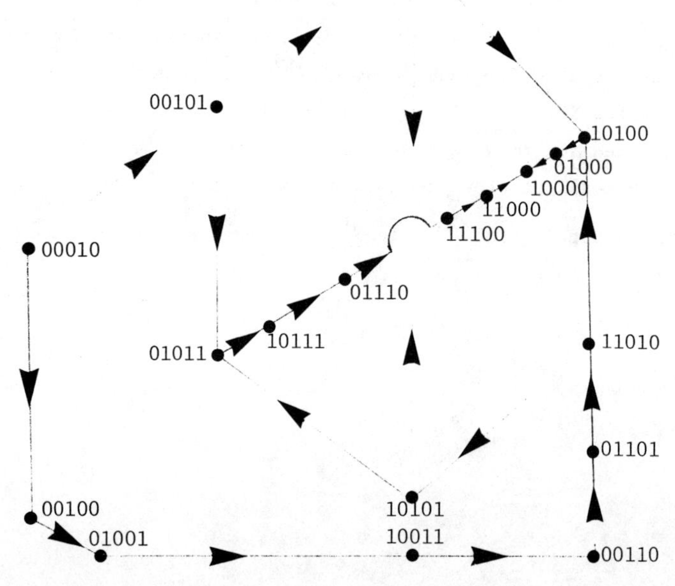

Figure 4.

We next distinguish two general cases viz. t even and t odd.

Case 1. t even $G_t^{(2)}$ $t \geq 6$, $t = 2u$.

Consider the following points;

$A = 0101\ldots01$, $B = 1010\ldots1011$, $C = 110101\ldots01$,
$D = 1010\ldots10$, $E = 0101\ldots0100$, $F = 001010\ldots10$,
$G = 0101\ldots0111$, $J = 1001\ldots0101$, $K = 10\ldots101111$,
$L = 01\ldots011111$

There is a unique directed path of length t-1 from E to L. Also, there is a unique directed path of length t-1 from J to C. It is readily verified that the paths from E to L and from J to C do not contain a common vertex nor do they contain any of the remaining lettered vertices. Figure 5 now shows a $K_{3,3}$ in $G_{2u}^{(2)}$.

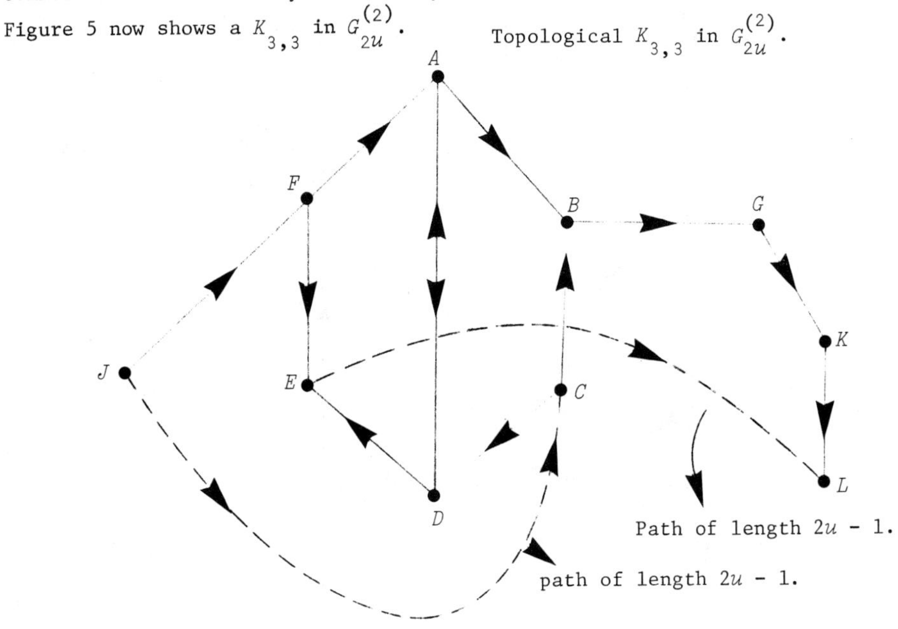

Figure 5.

Case 2. t odd, $G_t^{(2)}$ $t \geq 7$, $t = 2u + 1$.

Consider the following points;

$A = 0101\ldots010$, $B = 1010\ldots100$, $C = 11010\ldots10$
$D = 1010\ldots101$, $E = 0101\ldots011$, $F = 0010\ldots101$
$G = 01\ldots01000$, $H = 1010\ldots0000$, $K = 01101\ldots01$
$L = 00010\ldots10$

There is a unique directed path from L to K of length t and also a unique directed path from E to H of length t. Again, it is readily verified that the paths have no vertex in common nor any of the other lettered vertices. Figure 6 now shows the topological $K_{3,3}$ contained in $G^{(2)}_{2u+1}$.

Topological $K_{3,3}$ contained in $G^{(2)}_{2u+1}$.

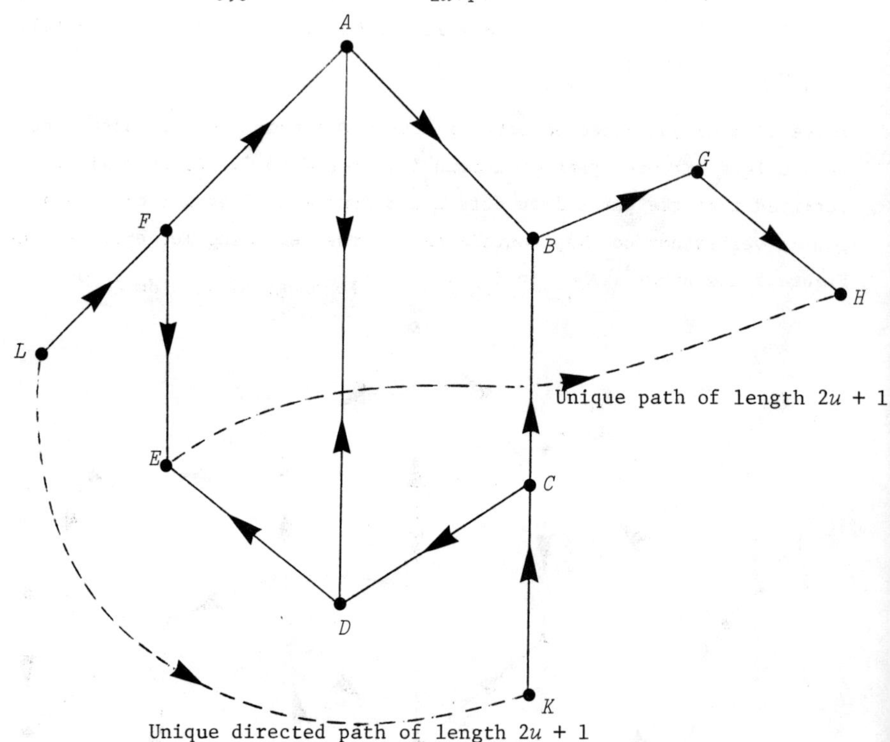

Figure 6.

Further Remarks.

The Good-de Bruijn graphs are examples of a more general class of graphs, namely, the directed graphs with the property that between any two of its vertices there is a unique path of length t. There has been a redundance in the proofs of non-planarity in the previous section. This is because we wanted as much as possible the proofs to go over to these more general graphs. In fact, the proof based on vertex and edge counts is valid for these more general graphs. We conjecture that all the results carry over. We further conjecture for the Good-de Bruijn graphs that in the case of non-planarity that every topological hexagon is part of a topological $K_{3,3}$.

References

1. de Bruijn, N.G., A Combinatorial Problem, Neder, Akad. Wetensch. Proc. 49, 758-764, *Indagationes Math.*, 8 (1946), 461-467.

2. Good, I.J., Normal Recurring Decimals, *J. London Math. Soc.*, 21 (1946), 167-169.

CHARACTERIZATION OF THE p-VECTOR OF A SELF-DUAL 3-POLYTOPE

Ernest Jucovič

P.J. Safarik University, Kosice, Czechoslovakia

A vector $p = (p_3, p_4, \ldots, p_m)$ is said to be the *p-vector of the 3-polytope M* if $p_i(M) = p_i$ for all i, where $p_i(M)$ denotes the number of i-gonal faces of M, and M doesn't have a k-gon with $k > m$. Analogically is defined the v-vector $v = (v_3, v_4, \ldots, v_n)$ for the numbers v_i of i-valent vertices of M. We call also the sequences $p = (p_3, p_5, \ldots, p_m)$ and $v = (v_3, v_5, \ldots, v_n)$ *realizable by the polytope M* if $p_i(M) = p_i$ and $v_j(M) = v_j$ for all i, j.

It is a difficult problem attacked in recent years especially by B. Grünbaum (cf. [1,2,3]) to characterize the p-vector and the v-vector of a general 3-polytope or a simple 3-polytope or a 4-valent 3-polytope. In this note a characterization of the p-vector and the v-vector of a self-dual 3-polytope is given. Self-dual is a 3-polytope if it is combinatorially isomorphic with its dual polytope. So in a self-dual polytope there exists a one-to-one mapping ω between the vertices and the faces preserving incidence.

<u>Theorem 1</u>. *Necessary and sufficient for the vector $p = (p_3, p_4, \ldots, p_n)$ and the vector $v = (v_3, v_4, \ldots, v_n)$ where p_i and v_j are non-negative integers, to be a p-vector and a v-vector of a self-dual 3-polytope is $p_i = v_i$ for all i and $\sum_{i=3}^{m} (4-i).p_i = 4$.*

The necessity of the conditions mentioned being obvious we shall construct, for every pair p, v of vectors satisfying these conditions, a self-dual 3-polytope whose p- or v-vector is p or v.

Every n-sided pyramid is self-dual. In the mapping ω described above correspond to each other the triangular side-faces and the trivalent vertices of the base-face σ and of course this face σ and the opposite vertex S. It is easy to verify that if we denote the vertices of the base-face by $A_1, \dots, A_n$ in one orientation and the side-faces successively by $\alpha_1, \dots, \alpha_n$ in the opposite orientation starting with any face and any vertex then in the mapping ω correspond A_i with α_i. So there exist such two indices j, k that A_j lies on α_j and A_k on α_k. Let e.g. the vertex A_1 lie on α_1. We dissect the face α_1 by new edges $A_nB_1, A_nB_2, \dots, A_nB_c$ where the new vertices B_i dissect the edge SA_1. (Fig. 1a). We choose the labelling so that the vertices form a sequence $S, B_1, \dots, B_c, A_1$. The polytope we get is self-dual with an n-gonal and a $(c+3)$-gonal face if denoting the faces $SA_2A_1B_c \dots B_1$ or SB_1A_n or $B_iB_{i+1}A_n$ or $B_cA_1A_n$ by α_n or α_1 or β_{m+1-i} or β_1 respectively and map in ω the vertices A_i or B_i or S upon the faces α_i or β_i or σ respectively. The described procedure may be repeated with the vertex B_{u+1} the edge B_uA_n and the faces β_{u+1}, β_u, $\left(u = \left|\frac{c+2}{2}\right|\right)$ etc. (Fig. 1b). The only rule which must be maintained is that the new 3-valent vertices and the new triangular faces which correspond with them in ω form sequences in opposite directions.

So we get to every $p = (p_3, p_4, \dots, p_n)$ and $v = (v_3, v_4, \dots, v_n)$ with $p_i = v_i$, a self-dual 3-polytope M with $p_i(M) = v_i(M) = p_i$, $i \geq 4$. Since M is self-dual, $p_3(M)$ satisfies $\sum_{i=3}^{n} (4-i)\cdot p_i(M) = 4$ and therefore $p_3(M) = p_3$. The proof is finished.

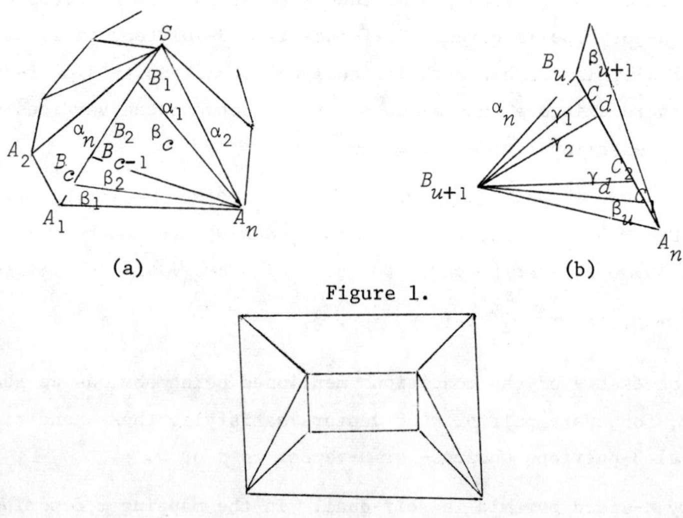

(a) (b)

Figure 1.

Figure 2.

Remark. Analogously, beginning with the self-dual centrally symmetric 3-polytope on Fig.2 we can prove:

Theorem 2. *Necessary and sufficient for two sequences* $p = (p_3, p_5, \ldots, p_n)$ *and* $v = (v_3, v_5, \ldots, v_n)$ *of non-negative integers to be realizable by a self-dual centrally symmetric 3-polytope is* $p_i = v_i = 2b_i$ *for all* i *and* $\sum_{i=3}^{n} (4-i).p_i = 4$.

References

1. Grünbaum, B., Convex Polytopes, J. Wiley, New York, 1967.

2. Grünbaum, B., Planar maps with prescribed types of vertices and faces (to appear).

3. Grünbaum, B., Some analogues of Eberhard's theorem on convex polytopes (to appear).

A VARIATION OF n-CONNECTEDNESS

H. A. Jung

Technische Universität, Berlin (West), Germany

Let $v(G)$ and $e(G)$ denote the numbers of vertices resp. edges of the graph G.

Definition. A graph G with $v(G) \geq 2n$ is *n-pair-connected* if for any n pairs (a_ν, b_ν) of vertices in G exist openly disjoint paths P_ν joining a_ν and b_ν $(1 \leq \nu \leq n)$.

A family of paths is defined to be openly disjoint if any two of these paths have at most terminal vertices in common.

It is a special case of the definition that every n-pair-connected graph G contains as subgraphs subdivisions of every graph with n edges; moreover the nodes or main vertices of the subdivision can be prescribed. So we have finally the

Remark. If G is $\binom{n}{2}$-pair-connected and if n different vertices $v_1, \ldots, v_n$ in G are given there exists a subdivision $U(K_n)$ of the complete graph K_n with n vertices in G such that the v_ν's are the nodes of this subdivision (i.e. $d_{G'}(v_\nu) \leq n$ where $G' = U(K_n)$).

An easy application of Menger's theorem gives

Theorem 1. (M. E. Watkins): *G n-pair-connected $\Rightarrow$ G (2n-1)-connected.*

On the other hand we have the following sufficient condition for a graph to be n-pair-connected.

Theorem 2. *G 2n-connected and $G \supseteq U(K_{3m})$ $\Rightarrow$ G n-pair-connected.*

For the case that all the a_ν and all the b_ν are different this result was also obtained by Larman and Manil(to appear in Proc. London Math. Soc.)

W. Mader proved the following

Theorem. *There exists a (minimal) function* $f(n)$ *such that* $\frac{2e(G)}{v(G)} \geq f(n)$ *implies* $G \supseteq U(K_n)$.

Combining Mader's result with Theorem 2 we get

Corollary 1. G $2n$-connected and $\frac{2e(G)}{v(G)} \geq f(3n) \Rightarrow G$ n-pair-connected.

Combining Mader's result with the remark at the beginning we get

Corollary 2. If G is $2\binom{n}{2}$-connected and if $\frac{2e(G)}{v(G)} \geq f(3\binom{n}{2})$ then for any n different vertices $v_1,\ldots,v_n$ of G exists a subdivision $U(K_n) = G' \subseteq G$ having $v_1,\ldots,v_n$ as nodes (i.e. $d_{G'}(v_\nu) = n$).

Since for every vertex v in the n-connected graph G the inequality $d_G(v) \geq n$ is true we get

Corollary 3. There exists a (minimal) function $g(n)$ such that every $g(n)$-connected graph is n-pair-connected.

Most probably the bounds for $f(n)$ and $g(n)$ known so far are far from being best possible. The determination of $f(n)$ and $g(n)$ in general seems to be a very difficult problem. Only in the cases $n = 2$ and $n = 3$ (the case $n = 1$ is trivial) we have more information.

Theorem 3. (M. E. Watkins): *Let G be 4-connected.*

(i) $G \supseteq U(K_5) \Rightarrow G$ *2-pair-connected.*

(ii) If G is planar then G is 2-pair-connected iff $3v(G) - 6 = e(G)$.

For (i) see Theorem 2. The condition in (ii) means that if G is embedded into the sphere then this embedding triangulates the sphere. In view of Theorem 3 and Kuratowski's theorem Watkins was lead to the conjecture that every 4-connected non-planar graph is 2-pair-connected.

This indeed is true. So theorem 3 can be sharpened to

Theorem 4. *Let G be 4-connected. Then G is 2-pair-connected iff G is non-planar or* $3v(G) - 6 = e(G)$.

This theorem was independently proved by P. Mani and me.

An immediate consequence of Theorem 4 is $g(2) = 6$. This result was conjectured by M. Rosenfeld, (see Grünbaum's lecture in this volume).

References

1. Mader, W., Homomorphieeigenschaften und Mittlere Kantendichte von Graphen, *Math. Analen*, 174 (1967).

2. Watkins, M. W., On the Existence of Certain Disjoint Arcs in Graphs. *Duke Math. J.* 35/2 (1968).

EMBEDDINGS AND ORIENTATIONS OF GRAPHS

Paul C. Kainen

Cornell University, Ithaca, NY, U.S.A.

Let G be a finite graph. Denote by $V(G)$ the set of vertices of G and by $E(G)$ the set of edges. Write $v \in e$ if e is incident with v, i.e., if v is an endpoint of e. An orientation ω of G is a function which assigns to every $v \in V(G)$ a cyclic ordering of the set $E_v(G) = \{e \in E(G) | v \in e\}$. Call two orientations ω and η equivalent, $\omega \sim \eta$, if for every $v \in V(G)$ $\omega(v)$ and $\eta(v)$ are either identical or exactly reversed. Thus any two orientations of a trivalent graph are equivalent.

All manifolds will be closed and 2-dimensional. For simplicity, we assume that all maps are piecewise-linear. If $f:G \to M$ is an embedding where M is an oriented manifold, then f induces a unique orientation ω_f of G. If M is not oriented, f induces a family of equivalent orientations by choosing local orientations at each point of M. For the purpose of this note, we define $cr_M(G) = 0$ if and only if there exists $f:G \to M$ an embedding. For M oriented, we define $cr_M(G,\omega) = 0$ if and only if there exists $f:G \to M$ an embedding such that $\omega_f = \omega$. J. Edmonds [1] has shown that given (G,ω) there exists a unique oriented $M = M(G,\omega)$ such that $cr_M(G,\omega) = 0$ and if genus $N <$ genus M, then $cr_N(G,\omega) \neq 0$. Now if $f:G \to M$ is an embedding with $\omega_f = \omega$, we can "inflate" the vertices $f(v)$ of $f(G)$ to obtain a trivalent graph $\tau_\omega G$ depending on (G,ω).

Alternatively, we can give the following combinatorial definition of $\tau_\omega G$: $V(\tau_\omega G) = \{(v,e) \in V(G) \times E(G) | v \in e\}$ and we say that (v,e), (v',e') are adjacent in $\tau_\omega G$ if and only if either (1) $e = e'$ and $\{v,v'\}$ = set of endpoints of e or (2) $v = v'$ and e' follows immediately after e in the cyclic ordering $\omega(v)$. If $v \in V(G)$, denote by $\overline{v}$ the collection of all edges of $\tau_\omega G$ of the form (2) above. Thus, for example, if v has degree 1, $\overline{v}$ is a loop, and in any event, $\overline{v}$ is a circle.

The relationship between $cr_M(G)$ and $cr_M(\tau_\omega G)$ is expressed in the following:

Theorem 1. *$cr_M(G) = 0$ if and only if there exists ω such that $cr_M(\tau_\omega G) = 0$.*

Proof: It is clear that $cr_M(G) = 0$ implies $cr_M(\tau_\omega G) = 0$ for some ω - namely, an orientation arising from an embedding of G in M. Now suppose $cr_M(\tau_\omega G) = 0$ for some ω. Let $g:\tau_\omega G \to M$ be an embedding. Call g "nice" at $v \in V(G)$ if $g(\overline{v})$ bounds a closed disc D_v in M such that $D_v \cap g(\tau_\omega G) = g(\overline{v})$. Note that if g is nice at every $v \in V(G)$, we may shrink the discs D_v to points, obtaining an embedding $f:G \to M$. In fact, $\omega_f \sim \omega$. Thus, to complete the proof of Theorem 1, we must show that given $v \in V(G)$ we can make g nice at v without making g unnice at any $v' \in V(G)$ at which it is already nice.

But this is easy; for given $v \in V(G)$, $g(\overline{v})$ is a circle and so has a neighborhood N_v in M homeomorphic to an annulus or to a Mobius band. Thus the pair $(N_v, N_v \cap g(\tau_\omega G))$ is homeomorphic to

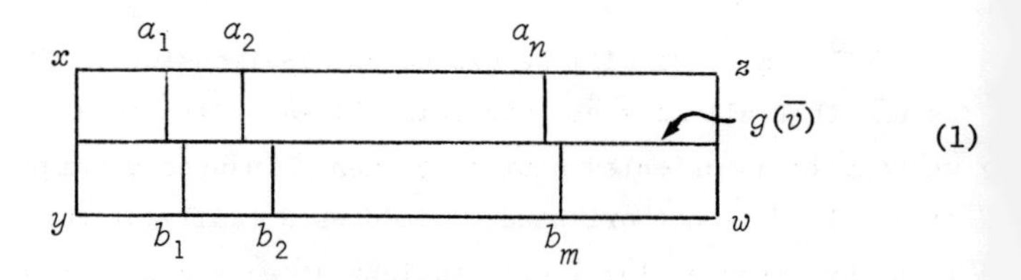

(1)

where xy is identified with zw or wz according to whether N_v is homeomorphic to an annulus or to a Mobius band. But now we may clearly change g so that $(N_v, N_v \cap g(\tau_\omega G))$ is homeomorphic to

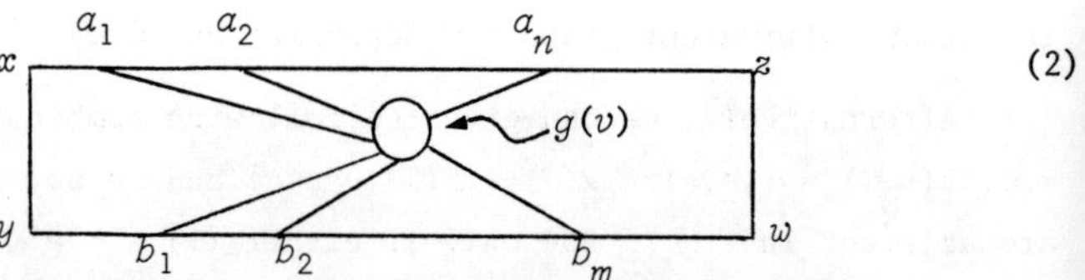

(2)

so that g is now nice at v and nothing has changed outside N_v.

Theorem 2. *Let G be a block and let cr denote cr_M when $M = S^2$. Then for any orientation ω of G, $cr(\tau_\omega G) = 0$ if and only if there exists*

$\eta \sim \omega$ *such that* $cr(G,\eta) = 0$.

Proof: Suppose there exists $\eta \sim \omega$ such that $cr(G,\eta) = 0$. Then $cr(\tau_\eta G) = 0$. But clearly $\eta \sim \omega$ implies that $\tau_\eta G$ is isomorphic to $\tau_\omega G$ so $cr(\tau_\omega G) = 0$.

Conversely, suppose $cr(\tau_\omega G) = 0$. Let $g:\tau_\omega G \to S^2$ be an embedding. It suffices, by the proof of Theorem 1, to show that g is nice at every $v \in V(G)$. So let $v \in V(G)$ be arbitrary. Then $g(\overline{v})$ is a circle and thus separates S^2 into two discs D_1, D_2. Since G is a block, $G-v$ is connected; and therefore so is $\tau_\omega G - \overline{v}$. Hence $g(\tau_\omega G - \overline{v})$ is contained in D_1 or D_2 and we see that g is nice at v.

Corollary A. G is planar if and only if there exists ω such that $\tau_\omega G$ contains no subgraph homeomorphic to $K_{3,3}$, the "three houses-three utilities" graph.

Corollary B. For any manifold M, there exists a trivalent G such that $cr_M(G) \neq 0$.

Corollary A uses Kuratowski's Theorem and the trivalence of $\tau_\omega G$. To prove Corollary B take any graph H such that $cr_M(H) \neq 0$ (for example, $H = K_m$ for some large m) and let $G = \tau_\omega H$ where $\omega = \omega_f$ for $f:H \to M'$ an embedding. One can also show this by taking G to be the dual of the 1-skeleton of a triangulation of N and applying a theorem of J.W.T. Youngs [3], where N is any manifold with lower Euler characteristic. Note that Theorem 1 and Corollary A are the geometric analogues to results of L. Neuwirth [2].

Conjecture. Let M be arbitrary and G trivalent. Then there exists a unique trivalent graph H such that $cr_M(G) = 0$ if and only if G does not contain a subgraph homeomorphic to H.

This conjecture, together with Theorem 1, would give a geometric criterion analogous to Corollary A for arbitrary M.

References

1. J. Edmonds, A combinatorial representation for polyhedral surfaces, *AMS Notices* 7 (1960), 646.

2. L. Neuwirth, Imbeddings in low dimensions, *Illinois J. Math.* 10 (1966), 470-478.

3. J. W. T. Youngs, Minimal imbeddings and the genus of a graph, *J. of Math. and Mech.* 12 (1963), 303-315.

A GENERALIZATION OF SOME GENERALIZATIONS OF SPERNER'S THEOREM

G. Katona

Mathematical Institute of the Hungarian Academy of Sciences, Budapest, Hungary

Sperner proved the following theorem [1]: Let $\mathcal{A} = \{A_1,\ldots,A_m\}$ be a family of subsets of a set S of n elements. If no two of them possess the property $A_i \subset A_j$ $(i \neq j)$, then $m \leq \binom{n}{\left[\frac{n}{2}\right]}$. Erdös answered the question which is the maximum of m if no h+1 different elements of the family form a chain $A_{i_1} \subset A_{i_2} \subset \ldots \subset A_{i_{h+1}}$. The answer [2] is the sum of the h largest binomial coefficients of order n. Kleitman [3] and Katona [4] independently proved a stronger form of Sperner's theorem: If $S = S_1 \cup S_2$, $S_1 \cap S_2 = \phi$ and $\mathcal{A} = \{A_1,\ldots,A_m\}$ is a family of subsets of S, such that no two different A_i, A_j satisfy the properties

$$A_i \cap S_1 = A_j \cap S_1 \quad \text{and} \quad A_i \cap S_2 \subset A_j \cap S_2$$

or

$$A_i \cap S_1 \subset A_j \cap S_1 \quad \text{and} \quad A_i \cap S_2 = A_j \cap S_2,$$

then $m \leq \binom{n}{\left[\frac{n}{2}\right]}$. De Bruijn, Tengbergen and Kruyswijk [5] generalized the original theorem of Sperner in the following manner: Let $f_1,\ldots,f_m$ be integer-valued functions defined on $S = \{x_1,\ldots,x_n\}$ such that $0 \leq f_i(x_k) \leq \alpha_k$, where α_k's are given positive integers. If no two different of them satisfy $f_i(x_k) \leq f_j(x_k)$ (for all k), then $m \leq M$, where M is the number of functions satisfying

$$\sum_{k=1}^{n} f(x_k) = \left[\frac{\sum_{k=1}^{n} \alpha_k}{2}\right].$$

Recently Schönheim [6] gave generalizations of both Erdös and Kleitman-Katona's results for integer-valued functions.

Now we give a common generalization of all these theorems in a more general language.

We say that the finite set G is a *partially ordered set* if a relation $<$ is defined on G and (a) at most one of the relations $g_1 < g_2$, $g_1 = g_2$, $g_2 < g_1$ holds; (b) the relation is transitive.

g_2 *covers* g_1 if $g_1 < g_2$ and there is no g_3 satisfying $g_1 < g_3 < g_2$, that is, if g_2 "immediately greater" than g_2. Assume there is a rank function $r(g)$ which corresponds a non-negative integer to every element of G, so that the statement g_2 covers g_1 results $r(g_2) = r(g_1) + 1$ and there is at least one element $g \in G$ for which $r(g) = 0$. We say in this case that G is a *partially ordered set with a rank function.*

A *chain of length* h is a sequence $g_1, \ldots, g_h \in G$, where g_h covers g_{h-1}, g_{h-1} covers $g_{h-2}, \ldots, g_2$ covers g_1. A chain is *symmetrical* if $r(g_1) + r(g_h) = n = \max_{g \in G} r(g)$.

We say that a partially ordered set is a *symmetric chain set* if we can split G into disjoint symmetrical chains. (It is defined in [7] under a different name. Instead of partially ordered sets graph terminology is used).

For example, the subsets of a finite set S of n elements form a partially ordered set if we order them by inclusion. A covers B if $A \supset B$ and $|A-B| = 1$. There is also a rank function $r(A) = |A|$, that is the number of elements of A.

More generally the integer valued functions f defined on $S = \{x_1, \ldots, x_n\}$ $(0 \leq f(x_k) \leq \alpha_k)$ form a partially ordered set by the ordering "$f \leq g$ but not $f = g$ for all x_k". The rank function is $r(f) = \sum_{k=1}^{n} f(x_k)$. It is proved in [5] that this partially ordered set is a symmetrical chain set (As a special case, the partially ordered set of all the subsets is a symmetric chain set, too).

If G and H are partially ordered sets, then the *direct sum* G+H is the set of ordered pairs (g,h) $g \in G$, $h \in H$ with the ordering $(g_1,h_1) < (g_2,h_2)$ iff $g_1 \leq g_2$ and $h_1 \leq h_2$ but equality can hold in at most one place. If the rank function of G and H is r and s, respectively then we can define a rank funcion on G+H as follows $t((g,h)) = r(g) + s(h)$

If G is the partially ordered set of the subsets of a set S_1 and H is the same for S_2 (S_1 and S_2 are disjoint), then G+H is the partially ordered set of the subsets of $S_1 \cup S_2$. The situation is similar in the case of the integer-valued functions;the direct sum of two sets of this type is again a partially ordered set of integer-valued functions defined on the union of the sets.

Now we can formulate the general

Theorem. *Let G and H be symmetrical chain sets. If we have a set* $(g_1,h_1),\dots,(g_m,h_m)$ *of the elements of G+H, such that*

no $h+1$ *different ones of them satisfy the conditions*

$$g_{i_1} = \dots = g_{i_w},$$

$$h_{i_1} < \dots < h_{i_w},$$

$$g_{i_w} < \dots < g_{i_{h+1}},$$

$$h_{i_w} = \dots = h_{i_{h+1}},$$

for some w $(1 \leq w \leq h+1)$

then $m \leq$ *the number of elements of G+H with ranks*

$$\left[\frac{n-h+1}{2}\right], \dots, \left[\frac{n-h+1}{2}\right] + h - 1, \text{ where } n = \max_{(g,h)=G+H} t((g,h)) .$$

The estimation is the best possible.

The proof is given in [7].

It is easy to see, that the theorem contains the theorems listed above as special cases.

Another interesting special case if G and H are totally ordered sets of p and q elements, respectively. $G+H$ is in this case a rectangle. Schönheim's generalization of Erdös's theorem would state in this special case that if we have a set of elements of this rectangle and no $h+1$ different ones form a configuration of type

(1)

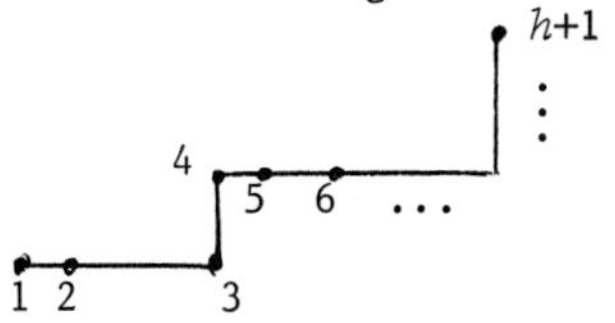

then the maximal system is the union of the h largest diagonal (*diagonal* is the set of elements of the rectangle with the same "coordinate-sum"). Our theorem say in this special case if we exclude only the configurations

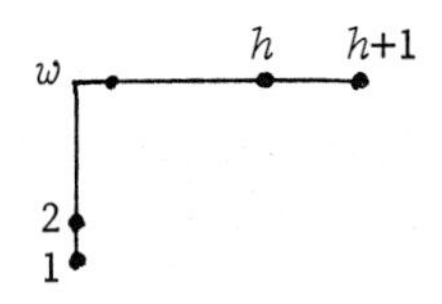

instead of (1), the maximal system is unchanged.

References

1. Sperner, E., Ein Satz über Untermengen einer endlichen Menge, *Math. Z.*, 27 (1928), 544-548.

2. Erdös, P., On a lemma of littlewood and Offord, *Bull. Amer. Math. Soc.*, 51 (1945), 8989-9002.

3. Kleitman, D., On a lemma of Littlewood and Offord on the distribution of certain sums, *Math. Z.*, 90 (1965), 251-259.

4. Katona, G., On a conjecture of Erdös and a stronger form of Sperner's theorem, *Studia. Sci. Math. Hungar.*, 1 (1966), 59-63.

5. De Bruijn, N.G., Van E. Tengbergen, C^A, and Kruiyswijk, D., On the set of divisors of a number, *Nieuw Arch. Wiskunde* (2), 23 (1949-51), 191-193.

6. Schönheim, J., A generalization of results of P. Erdös, G. Katona and D. Kleitman concerning Sperner's theorem, *J. Combinatorial Theory* (to appear).

7. Katona, G., A generalization of some generalizations of Sperner's theorem, *J. Combinatorial Theory* (to appear).

COMBINATORIAL INEQUALITIES

John B. Kelly

Arizona State University, Tempe, AZ, U.S.A.

1. Introduction. In many combinatorial investigations one seeks to decide whether, for a given integral, symmetric matrix $B = (b_{ij})$, $1 \leqslant i, j \leqslant n$, there exists a family of n subsets of a finite set, X, such that $|S_i \cap S_j| = b_{ij}$. The still unanswered question of the existence of finite geometries of prescribed orders is an example. Clearly, an equivalent problem is to decide whether there exists a zero-one matrix A such that $AA^T = B$. We have discussed this problem in [1].

Instead of dealing with intersections, one may deal with symmetric differences, since the intersection cardinalities b_{ij} determine the symmetric difference cardinalities d_{ij} by means of the formula

$$d_{ij} = b_{ii} + b_{jj} - 2b_{ij}, \qquad 1 \leqslant i,\ j \leqslant n.$$

The matrix $D' = (d_{ij})$ $1 \leqslant i, j \leqslant n$ does not contain as much information as B because its principal diagonal necessarily consists of zeroes. However, if one adjoins the empty set $\phi = S_{n+1}$ to the original family, one has, additionally

$$d_{i,n+1} = |S_i \,\Delta\, S_{n+1}| = b_{ii} \qquad (\Delta \text{ denotes symmetric difference.})$$

Then the matrix $D = (d_{ij})$, $1 \leqslant i, j \leqslant n+1$ contains as much information as B. B may be recovered from D by means of the formulae

$$b_{ii} = d_{i,n+1}, \qquad 1 \leqslant i \leqslant n$$

$$b_{ij} = \frac{1}{2}(d_{i,n+1} + d_{j,n+1} - d_{i,j}).$$

It is well known (cf. [3]) that the space 2^X of subsets of X is a metric space with metric $\rho(S_i, S_j) = |S_i \,\Delta\, S_j| = d_{ij}$. Hence our intersection problem has a solution if and only if D is the distance matrix for some subset, $\{S_i\}$, $1 \leqslant i \leqslant n+1$, of 2^X with $S_{n+1} = \phi$. However, this last restriction may be removed. In fact, if there is a family $\{T_i\}$, $1 \leq i \leq n+1$ of subsets of X with $|T_i \,\Delta\, T_j| = d_{ij}$, $1 \leq i,j \leq n+1$, then the family $\{S_i\}$, where $S_i = T_i \,\Delta\, T_{n+1}$ also satisfies the same conditions since the mapping $S \mapsto S \,\Delta\, T_{n+1}$ is an isometry of 2^X. Clearly $S_{n+1} = \phi$.

A study of the geometry of 2^X yields necessary conditions for the realizability of D. Certain linear metric inequalities which generalize the triangle inequality are satisfied by the metric, ρ, of 2^X. It turns out that these hypermetric inequalities hold in many of the important metric spaces of analysis and geometry. This observation has interesting consequences. For example, we are able to prove a conjecture of Fejes-Toth concerning the maximum sum of the mutual distances between m points on a sphere. It is perhaps surprising that this result is essentially combinatorial.

2. Hypermetric spaces. Let (M,ρ) be a metric space and let P_i and P_j be points of M. For brevity we shall frequently put $\rho(P_i,P_j) = \rho_{ij}$.

A metric space (M,ρ) is n-hypermetric provided that

$$\sum_{1\leq i<j\leq n} \rho_{ij}x_i x_j \leq 0 \tag{2.1}$$

for all sets of n points $\{P_1, P_2, \ldots, P_n\}$ of M and all sets of *integers* $\{x_1, x_2, \ldots, x_n\}$ such that

$$\sum_{i=1}^{n} x_i = 1. \tag{2.2}$$

Note that setting $n = 3$ and $x_1 = x_2 = 1$, $x_3 = -1$ gives $\rho_{12} - \rho_{13} - \rho_{23} \leq 0$ the triangle inequality. (M,ρ) will be said to be hypermetric if it is n-hypermetric for every positive integer n.

In [2], we discussed hypermetric spaces using an apparently weaker definition. That is, we called a metric space k-hypermetric if (2.1) held with $n = 2k + 1$ and $x_1 = x_2 = \ldots = x_k = x_{k+1} = 1$, $x_{k+2} = x_{k+3} = \ldots = x_n = -1$. But (2.1) may be obtained in the general case by applying the restricted definition to the set $\{P_1, P_2, \ldots, P_n\}$ modified by repeating each point, P_i, $|x_i|$ times.

That many metric spaces, in particular, 2^X, are hypermetric is a consequence of the following theorem.

Theorem 2.1. *Let (M,ρ) be a metric space and let (W,m) be a measure space. Suppose that with each point P of M there is associated a measurable subset S of W in such a way that $\rho(P_i,P_j) = \rho_{ij} = m(S_i \,\Delta\, S_j)$. Then (M,ρ) is hypermetric.*

Proof. Let $\Omega_n = \{1, 2, 3, \ldots, n\}$ and let $\tau \subset \Omega_n$. Let $R_\tau = \bigcap_{i\in\tau} S_i \cap \bigcap_{j\notin\tau} S'_j$, where S'_j is the complement of S_j in W. Thus R_τ is the (possibly empty) subset of W consisting of points belonging to those S_i for which $i \in \tau$ and to no other S_i. The sets R_τ are mutually disjoint measurable subsets of W (except perhaps for R_ϕ) and their union is W. Put $m_\tau = m(R_\tau)$ and $x_\tau = \sum_{i\in\tau} x_i$. Our proof of Theorem 2.1 is based upon the identity

$$\sum_{1\leq i<j\leq n} \rho_{ij}x_i x_j = \sum_{\tau\subset\Omega_n} m_\tau x_\tau x_{\tau'}. \tag{2.3}$$

To prove (2.3) note that

$$\rho_{ij} = m(S_i \,\Delta\, S_j) = m\left(\bigcup_{\substack{i\in\tau\\ j\notin\tau}} R_\tau \cup \bigcup_{\substack{i\notin\tau\\ j\in\tau}} R_\tau\right) = \sum_{\substack{i\in\tau\\ j\notin\tau}} m_\tau + \sum_{\substack{i\notin\tau\\ j\in\tau}} m_\tau .$$

Thus

$$\sum_{1\leq i<j\leq n} \rho_{ij}x_i x_j = \sum_{1\leq i<j\leq n}\left(\sum_{\substack{i\in\tau\\ j\notin\tau}} m_\tau + \sum_{\substack{i\notin\tau\\ j\in\tau}} m_\tau\right)x_i x_j = \sum_{\tau\subset\Omega_n} m_\tau \sum_{\substack{i\in\tau\\ j\notin\tau}} x_i x_j = \sum_{\tau\subset\Omega_n} m_\tau x_\tau x_{\tau'}.$$

If $\sum_{i=1}^{n} x_i = 1$, then $x_{\tau'} = 1 - x_\tau$ and $x_\tau x_{\tau'} = x_\tau(1-x_\tau)$. Since $m_\tau \geq 0$ and the function $t(1-t)$ is not positive for integral t, (2.1) follows immediately from (2.3). Hence (M,ρ) is n-hypermetric. Since n was arbitrary, (M,ρ) is hypermetric.

In [2] we proved this theorem using our earlier, but equivalent, definition of hypermetricity. We have reproved it here because we believe that the identity (2.3) is of independent interest.

Note that from (2.3) we can also deduce that (2.1) holds whenever $x_1, x_2, \ldots, x_n$ are real, not necessarily integral numbers whose sum is zero. However, it is possible to show that this is a weaker property of metric spaces, i.e., it is implied by hypermetricity.

If $M = 2^X$, one may take $W = 2^X$, $S_i = P_i$, and $m(S_i) = |S_i|$. Theorem (2.1) yields

Corollary (2.1). 2^X, under the symmetric difference metric, is hypermetric.

Corollary (2.2). The real line, (E_1,ρ) where $\rho(x,y) = |x-y|$, is hypermetric.

Proof. Let $W = E_1$ and associate with each real number x the interval $(0,x)$. Then $|x-y| = m((0,x) \,\Delta\, (0,y))$, so that E_1 is hypermetric.

From Corollary (2.2), one may deduce, as in [2], that inner product spaces are hypermetric under the metric induced by the norm. In particular, E_n, (Euclidean n-space) and H (Hilbert space) are hypermetric.

On a finite connected graph, Γ, one may define the distance $\rho(P,Q)$ between any two distinct nodes as the number of edges in a shortest path joining them.

Corollary 2.3. If Γ is a tree, then (Γ,ρ) is hypermetric.

Proof. Fix a node, O, of Γ. Associate with any node P the collection, E_{OP}, of edges in the unique path from O to P. Let the measure of any set of edges be its cardinality. It is readily seen that $\rho(P,Q) = m(E_{OP} \Delta E_{OQ})$. Theorem 2.1 now gives the corollary. The result still holds if the edges are given arbitrary non-negative weights.

Other graphs are hypermetric, e.g., circuits, complete graphs, the graphs of the 5 regular solids, but not all. A non-hypermetric graph is shown in Figure 2.1. Here $\rho_{12} = \rho_{13} = \rho_{23} = \rho_{45} = 2$,

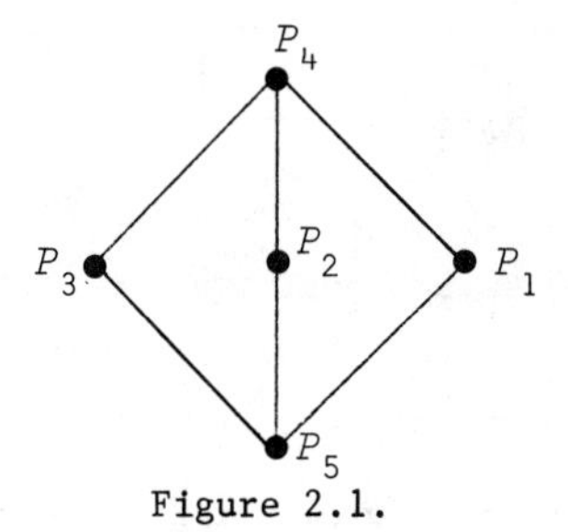

Figure 2.1.

$\rho_{14} = \rho_{24} = \rho_{34} = \rho_{15} = \rho_{25} = \rho_{35} = 1$. If one puts $x_1 = x_2 = x_3 = 1$, $x_4 = x_5 = -1$, one readily checks that (2.2) is satisfied but (2.1) is not. Hypermetricity is a necessary, but, as examples show, not sufficient condition that a given matrix be the distance matrix of a tree.

In [2] it is shown that the function spaces $L_1[0,1]$ and $L_2[0,1]$ are hypermetric but that $L_\infty[0,1]$ (the space of bounded functions on [0,1] with sup norm) is not hypermetric. All normed linear spaces of dimensions one and two are hypermetric under the metric induced by the norm, but there are spaces of dimension three which are not hypermetric. A normed lattice is always modular, but will be a hypermetric space under the induced metric if and only if it is distributive. Cases of equality in (2.1) are examined in [2].

3. The sphere. Another consequence of Theorem 2.1 is that the n-sphere, provided with the usual angular or great circle metric, is hypermetric. For convenience we assume that the radius is 1.

To see that the 1-sphere, i.e., the circle, with the arc length metric is hypermetric we associate with each point, P, the semi-circle symmetric about P. It is easy to see that if $PQ = \theta$, the angular measure of the symmetric difference of the corresponding semicircles is 2θ. (See Figure 3.1.) Since (2.1) is linear in the distances ρ_{ij}, it follows that the 1-sphere is hypermetric.

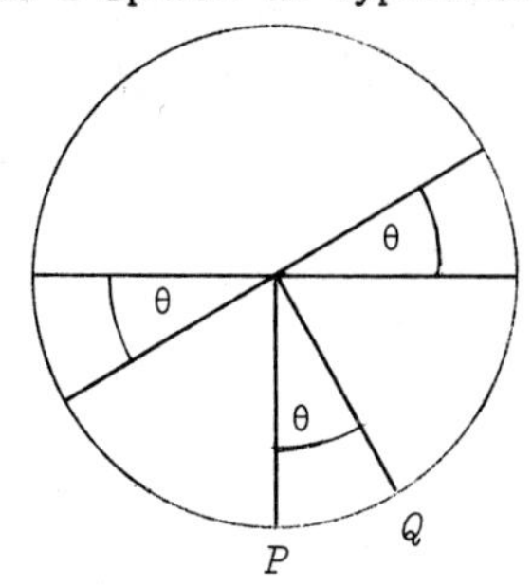

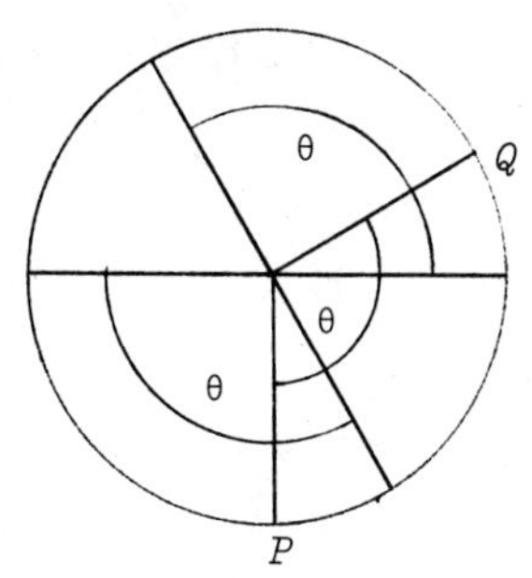

Figure 3.1.

The same idea works for the n-sphere, S_n. We let $W = S_n$ and associate with each point P of the n-sphere the hemisphere H_P symmetric about P. That is, $H_P = \{Q | Q \in S_n, \rho(P,Q) \leq \frac{\pi}{2}\}$. Define $m(J)$ to be the n-dimensional hypersurface area of the set J, considered as a subset of E_{n+1}. The hypermetricity of the n-sphere will be a consequence of the following lemma.

<u>Lemma 3.1</u>. Let P and Q be points of S_n whose angular distance is θ. Then $m(H_P \Delta H_Q) = K_n\theta$ where K_n is a constant depending only on n.

<u>Proof</u>. We proceed by induction on n. We have already treated the case $n = 1$.

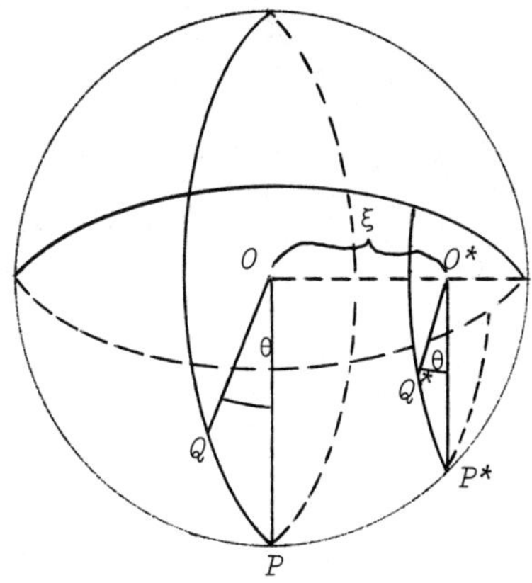

Considering P and Q as points of E_{n+1}, we may, without loss of generality, assume that $P = (1, 0, 0, \ldots, 0)$, $Q = (\cos\theta, \sin\theta, 0, \ldots, 0)$. Denote the coordinates in E_{n+1} by $\xi_1, \xi_2, \ldots, \xi_{n+1}$ and let Π_ξ be the hyperplane $\xi_{n+1} = \xi$, where $-1 \leq \xi \leq 1$. It is easy to see that $H_n(P) \cap \Pi_\xi$ is an $(n-1)$-dimensional hemisphere $H_{n-1}(P^*)$ of radius $\sqrt{1-\xi^2}$ symmetric about a point P^* with coordinates $(\sqrt{1-\xi^2}, 0, 0, \ldots, \xi)$. Similarly, $H_n(Q) \cap \Pi_\xi$ is an $(n-1)$-dimensional hemisphere $H_{n-1}(P^*)$ symmetric about a point $Q^* = (\sqrt{1-\xi^2}\cos\theta, \sqrt{1-\xi^2}\sin\theta, 0, 0, \ldots, \xi)$. (See figure 3.2 for an elucidation of the two-dimensional case.) Clearly

$$(H_n(P) \,\Delta\, H_n(Q)) \cap \Pi_\xi = (H_n(P) \cap \Pi_\xi) \,\Delta\, (H_n(Q) \cap \Pi_\xi) = H_{n-1}(P^*) \,\Delta\, H_{n-1}(Q^*).$$

The point $O^* = (0, 0, \ldots, \xi)$ is the common center of $H_{n-1}(P^*)$ and $H_{n-1}(Q^*)$ and P^* and Q^* subtend an angle θ at O^*. By our inductive hypothesis, the $(n-1)$-dimensional measure of $H_{n-1}(P^*) \,\Delta\, H_{n-1}(Q^*)$ is $K_{n-1}\theta(\sqrt{1-\xi^2})^{n-1}$. Thus

$$m(H_n(P) \,\Delta\, H_n(Q)) = \int_{-1}^{1} \frac{K_{n-1}\theta(\sqrt{1-\xi^2})^{n-1}}{\sqrt{1-\xi^2}}\, d\xi = K_n\theta.$$

Theorem 3.1. *The n-sphere with the great circle metric is hypermetric.*

Proof. Lemma 3.1 and Theorem 2.1.

Theorem 3.2. *Let $P_1, P_2, \ldots, P_m$ be points on the n-sphere, S_n, of unit radius. Then, if m is odd,*

$$\sum_{1 \leq i < j \leq m} \rho_{ij} \leq \frac{m^2-1}{4}\pi \tag{3.1}$$

while if m is even

$$\sum_{1 \leq i < j \leq m} \rho_{ij} \leq \frac{m^2}{4}\pi. \tag{3.2}$$

Proof. Let m be odd and set $m = 2k+1$. Put $x_i = 1$, $1 \leq i \leq k+1$, $x_i = -1$, $k+2 \leq i \leq m$, so that $\sum_{i=1}^{m} x_i = 1$. Consider the points $P_1, P_2, \ldots, P_{k+1}, P'_{k+1}, \ldots, P'_m$ where P'_j is the point on S_n antipodal to P_j, $k+2 \leq j \leq m$. Then, since S_n is hypermetric, we have, from (2.1),

$$\sum_{1 \leq i < j \leq k+1} \rho(P_i,P_j) + \sum_{k+2 \leq i < j \leq m} \rho(P'_i,P'_j) - \sum_{i=1}^{k+1} \sum_{j=k+2}^{m} \rho(P_i,P'_j) \leq 0. \tag{3.3}$$

Using $\rho(P'_i,P'_j) = \rho(P_i,P_j)$ and $\rho(P_i,P'_j) + \rho(P_i,P_j) = \pi$, we obtain (3.1) immediately from (3.3). (3.2) is obtained similarly. Here we do not use directly the fact that S_n is hypermetric but rather the weaker property mentioned at the conclusion of the proof of Theorem (2.1). (3.2) may also be derived directly from the stronger inequality (3.1).

It is easy to see that Theorem (3.2) is best possible. Equality holds in 3.1 if $P_1 = P_2 = \ldots = P_{k+1} = P'_{k+2} = \ldots = P'_m$, and in 3.2 if $P_1 = P_2 = \ldots = P_k = P'_{k+1} = \ldots = P_m$ (with $m = 2k$).

Theorem 3.2 was conjectured by Fejes-Toth and the case of even m was previously handled by Sperling [4].

4. Conclusion. Although our purpose in introducing the concept of hypermetricity was to produce a tool for dealing with combinatorial problems, we have not had great success in using it to tackle the most delicate questions. Essentially the reason for this is that our proof of Theorem 2.1 exploited only the fact that the quantities m_τ which occur in the fundamental identity (2.3) are non-negative whereas in most combinatorial problems one knows additionally that the quantities m_τ are integers. In some instances it is possible to apply the Hasse-Minkowski theory of the congruence of quadratic forms to (2.3) so that this powerful weapon, one of the principal assets of the matric approach to combinatorial problems is still available in our metric approach.

Our work has implications for coding theory. It turns out that the space of q-ary codes, under either the Lee or Hamming metric, is hypermetric. The consequences of this fact will be developed elsewhere.

References

1. Kelly, J. B., Products of Zero-One Matrices, *Can. J. Math.* 20, 1968, 298-329.

2. Kelly, J. B., Metric Inequalities and Symmetric Differences, Proceedings of Second Symposium on Inequalities (Colorado Springs, 1967), *Academic Press*, to appear.

3. Silverman, R., A Metrization for Power Sets with Applications of Combinatorial Analysis, *Can. J. Math.*, 12 (1960) 153-176.

4. Sperling, G., Lösung einer elementärgeometrischen Frage von Fejes-Toth, *Arch. Math.*, 11 (1960) 69-71.

THE NUMBER OF TOURNAMENT SCORE SEQUENCES FOR A LARGE NUMBER OF PLAYERS

D. Kleitman

Massachusetts Institute of Technology
Cambridge, MA, U.S.A.

Suppose n people play in a round robin tournament, playing each other participant once, one point being credited to the winner in each game. We order the players in increasing order of their total scores, and associate to each tournament outcome the sequence of scores obtained by the various players in it. This sequence will be an increasing sequence consisting of n integers; we denote as $\{S_1,\ldots, S_n\}$.

In this paper we consider the question: How many distinct sequences of this kind are there for n players? We show that the answer, called $T_{012}(n)$ below, satisfies $c_1 \dfrac{2^{2n}}{n^{5/2}} < T_{012}(n) < c_2 \dfrac{2^{2n}}{n^{5/2}}$.

This result was conjectured by L. Moser and P. Erdös who have obtained weaker bounds; a full discussion of their work appears in Moon's book on tournaments.

The problem can be reformulated in several ways. It is well known (see Moon) that necessary and sufficient conditions on increasing sequences of this kind are:

$$1^\circ \qquad \sum_{i=1}^{n} S_i = \binom{n}{2}$$

$$2^\circ \qquad \sum_{i=1}^{k} S_i \geq \binom{n}{2} \text{ for all } k,\ k \leq n.$$

The former condition reflects the fact that exactly one point must be scored for each game played. The second condition involves the fact that the weakest k players play $\binom{k}{2}$ games among themselves, and hence must score at least $\binom{k}{2}$ points.

Another formulation (used by Erdös and Moser) is as a random walk on the plane in which the following restrictions apply.

We start from the origin and each step is of unit length directed either up or to the right.

(0) There are n steps up and n to the right,

(1) The area between the resulting path and the horizontal axis is $\binom{n}{2}$.

(2) The similar area restricted to the first k columns is at least $\binom{k}{2}$.

Under these circumstances we can interpret the height of the j-th column as S_j which gives us a direct correspondence between such walks and tournament score sequences. Notice that the j-th horizontal line is the $(S_j + j)$-th step in the random walk.

For convenience we let $T_i(n)$, $T_{ij}(n)$ and $T_{ijk}(n)$ be the number of random walks of $2n$ steps satisfying conditions i; i, j and i, j, k respectively.

We can associate a monomial $\left(\prod_{k=1}^{2n} w_k\right)$ to a random walk by setting $w_k = xy^k$ if the k-th step is to the right, and $w_k = 1$ otherwise.

The various monomials so obtained each correspond to a unique term in the polynomial

$$(1+xy)\ (1+xy^2)\ \ldots\ (1+xy^{2n}) \simeq \prod_{k=1}^{2n} (1+xy^k)$$

The coefficient of $x^n y^{n^2}$ in this polynomial counts those monomials corresponding to walks satisfying the conditions:

there are exactly n steps to the right, and

the area between the walk and the horizontal axis is $\binom{n}{2}$.

The last fact follows since the j-th horizontal step will give rise to a factor xy^{S_j+j}; the contribution from n horizontal steps is therefore $x^n y^{\left(\sum_{j=1}^{n} j + \sum_{j=1}^{n} S_j\right)}$, or $x^n y^{n(n+1)/2 + \sum_{i=1}^{n} S_j}$.

Extraction of the coefficient of $x^n y^{n^2}$ in this polynomial is a straightforward exercise; the results are

$$T_{10}(n) = \left(\frac{1}{2\pi i}\right)^2 \oint \frac{dx}{x^{n+1}} \oint \frac{dy}{y^{n^2+1}} \prod_{y=1}^{n} (1+xy^j)$$

$$= \frac{2^{2n}}{(2\pi)^2} \int_0^{2\pi} d\theta \int_0^{2\pi} d\theta\, e^{in\phi} \prod_{y=1}^{n} \cos(\theta+y\phi)$$

$$= c_{12} \frac{2^{2n}}{n^2} (1+0(1)) \equiv T_{01}^*(n)$$

The estimate $T_{01}^*(n)$ here obtained for $T_{01}(n)$ is also an upper bound on the number of random walks having any other specified area under them.

A tournament in which each player beats all weaker players and loses to all others will have score sequence $(0,1,2,\ldots,n-1)$ which corresponds to a staircase walk path which we call S.

We can classify our random paths P according to the number, $k(P) + 1$, of connected components of their intersection with S. Let $T_{01}(n,k)$ (similarly $T_0(n,k)$) be the number of paths having $(k + 1)$ such components and satisfying conditions (0) and (1). If P is any path let the area between P and S (with negative sign if P lies below S) lying between the l-th and $(l+1)$-th component of $P \cap S$ be denoted by $\sigma_i(P)$. Notice that if P satisfies condition (1) it must satisfy $\Sigma\sigma_i(P) = 0$, as the total area under it and S must be the same.

The following two statements allow us to bound $T_{012}(n)$.

1. To verify that P satisfies condition (2) only requires us to verify that for each $l \leqslant k(P)$

$$\sum_{j=1}^{l} \sigma j(P) \geqslant 0$$

i.e., that all partial sums of the sequence $\{\sigma j\}$ are non-negative.

2. If $\sigma_1,\ldots,\sigma_k$ is any sequence of non-zero real numbers satisfying $\sum_{i=1}^{k} \sigma_i = 0$ then, among the $n!$ permutation of the indices $L \to \pi(i)$ a proportion Ω with $\frac{1}{k} \leqslant \Omega \leqslant \frac{c}{k}$, $c < 4$ satisfy $\sum_{L=1}^{l} \sigma_{\pi(i)} \geqslant 0$ for all l.

Proof of these remarks will be published elsewhere. They would have occurred in an appendix which space limitations here cause to be ruptured. To prove the lower bound here one need only notice that cyclic permutation of the k σ_y's must give rise to at least one ordering for which all partial sums are positive; namely we can start after the minimal partial sum in the original ordering.

These remarks allow us to deduce the inequalities

$$c \sum \frac{T_{01}(n,k)}{k} \geq T_{012}(n) \geq \sum \frac{T_{01}(n,k)}{k}$$

All that remains to give us a bound $T_{012}(n)$ are bounds on $\sum \frac{T_{01}(n,k)}{k}$. Speaking crudely, paths satisfying conditions (0) and (1) mostly have $k(P)$ values of the order of $n^{1/2}$ so that $\sum_k \frac{T_{01}(n,k)}{k}$ is of the order of $\frac{c^1}{n^{1/2}} T_{10}(n)$ which leads to the behaviour $C^{11}\ 2^{2n}/n^{5/2}$ for $T_{012}(n)$.

We can deduce limits and probably even obtain an asymptotic formula for $\sum(1/k)T_{01}(n,k)$ by utilizing the fact that random walks satisfying condition (0) represent a classical situation - coin tossing which breaks even after 2 tosses. The properties of such walks can be determined in great detail by standard methods using such concepts as "recurrent events"; and th central limit theorem. Thus $T_0(n,k)$, the distribution of intervals between successive break even points, and area distributions under walks counted by $T_0(n,k)$, can be determined asymptotically. Such expressions allow asymptotic determination of $T_0(n,k)$ and $\frac{T_{01}(n,k)}{T_0(n,k)}$ and hence of $\sum \frac{T_0(n,k)}{k} \left(\frac{T_{01}(n,k)}{T_0(n,k)}\right)$ which is $\sum \frac{T_{01}(n,k)}{k}$. The detailed determinations will not be presented here.

It is relatively easy to obtain an upper bound on this sum. By noticing the meaning of $\Sigma(k+1)T_{01}(n,k)$ we can find an upper bound θ in this quantity. Since we know $\Sigma T_{01}(n,k)$ and since the produce of the average value of the quantity and of its reciprocal always exceeds one, we obtain $\sum \frac{T_{01}(n,k)}{k} \geq \frac{T_{01}(n)^2}{\theta}$.

The sum $\Sigma(k+1)T_{01}(n,k)$ counts the number of connected components of $P \cap S$ for all walks P satisfying conditions (0) and (1). These can be overestimated if we sum over all j of the number (N_j) of paths of the desired kind for which $P \cap S$ contains (j,j). For $j \geq \frac{n}{2}$ (and otherwise interchange j and $n-j$) we obtain $N_j \leq T^*_{01}(j)\ T_0(n-j)$ which yields

$$\Sigma\, k\, T_{01}(n,k) \leq 2 \sum_{j=n}^{n} T_{01}{*}(j)\, T_0(n-j) = \theta$$

which yields a bound of the required kind.

Some additional conjectures related to this problem are

1° given k real numbers σ_i with $\sum_{i=1}^{k} \sigma_i = 0$.
Then the proportion of their orderings which make all partial sums positive cannot exceed $\frac{2}{k+2}$ (which bound is realized if all σ_i are of equal magnitude).

2° with the σ_i distributed as they are in our problem, that the proportion of orderings making all partial sums positive is $\frac{1}{k}$ (1 + 0(1)) for large values of k and asymptotically all random walks.

3° that a similar analysis can be applied to tournaments in which each player plays l games with each opponent, $l > 1$.

4° that the remarks in the last pages can be utilized to obtain an explicit asymptotic formula for $\Sigma T_{01}(n,k)\ /\ k$ rather than only a lower bound.

Proof of conjectures 2° and 4° would yield an asymptotic formula for $T_{012}(n)$.

References

1. Moon, J.W., *Topics on tournaments*, New York, Hold, 1968.
2. Feller, W., *An introduction to probability theory and its applications*, New York, John Wiley and Sons, 1967.

GROUPOIDS AND PARTITIONS OF COMPLETE GRAPHS

Anton Kotzig

The University of Calgary, Alta., Canada

In this lecture every set is understood to be finite. We understand as usually under a binary operation on the set S a mapping of the cartesian product $S \otimes S$ into S. If V is a set and $\otimes$ a binary operation on V then we call the pair $\{V,\otimes\}$ a groupoid. $\{V,\otimes\}$ is a P-groupoid, if it is a groupoid and has the following three properties:

(1) $a \otimes a = a$ for every $a \in V$;

(2) $[a \neq b] \Longrightarrow [a \neq a \times b \neq b]$ for every $a, b \in V$;

(3) $[a \otimes b = c] \Longleftrightarrow [c \otimes b = a]$ for every $a, b, c \in V$.

Lemma 1. Let $\{V,\otimes\}$ be a P-groupoid and let a, b be two arbitrary elements of V then the equation $x \otimes a = b$ has one and only one solution in V.

The proof is clear: $[x \otimes a = b] \Longrightarrow [b \otimes a = x]$, $[y \otimes a = b] \Longrightarrow [b \otimes a = y]$ and therefore $x = y$. The equation has at most one solution. This solution is obviously $x = b \otimes a$.

The groupoid $\{V,\otimes\}$ will be called a quasigroup if the equation $a \otimes x = b$ (and similarly the equation $x \otimes a = b$) has exactly one solution for every $a, b \in V$.

There exist P-groupoids, which are not quasigroups. Let us show an example: $V = \{1,2,3,4,5\}$; the binary operation $\otimes$ is described by the following table of multiplication:

⊗	1	2	3	4	5
1	1	5	5	5	3
2	4	2	4	3	4
3	5	4	3	2	1
4	2	3	2	4	2
5	3	1	1	1	5

Table of multiplication

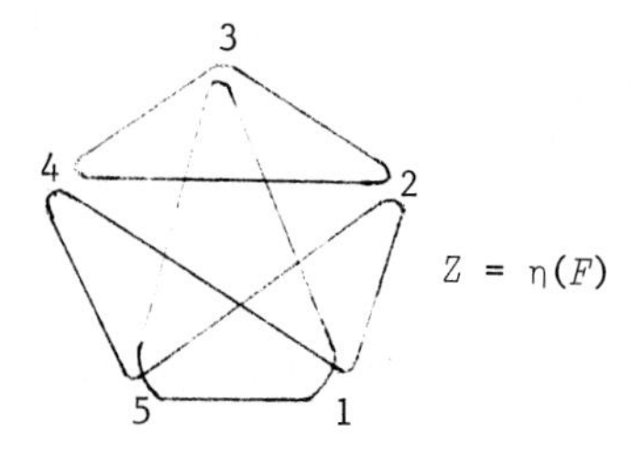

Fig. 1

We can easily see, that in this case $\{V,\otimes\} = F$ has the properties (1), (2), (3).

Let us recollect some definitions of the notions of graph theory: Let G be a graph. A sequence $S = \{x_1,y_1,x_2,y_2,\ldots,y_p,x_{p+1}\}$ of elements G (where $p > 0$; x_i are vertices; y_j are edges and y_j joins in G the vertices x_j, x_{j+1}) is called a line of G if an arbitrary edge of G appears in S at most once.

A line is closed (or open) if $x_{p+1} = x_1$ (or $x_{p+1} \neq x_1$). Lines obtained from S by the reversion of its elements and by a translation of its elements (if S is closed) are not considered as different of S. By a transition of S through x_i we mean a triple of elements $\{y_i,x_i,y_{i+1}\}$. In the case where S is a closed line we consider as a transition through $x_1 = x_{p+1}$ also the triple $\{y_p,x_1,y_1\}$

Under a partition of the graph G into closed lines (or briefly: under a ξ-partition) we understand a set Z of closed lines of G such that an arbitrary edge of G belongs to exactly one closed line of Z. If the ξ-partition of G contains only one element then we call this closed line an Eulerian line of G.

Let Z be a ξ-partition of the graph G and $V = \{v_1,v_2,\ldots,v_n\}$ the set of vertices of G. We denote by $H(i)$ the set of all edges of G incident to $v_i \in V$. Let d_i be the degree of v_i. We know very well that a ξ-partition of G exists if and only if d_i is an even number for every $i \in \{1,2,\ldots,n\}$. (We put $d_i = 2c_i$; c_i is an integer). In the closed lines of Z we evidently find c_i different transitions $\{e_i(k),v_i,f_i(k)\}$ (where $k = 1,2,\ldots,c_i$) through v_i. Obviously we have:

$$\bigcup_{k=1}^{c_i} \{e_i(k),f_i(k)\} = H(i);\ [r\neq s] \Longrightarrow [\{e_i(r),f_i(r)\} \cap \{e_i(s),f_i(s)\} = \emptyset]$$

The partition $D_i = \{\{e_i(1),f_i(1)\}, \{e_i(2),f_i(2)\},\ldots, \{e_i(c_i),f_i(c_i)\}\}$ of

$H(i)$ into pairs of edges is called a δ-partition in v_i formed by Z and the system $D = \{D_1, D_2, \ldots, D_n\}$ is called a δ-system formed in G by Z. We denote the fact that D is the δ-system formed by Z thus: $D = \delta(Z)$.

The following lemma is well known:

Lemma 2. Every ξ-partition of G forms exactly one δ-system in G and to every system $D = \{D_1, D_2, \ldots, D_n\}$ of partitions of the sets $H(i)$ into pairs of edges there exists exactly one ξ-partition Z of G so that $D = \delta(Z)$.

Let n be a natural number. We denote by $<n>$ the complete graph with vertices $v_1, v_2, \ldots, v_n$. Let $\{V, \otimes\} = G$ (where $V = \{v_1, v_2, \ldots, v_n\}$) be a P-groupoid.

We construct the partition D_i of the set $H(i)$ $(i = 1, 2, \ldots, n)$ of all edges in $<n>$ incident with v_i as follows: the edge $e_{i,j}$ joining in $<n>$ the vertices v_i, v_j belongs to the same class of D_i as the edge $e_{i,k}$ (which joins v_i, v_k) just when $v_j \otimes v_i = v_k$ (then we have also $v_k \otimes v_i = v_j$ - see the property (3) of P-groupoid). From the definition of P-groupoid it is clear: every class of D_i contains just two edges and D_i is a partition of $H(i)$ into pairs of edges. Therefore: $D = \{D_1, D_2, \ldots, D_n\}$ is a δ-system in $<n>$ and we have:

Lemma 3. Every P-groupoid of $G = \{V, \otimes\}$ forms in the way described a δ-system in $<n>$ and to every P-groupoid corresponds one and only one ξ-partition Z of $<n>$ so that $D = \delta(Z)$.

We denote the fact that Z corresponds in the way described to G thus: $Z = \eta(G)$. (See e.g. Fig. 1).

We obtain immediately:

Lemma 4. Every P-groupoid contains an odd number of elements. Let $Z = \{Z_1, Z_2, \ldots, Z_p\}$ be some ξ-partition of $<n>$ into closed lines and let $V = \{v_1, v_2, \ldots, v_n\}$ be the set of vertices of $<n>$. Let us define the binary operation $\otimes$ on V as follows: (i) $v_i \otimes v_i = v_i$ for every $i \in \{1, 2, \ldots, n\}$ (ii) Let v_i, v_j be two different vertices $\in V$ and let e be the edge of $<n>$ joining the vertices v_i, v_j. We denote by Z_r the line of Z which contains e. Let $v_i, v_j, v_k, \ldots$ be the order in which we travel through the vertices of Z_r starting from v_i through the edge e. Then we put $v_i \otimes v_j = v_k$.

Z is a ξ-partition of a complete graph. We obviously have $\otimes$ is a binary operation on V and $\{V, \otimes\} = G$ is a P-groupoid where $Z = \eta(G)$.

The validity of the following theorem is clear from this:

Theorem 1. *Let V be a given set with n elements (n is odd). There exists a one to one mapping of the set of all P-groupoids* $\{V,\otimes\}$ *onto the set of all partitions of a complete graph with n vertices into closed lines.*

Let us consider the following: A set with an even number $2k$ of elements can be parted into k pairs in $1.3. \ldots . (2k-3).(2k-1) = \phi(k)$ different ways. This means, that there exist exactly $\phi(k)$ different δ-partitions of the set of all edges incident with a given vertex of $\langle 2k+1\rangle$ and that the number of different δ-systems in this graph is

$$\theta(k) = \prod_{i=1}^{k} [(2i-1)^{2k+1}].$$

The number $\theta(k)$ means also the number of different ξ-partitions of $\langle 2k+1\rangle$ and (see Theorem 1) $\theta(k)$ is the number of different P-groupoids with $\langle 2k+1\rangle$ elements (two isomorp P-groupoids are considered as different).

It seems to me that an investigation of the isomorphisms in the set of all P-groupoids with a given number of vertices is full of hopes. An investigation of special P-groupoids, which correspond in the above mentioned sense to the partitions of a complete graph into closed lines of the fixed form (e.g.: into circuits, quadratic factors, r-angles, into closed lines which all are isomorph and so on) is also promising. I am not going to deal with these questions in this lecture. I wish to concentrate your attentions in an other direction: If the equation $a \otimes x = b$ has in the groupoid $G = \{V,\otimes\}$ a solution for every $a, b \in V$, then this equation has in G for every $a, b \in V$ exactly one solution and G is a quasigroup (see Lemma 1). We shall call this quasigroup a P-quasigroup. As the example of such a quasigroup we introduce the following: If Z is the partition of a complete graph into triangles and $Z = \eta(G)$, then G is a P-quasigroup.

Let $G = \{V,\otimes\}$ be a given P-quasigroup. With the help of G we define the binary operation $\odot$ on the set V as follows: for every $a, b \in V$,

$$a \otimes (a \odot b) = b \text{ holds.}$$

Remark. Using the above mentioned method it is possible to define a binary operation $\odot$ when and only when for every a, b the equation $a \otimes x = b$ has only one solution, so: only when $\{V,\otimes\}$ is a quasigroup. If the mentioned equation would have more solutions or not solution at all, then the term $a \otimes b$ would not be defined.

The following theorem holds:

Theorem 2. *If* $\{V,\otimes\}$ *is a P-quasigroup then* $\{V\odot\}$ *is a commutative quasigroup in which* $a \odot a = a$ *for every* $a \in V$. *Conversely: If* $\{V,\odot\}$ *is a commutative quasigroup in which* $a \odot a = a$ *for every* $a \in V$, *then there exists just one P-quasigroup* $\{V,\otimes\}$ *such, that for every* $a,b,c \in V$ *the following holds:*

$$[a \otimes b = c] \Longleftrightarrow [a \odot c = b].$$

We call the commutative quasigroup $\{V,\odot\}$ in which $a \odot a = a$ for every $a \in V$ a K-quasigroup.

It follows from our considerations, that the class of all K-quasigroups corresponds (in a one to one mapping) to a special class of partitions of a complete graph into closed lines.

This problem, which has for the theory of K-quasigroups an essential importance is not yet solved in a satisfactory way: Which are properties (from the point of view of the theory of graphs) of the partition Z of $\langle n \rangle$ into closed lines, in order to the binary operation $\otimes$ defined with the help of Z forms on the set of vertices of $\langle n \rangle$ a P-quasigroup (and not only a P-groupoid). For instance, it is known that if Z is a partition of $\langle n \rangle$ into triangles, then $\{V,\otimes\}$ is surely a P-quasigroup. Let us remark that in this case and only in this case $\{V,\otimes\} = \{V,\odot\}$ holds. Some attentions are paid to the theory of these partitions. This can be helpful to the theory of K-quasigroups.

The following case is also very interesting: Z is a partition of $\langle n \rangle$ into the circuits of length 5 and $\{V,\otimes\} = G$ (where $Z = \eta(G)$) is a P-quasigroup. If we replace in Z every circuit by its complement in the corresponding complete graph with 5 vertices, we obviously obtain a new partition Z' of $\langle n \rangle$ into circuits of length 5 and this partition again forms some P-quasigroup G'. Both quasigroups G, G' are conjugated: if we made the mentioned arrangement in Z' we obtain so the partition Z. I think that the research of such conjugated pairs of P-quasigroups (and also K-quasigroups) can give very interesting results.

Another interesting domain of such problems is the following: Every Eulerian line of a graph can be considered as an only element of a certain ξ-partition. The following question arises: which of the Eulerian lines define such binary operation $\otimes$ on the set V of all vertices of $\langle n \rangle$, that $\{V,\otimes\}$ is a quasigroup? Do such Eulerian lines exist? In this matter we can give at once a definite answer as far as $n < 9$. We know that n must be odd, therefore n belonging to $\{2,4,6,8\}$ is not taken into consideration. For the case $n = 1$ this question has no sense at all. For $n = 3$ the answer is trivially affirmative, for $n = 5$ Eulerian line with the expected properties does not exist and for $n = 7$ there exists (up to the isomorphism) exactly one such Eulerian line (see Figure 2).

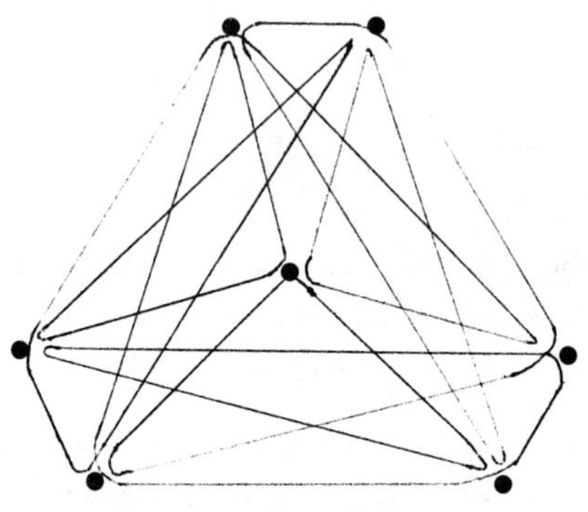

Fig. 2

I suppose even in a smaller team the answer for some greater n could be found. It is very likely that in this process some general results could be reached.

Very interesting is also the connection between the K-quasigroups and the partitions of n into nearly linear factors. Under the nearly linear factor of a graph we understand such its factor, in which one vertex is isolated and the remaining vertices are of first degree.

Let $G = \{V, \odot\}$ be a K-quasigroup with $2k+1$ elements and c be its arbitrary element. It follows directly from the definition of K-quasigroups, that the elements of V other than c can be divided into k pairs $\{x_1, x_2\}$, $\{x_3, x_4\}$, ..., $\{x_{2k-1}, x_{2k}\}$ so, that $x_{2i-1} \odot x_{2i} = c$ for every i. It follows from this, that the elements of the graph $\langle 2k+1 \rangle$ can be denoted as follows: the vertices of $\langle 2k+1 \rangle$ are elements of V and the edge joining the vertices x, y is denoted by the sign $z = x \odot y$. It is clear that the edges denoted by the same sign form the set of edges of some nearly linear factor of $\langle 2k+1 \rangle$ and that G forms by this method a partition of $\langle 2k+1 \rangle$ into nearly linear factors.

It is also evident that to every partition of $\langle 2k+1 \rangle$ into nearly linear factors corresponds unambigously by some K-quasigroup and that to different such partitions correspond different K-quasigroups.

So to every K-quasigroup $G = \{V,\odot\}$ with n elements corresponds (in a one to one mapping) a certain partition of the graph $\langle n \rangle$ into nearly linear factors and also some partition of the graph into closed lines $\eta(\{V,\otimes\})$, where $\{V,\otimes\}$ and $\{V,\odot\}$ are quasigroups connected by the relations $a \otimes (a \odot b) = b$.

The fact that every K-quasigroup has two mentioned different interpretations in the theory of graphs facilitates very substantially the finding of the isomorphisms in the class of all K-quasigroups. I used this fact and I found all different (non isomorph) K-quasigroups with the number of less than 9 elements. Figure 3 describes my results.

$\underline{1}$

$\underline{1}$	3	2
3	$\underline{2}$	1
2	1	$\underline{3}$

$\underline{1}$	4	2	5	3
4	$\underline{2}$	5	3	1
2	5	$\underline{3}$	1	4
5	3	1	$\underline{4}$	2
3	1	4	2	$\underline{5}$

$\underline{1}$	5	2	6	3	7	4
5	$\underline{2}$	6	3	7	4	1
2	6	$\underline{3}$	7	4	1	5
6	3	7	$\underline{4}$	1	5	2
3	7	4	1	$\underline{5}$	2	6
7	4	1	5	2	$\underline{6}$	3
4	1	5	2	6	3	$\underline{7}$

$\underline{1}$	3	5	2	6	7	4
3	$\underline{2}$	6	5	7	4	1
5	6	$\underline{3}$	7	4	1	2
2	5	7	$\underline{4}$	1	3	6
6	7	4	1	$\underline{5}$	2	3
7	4	1	3	2	$\underline{6}$	5
4	1	2	6	3	5	$\underline{7}$

$\underline{1}$	7	6	5	4	3	2
7	$\underline{2}$	5	6	3	4	1
6	5	$\underline{3}$	7	2	1	4
5	6	7	$\underline{4}$	1	2	3
4	3	2	1	$\underline{5}$	7	6
3	4	1	2	7	$\underline{6}$	5
2	1	4	3	6	5	$\underline{7}$

$\underline{1}$	5	6	7	2	3	4
5	$\underline{2}$	1	6	7	4	3
6	1	$\underline{3}$	5	4	7	2
7	6	5	$\underline{4}$	3	2	1
2	7	4	3	$\underline{5}$	1	6
3	4	7	2	1	$\underline{6}$	5
4	3	2	1	6	5	$\underline{7}$

$\underline{1}$	7	4	3	2	5	6
7	$\underline{2}$	5	6	3	4	1
4	5	$\underline{3}$	7	6	1	2
3	6	7	$\underline{4}$	1	2	5
2	3	6	1	$\underline{5}$	7	4
5	4	1	2	7	$\underline{6}$	3
6	1	2	5	4	3	$\underline{7}$

$\underline{1}$	3	4	7	2	5	6
3	$\underline{2}$	1	6	7	4	5
4	1	$\underline{3}$	5	6	7	2
7	6	5	$\underline{4}$	3	2	1
2	7	6	3	$\underline{5}$	1	4
5	4	7	2	1	$\underline{6}$	3
6	5	2	1	4	3	$\underline{7}$

$\underline{1}$	3	2	5	7	4	6
3	$\underline{2}$	7	6	4	1	5
2	7	$\underline{3}$	1	6	5	4
5	6	1	$\underline{4}$	3	7	2
7	4	6	3	$\underline{5}$	2	1
4	1	5	7	2	$\underline{6}$	3
6	5	4	2	1	3	$\underline{7}$

Fig. 3. The K-quasigroups with the number of elements < 9.

There is no doubt that the shifting of the border over 9 can give us new and interesting knowledge. It can be that there will be more light even in the following conjecture:

Conjecture. In every complete graph with an odd number $n = 2k-1$ of vertices there exists such a partition into $2k-1$ nearly linear factors such that the composition of every two of them is a Hamiltonian path of the graph.

I proved a long time ago the validity of the mentioned assertion for the case when n is a prime and I can prove this also for the case when k is a prime. The remaining cases wait still patiently for a discoverer.

COUNTING PROBLEMS IN DENDROIDS

G. Kreweras

University of Paris, France

We define a *graphoid* (A,B) by

1. a (finite) set A of *vertices*

2. a set B of subsets of A (or "*bonds*") subject to the following conditions:

   a. For each bond j, card $j \geq 2$ (if card $j = 2$ the bond will be called an *edge*).

   b. For any two bonds j_1 and j_2, card$(j_1 \cap j_2) \leq 1$.

If all the bonds are edges, the graphoid is a *graph*.

A *chain* is any sequence $a_1b_1a_2b_2 \ldots a_{n-1}b_{n-1}a_n$ the terms of which are alternately vertices and bonds, beginning and ending with vertices, and such that

i. the $2n$-1 terms are distinct

ii. for each b_i, a_i and a_{i+1} belong to b_i.

If the only non-distinct terms are $a_1 = a_n$, the sequence becomes a *cycle*. A graphoid is *connected* if any two distinct vertices belong to a chain. A connected graphoid without cycles will be called a *dendroid*.

In order to solve some counting problems (and also for other possible purposes) it is convenient to introduce the *associate tree* $T(D)$ of a given dendroid $D = (A,B)$. We shall so call the graph (X,Y) in which

i. the set of vertices X is $X = A + B$ (cartesian sum)

ii. the set of edges Y is defined by all the incidence-relations (ε-relations) between vertices and bonds of D.

It follows clearly from the definitions that $T(D) = (X,Y)$ is in fact a tree. Moreover $\{A,B\}$ is the "even-odd" partition (or 2-colouring partition) of X; we shall speak of the A-vertices and B-vertices of $T(D)$.

If we write card $A = a$ and card $B = b$, each of the $a + b$ vertices of $T(D)$ has its own degree. Since the total number of edges of $T(D)$ is $a + b - 1$, the same number is the sum of the degrees of the A-vertices as well as of the B-vertices; thus, if d_i and d_j are the degrees of vertices i and j,

$$\sum_{i\varepsilon A} (d_i - 1) = b - 1 \quad \text{and} \quad \sum_{j\varepsilon B} (d_j - 1) = a - 1. \tag{1}$$

Conversely, consider any tree T that can be 2-colored with the same color on all the terminal vertices. If $T = (X,Y)$, the "even-odd" partition of X is $\{A,B\}$, where A includes the terminal vertices; the sums (1) hold for the degrees.

T is the associate tree of a dendroid D, as may be seen from the incidence-matrix of T, with lines for A-vertices and columns for B-vertices. D will have the A-vertices as vertices, and each j of B will define a bond as the set of A-vertices with which j is incident; hence $D = (A,B)$.

Lemma. *Given two finite sets A and B, with cardinal numbers a and b, and, assigned to each of their elements i or j, is a positive integer d_i or d_j such that equalities* (1) *hold, the number of distinct trees having $\{A,B\}$ as "even-odd" partition and the given integers as degrees is the following product of two multinomial numbers:*

$$\frac{(b-1)!}{\prod_{i\varepsilon A} (d_i-1)!} \times \frac{(a-1)!}{\prod_{j\varepsilon B} (d_j-1)!}. \tag{2}$$

This lemma is easy to prove, e.g. by induction with respect to the sum $a + b$.

It holds of course in the particular case when all the d_j's are $\geqslant 2$, which implies $b \leqslant a - 1$.

A number of consequences can be obtained from this lemma by considering all possible combinations of the following assumptions for A or B or both:

i. vertices labeled or not

ii. degrees completely assigned, or specified only by their *type* (i.e. the sequence $s_1 s_2 \ldots s_k \ldots$, s_k being the number of degrees equal to k, or not specified at all.

Let us for instance consider the following sequence of cases.

1. A-vertices labeled, with assigned degrees d_i; B-vertices also

labeled, with given degree assignment including no degrees equal to 1 (s_1=0). If we call $(t_1 t_2 \dots t_k \dots)$ the *type* of the corresponding (d_j-1) assignment, we have

$$t_1 + t_2 + \dots + t_k + \dots = b$$

$$t_1 + 2t_2 + \dots + kt_k + \dots = a - 1,$$

so that the second factor of (2) may be written

$$\frac{(a-1)!}{(1!)^{t_1}(2!)^{t_2}\dots(k!)^{t_k}\dots}. \tag{3}$$

2. Nothing changed for A; B-vertices still labeled, no given degree assignment but *given type* $(t_1 t_2 \dots t_k \dots)$ for the numbers d_j-1. The factor (3) must be repeated a number of times, namely

$$\frac{b!}{t_1!t_2!\dots t_k!\dots}, \tag{4}$$

i.e. *multiplied by* (4).

3. Still nothing changed for A; still the same given type for the (d_j-1)-assignment, but B-vertices *unlabeled*. The previously calculated number has to be divided by $b!$, which now replaces the second factor of (2) by

$$\frac{(a-1)!}{(1!)^{t_1}t_1!(2!)^{t_2}t_2!\dots(k!)^{t_k}t_k!\dots}; \tag{5}$$

this number is known to be the number of ways to partition a given set of a-1 elements into subsets, t_k of which have cardinal number k (k=1,2,...).

4. Type of (d_j-1)-assignment *no longer specified* except that there are still no degrees =1; nothing else changed. Factor (5) has to be summed for all possible types $(t_1 t_2 \dots t_k \dots)$, which gives the total number of ways to partition a given set of $a - 1$ elements into b non-empty subsets, i.e. the Stirling number of the second kind $S(a-1,b)$.

5. Same conditions as in 4. for the B-vertices, but *no specification of the degrees of the (still labeled) A-vertices*. We have then to sum the *first* factors of (2) for all possible (d_i-1)-assignments of a non-negative integers which add up to b-1. The sum is equal to a^{b-1}, as may be seen from the multinomial development of $(1+1+\dots+1)^{b-1}$.

The last result provides a reply to the following question: how many distinct dendroids with b bonds may be defined on a given set of a (labeled) vertices? Replacing the problem by its equivalent in terms of associate trees, it turns out that there are $a^{b-1}S(a-1,b)$ such dendroids. In case b has its greatest possible value $a-1$, the dendroids become trees themselves; the Stirling number $S(a-1,a-1)$ is equal to 1 and the result goes back to the well known Cayley's expression a^{a-2}.

THE POINT ARBORICITY OF S_n

Hudson V. Kronk

State University of New York at Binghamton, NY, U.S.A.

This note is a summary of the results of references [2,5] given below. The *point-arboricity* $\rho(G)$ of a graph G has been defined [3] as the minimum number of subsets into which the point set of G may be partitioned so that each subset induces an acyclic subgraph. (A subgraph H of G is called *induced* if every line of G which joins two points of H is a line of H.) Equivalently, $\rho(G)$ is the minimum number of colors needed to color the points of G so that no cycle of G has all its points colored the same. The *chromatic number* $\chi(G)$ is the minimum number of subsets in a partition of the point set of G such that each subset induces a subgraph containing no lines. Each such subset is called a *color class*. Since each color class is acyclic and every acyclic graph has chromatic number not exceeding 2, it follows that

$$\rho(G) \leq \chi(G) \leq 2\rho(G) \tag{1}$$

for every graph G. If $\delta(G)$ denotes the minimum degree of the points of G, then it can be shown that [2]

$$\rho(G) \leq 1 + \left[\frac{\max\ \delta(G')}{2}\right] , \tag{2}$$

where the maximum is taken over all induced subgraphs G' of G. This upper bound is analogous to the upper bound for $\chi(G)$ given by Szekeres and Wilf [7]. It follows immediately from (2) that every planar graph has point-arboricity not exceeding 3. After considering several examples, one may very well be led to conjecture that this upper bound can be reduced to 2. If this were the case, then, by (1), the four-color theorem would follow. In [2], however, it is shown that there does exist a planar graph having point-arboricity 3. In fact, the dual G^* of the graph G pictured in Figure 2 of [8] has point-arboricity 3. It might be thought that the point-arboricity of the graph G^* might depend upon the fact that G is a non-hamiltonian graph of a cubic convex polyhedron. However, this is not the case since another graph with these same properties has recently been found (see [1, Fig. 13]), whose dual has point-arboricity 2.

The *chromatic number* of a surface (closed orientable 2-manifold) S_n of genus n, denoted by $\chi(S_n)$, is the maximum chromatic number among all graphs which can be embedded on S_n. Heawood [4] showed that

$$\chi(S_n) \leq \left[\frac{7 + \sqrt{1 + 48n}}{2}\right] = H(n)$$

for all $n > 0$. The statement that $\chi(S_n) = H(n)$ for $n > 0$ eventually came to be known as the Heawood Map-Coloring Conjecture. In 1968, Ringel and Youngs announced an affirmative solution to the conjecture. Defining $\rho(S_n)$ in the obvious manner, it is natural to seek a corresponding result for $\rho(S_n)$. In [5], it is shown that

$$\rho(S_n) = \left[\frac{9 + \sqrt{1 + 48n}}{4}\right] = K(n) \qquad (3)$$

for all $n > 0$. The inequality $\rho(S_n) \leq K(n)$ was established by using the concept of a graph being critical with respect to point-arboricity. A graph G is called *critical with respect to point-arboricity* or simply *critical* if $\rho(G - v) < \rho(G)$ for all points v of G. An *n-critical* graph is a critical graph G having $\rho(G) = n$. It is easy to show that if G is n-critical, $n \geq 2$, then $\delta(G) > 2(n - 1)$. The inequality $\rho(S_n) \geq K(n)$ follows from the result of Ringel and Youngs [6] that the complete graph on p points has genus $\{\frac{(p-3)(p-4)}{12}\}$, where $\{x\} = -[-x]$.

In closing, we note that since the sphere S_0 has point-arboricity 3, formula (3) above doesn't hold when $n = 0$.

References

1. Bosák. J., Hamiltonian lines in cubic graphs, *Theory of graphs international symposium*, Rome, 1966 (New York: Gordon and Breach, pp. 35-46).

2. Chartrand, G. and Kronk, H. V., The point-arboricity of planar graphs *J. London Math. Soc.*, 44(1969). 612-616.

3. Chartrand. G., Kronk, H. V., and Wall, C. E., The point-arboricity of a graph, *Israel J. Math.*, 6(1968), 169-175.

4. Heawood, P. J., Map colour theorem. *Quart. J. Math.*, 24(1890), 332-338.

5. Kronk, H. V., An analogue to the Heawood map-coloring problem, *J. London Math. Soc.*, (to appear).

6. Ringel, G. and Youngs, J. W. T., Solution of the Heawood map-coloring problem, *Proc. Nat. Acad. Sci., USA*, 60(1968), 438-445.

7. Szekeres, G. and Wilf, H. S., An inequality for the chromatic number of a graph, *J. Combinatorial Theory*, 4(1968), 1-3.

8. Tutte, W. T., On hamiltonian circuits, *J. London Math. Soc.*, 21(1946), 98-101.

FRAÏSSÉ'S CONJECTURE ON ORDER TYPES

Richard Laver

University of California, Berkeley, CA, U.S.A.

For order types (isomorphism types of linear orderings) ϕ and ψ, let $\phi \leqslant \psi$ mean there is a 1-1 order preserving function from ϕ into ψ. $\phi < \psi$ iff $\phi \leqslant \psi$ and $\psi \nleqslant \phi$. Call a collection R of order types *well-quasi-ordered* iff R is well founded (under $<$) and there is no infinite set of mutually incomparable (under $\leqslant$) members of R. Fraïssé (C. R. Acad. Sci., 1948) conjectured that the set of countable order types is well-quasi-ordered. An order type ϕ is called *scattered* iff $\eta \nleqslant \phi$ (η the order type of the rationals). Let M be the class of all order types ϕ such that a linearly ordered set of type ϕ can be partitioned into $\leqslant \aleph_0$ subsets, each of which is scattered.

<u>Theorem</u>: *M is well-quasi ordered (Fraïssé's conjecture is of course a corollary).*

The proof uses the following generalization of Nash-Williams' infinite tree theorem (Proc. Camb. Phil. Soc., 1965). For Q a quasi-ordering, define a Q-tree to be a (graph theoretic) tree whose nodes are labelled by elements of Q. Quasi-order the class of Q-trees as follows: $T_1 \leqslant T_2$ iff there is a homeomorphism f: $T_1 \to$ some subtree of T_2, such that for all nodes x in T_1, the label of $x \leqslant$ the label of fx.

<u>Theorem</u>: *If Q is better-quasi-ordered then the class of all Q-trees is better-quasi-ordered.*

In extending the result that the scattered order types are well-quasi-ordered to include the whole class M, a Hausdorff-like classification of M due to Galvin is used. Details of all these results will appear elsewhere.

OPTIMAL MATROID INTERSECTIONS*

Eugene L. Lawler

University of California, Berkeley
and
University of Michigan, Ann Arbor, U.S.A.

Given two matroids M_1 and M_2 over the same set of elements E, we say that $I \subseteq E$ is an *intersection* of the two matroids if I is independent in both M_1 and M_2. In this paper, our principal concern is with the problem of constructing an intersection with the largest possible membership (the "maximum cardinality intersection problem"), and with the problem of constructing a maximum weighted intersection, for an arbitrary weighting of the elements of E (the "maximum weighted intersection problem").

Computationally efficient algorithms are presented for both problems. The proposed algorithms are generalizations of well-known network flow techniques. In particular, the algorithm for solving the weighted intersection problem is a generalization of primal-dual methods for the assignment or bipartite matching problem. This is one manifestation of the fact that matroid intersection problems are proper generalizations of network flow problems.

The cardinality intersection algorithm provides a constructive proof for a generalization of the Konig-Egervary theorem due to Edmonds. Namely, the maximum cardinality of an intersection of two matroids is equal to the minimum rank of a "covering" of their elements. The weighted intersection algorithm provides a proof of a theorem of Edmonds to the effect that the intersection of two matroid polyhedra (convex polyhedra determined by inequalities based on span-rank relations of the matroids) has integer vertices which are in one-one correspondence with the intersections of the matroids.

It is shown that the cardinality intersection algorithm can be applied to solve what would appear to be a rather difficult general problem. That is, given a matroid M over E, and an arbitrary function

*Abstract

$h:E \rightarrow E'$, how does one determine, for a given set $I' \subseteq E'$, whether I' is the image of some independent set I of M? (A theorem of Nash-Williams shows that the family of all such sets I' forms a matroid M' over E'. Therefore, this is a problem of testing for independence in one matroid M' in terms of independence in another matroid M.) Given k matroids $M_1, M_2, \ldots, M_k$ the problem of determining whether a given set I' can be partitioned into k blocks, where the i-th block is independent in matroid M_i, can be shown to be a special case of this problem. The partitioning algorithm of Edmonds, which was designed to solve this problem, can be used to solve any problem solvable by the cardinality intersection algorithm, and vice versa.

Some applications of the matroid intersection theorems and the computational algorithms are discussed, and some areas of inapplicability are indicated, *e.g.*, the traveling salesman problem.

THE CATEGORIES OF COMBINATORICS

K. Leeb

University of Alberta, Edmonton, Alta., Canada

In this note I want to show how to bridge the gap between Algebra in the sense of Birkhoff on one side and Combinatorics on the other. Those who doubt the existence of such a gap I refer to a universal algebraist who calls Bourbaki's remedy "ugly", and to the situation in combinatorics, where a standard notation for many standard concepts is lacking mainly because structures are not properly identified. Those who doubt the need for filling this gap I hope to convince by exhibiting some instances in which the two fields meet.

Let us define first what we mean by homomorphisms of general algebras, a concept related to Bourbaki's espèces de structures, Wyler's operational categories and the weak and strong homomorphisms of Hedrlin and Pultr.

Let Ens denote the category of functions, Rt the category of weak (in the sense of Hedrlin and Pultr) homomorphisms of relational structures with one reflexive transitive binary relation (usually denoted by $\leqslant$). Let further $F:S \to Ens$ and $G:S \to Rt$ be functors of any variances from some category S. Then an (F,G)-algebra is a pair $(A,+)$ where A is an object of S and the operation $+:F(A) \to G(A)$ is a function (notice that we dropped the forgetful functor $U:Rt \to Ens$ in $UG(A)$!). An (F,G)-homomorphism $f:(A,+) \to (B,+)$ is a morphism $f:A \to B$ in S satisfying: The diagram

$$\begin{array}{ccccc} & F(A) & \overset{+}{\to} & G(A) & \\ F(f) & | & & | & G(f) \\ & F(B) & \overset{+}{\to} & G(B) & \end{array}$$

$\leqslant$-commutes in the corner which is target for two arrows according to the variance of G, with the convention that the horizontal arrow hits "higher" than the vertical one. For example if F is co- and G is contravariant this means:

$$\forall\, a \in F(A) \;\; + a \geqslant G(f) + F(f)\, a$$

Before we turn to combinatorics we give some algebraic examples which should justify the adoption of the above definition:

An algebra with a k-ary operation is a pair $(X,+)$ where X is a set and $+: {}^{k}X \to X$. So choosing S to be *Ens*, F to be the covariant $\lambda X \cdot {}^{k}X$ and G to be the functor which associates to each set X the relational structure $(X,=)$, we regain the usual concept of homomorphism.

A relational structure (X,R) is defined by $R: {}^{k}X \to 2$. Hence S is as above, F is too, but G can be chosen to be the constant functor with value $(2,\leqslant)$ or $(2,=)$ resulting in weak or strong homomorphisms respectively.

Let us call Hartmanis' generalized partitions for any r geometries, as he does in the case $r = 2$, so that we can call their morphisms subspaces. In this example S is the category of all 1-1 - functions and F the functor $\binom{\ }{r}$ assigning to each set the set of its r-subsets. Giving ${}^{X}2$ the structure $({}^{X}2,=)$ defines the functor G. Then geometries are just those operations $g: \binom{A}{r} \to {}^{A}2$ satisfying $g_{-1}(gu) = \binom{gu}{r}$ and subspaces are the homomorphisms of the general algebras so defined.

There is no space to list the many categories and functors relevant to the van der Waerden - Ramsey - like theorems. I hope to do this somewhere else. Here I only define some Galois connexions arising from categorical concepts which induce well and less wellknown closures and finish with an analysis of Nash-Williams' concept of better quasiorder.

The subobjects of a set are its subsets and the quotient objects its partitions. We can connect subobjects and quotient objects by requiring that their composite does not factor through 1, is monic, is epic, respectively. Much attention has been paid to the closure induced by the first connexion (chromatic number), but little to the two others.

Better quasiorder is the following property: Take $(\overset{\rightarrowtail\!\!\twoheadrightarrow}{\omega},\omega)$, the set of all strictly ascending sequences. This is a contravariant functor in the first argument, hence by the operation of successor $S:\omega \to \omega$ we ca induce a map $(\overset{\rightarrowtail\!\!\twoheadrightarrow}{S},\omega)$, which we finally write as a relation $S: {}^{2}(\overset{\rightarrowtail\!\!\twoheadrightarrow}{\omega},\omega) \to 2$. quasiorder B is better iff for all pairs of functions $(f,g):(\overset{\rightarrowtail\!\!\twoheadrightarrow}{\omega},\omega) \to (\omega\backslash 1$ such that $\forall\, a\ a|fa{=}b|fb \Rightarrow ga = gb$ there is a successor homomorphism $(2,S) \to ((\overset{\rightarrowtail\!\!\twoheadrightarrow}{\omega},\omega),S)$ such that its composite with (f,g) is still an order homomorphism, where the product has the cardinal order.

One ought to study the interrelation with other ordering concepts defined by different orders and different restrictions on the left and right hand homomorphisms. (Exercise: Characterize wellorder!)

I wish to acknowledge the hospitality and financial support extended to me by the Department of Electrical Engineering, University of Alberta, Edmonton.

ON EXTREMAL FACTORS OF THE DE BRUIJN GRAPH

Abraham Lempel

Sperry Rand Research Center, Sudbury, MA, U.S.A.

The n-th order de Bruijn graph G_n is a directed graph with 2^n vertices, labeled by the binary n-tuples $x = (x_1, \ldots, x_n)$, $x_i \in \{0,1\}$, and 2^{n+1} arcs. The vertices $x = (x_1, \ldots, x_n)$ and $y = (y_1, \ldots, y_n)$ of G_n are joined by an arc, directed from x to y, if and only if

$$y_i = x_{i+1}, \qquad i = 1, \ldots, n-1. \tag{1}$$

A factor of G_n is a subgraph formed by a set of vertex-disjoint cycles (directed circuits) which, together, include all the vertices of G_n. A factor which contains the maximum possible number of cycles will be referred to as an *extremal* factor.

An outstanding problem in combinatorial theory is to determine the number of cycles in an extremal factor of G_n. Golomb [1] conjectured:

Conjecture 1. The number of cycles $Z(n)$ in an extremal factor of G_n is given by:

$$Z(n) = \frac{1}{n} \sum_{d|n} \phi(d) 2^{\frac{n}{d}} \tag{2}$$

where the summation extends over all positive divisors d of n and ϕ is the Euler ϕ-function. The first ten values of $Z(n)$ are listed in Table 1. It is well known [1], that for each $n \geq 1$ there exists a factor of G_n with $Z(n)$ cycles, but no promising method of proving that this number cannot be exceeded has yet been discovered. Various aspects of the problem are discussed in recent publication [2], where it is also shown that the conjecture is true for all $n \leq 6$.

In this paper, the problem is approached by proposing and discussing yet another pair of conjectures.

Conjecture 2. The number of cycles in an extremal factor of G_n is equal to the minimum number of vertices which, if removed from G_n will leave a graph with no cycles.

Although this conjecture need not be true for a general directed graph, it is believed to hold for the de Bruijn graphs.

Conjectures 1 and 2 are independent of each other, but both are weaker than and implied by:

Conjecture 3. The minimum number of vertices which, if removed from G_n will leave a graph with no cycles is $Z(n)$.

Of all three conjectures, the last one seems to be the easiest to verify for a given n. A simple procedure for verifying Conjecture 3 is derived and its validity is established for all $n \leqslant 8$. Also, some related results are obtained which, it is hoped, will lead to a confirmative proof of all three conjectures.

CONJECTURE ON A THEOREM SIMILAR TO SPERNER'S

Bernt Lindström

University of Stockholm, Sweden

Let P_n be the family of all subsets of the set of integers $S_n = \{1,2,\ldots n\}$. If I and J are nonzero subsets of S_n, we shall write $I \leqslant J$ if and only if there is a function f, which is defined on I and maps I into J strictly increasingly and such that $f(i) \geqslant i$ for every $i \in I$. For the zero subset ϕ and any I we write $\phi \leqslant I$. The relation $I \leqslant J$ is a partial order in P_n. We have the rank function

$$r(I) = \sum_{i \in I} i \ (I \neq \phi), \ r(\phi) = 0.$$

A family F: $I_1, I_2, \ldots, I_r$ of subsets of S_n is independent if $I_v \leqslant I_\mu$ does not hold for any $v \neq \mu$, $1 \leqslant v, \mu \leqslant r$.

Conjecture. The number of sets in an independent family F never exceeds the maximum number of sets I with the same rank $r(I)$.

Corollary.
If $f(b, a_1, a_2 \ \ldots, a_n)$ is the number of solutions in $\{0, 1\}$ of the equation

$$a_1 x_1 + a_2 x_2 + \ldots + a_n x_n = b, \quad 0 < a_1 < a_2 < \ldots < a_n,$$

then

$$f(b, a_1, a_2, \ldots a_n) \leqslant \max_m f(m, 1, 2, \ldots, n).$$

The conjecture is true when $n \leqslant 7$. This has been proved by decomposition of P_n into disjoint maximal chains $J_1 < J_2 < \ldots < J_s$ for which $r(J_1) + r(J_s) = r(S_n)$. (This suggests a stronger conjecture.)

THE FACTORIZATION OF GRAPHS

L. Lovász

Eötvös L. University, Budapest, Hungary

In this lecture we are giving a structural description of graphs, from the point of view of their factorization. The criteriums for having a certain factor, given by Tutte and Ore, and the description of maximal independent edge-systems given by Gallai can be deduced from our results easily.

Throughout this lecture we consider a fixed, finite undirected graph without loops. Vertex, subgraph, etc. means vertex, subgraph, etc. of this graph. If G is a subgraph and x is a vertex, then we denote by $\phi_G(x)$ and $\phi(x)$ the valency of x in G and in the original graph, respectively. We associate with every vertex x a set $H(x)$ of integers and ask, when does a subgraph exist satisfying

$$\phi_G(x) \in H(x) \tag{1}$$

in every vertex x. Furthermore, if such a subgraph does not exist, which are "optimally approaching" subgraphs and what is the measure of this "approach"?

If G is any subgraph, then it is natural to define its deviation from condition (1) by the value

$$\delta_G(H) = \Sigma\delta_G(H;x)$$

where $\delta_G(H;x)$ denotes the distance of $\phi_G(x)$ and $H(x)$. Thus, the solvability of (1) can be characterized by

$$\delta(H) = \min_G\{\delta_G(H)\}.$$

The subgraphs G giving here the minimum are called *optimal*. We are going to describe the optimal subgraphs. This succeeded only besides two further suppositions:

i. $H(x) \cap [0, \phi(x)] \neq \emptyset$,

ii. If w is an integer such that $w \notin H(x)$, $w+1 \notin H(x)$ then either $(-\infty, w+1] \cap H(x) = \emptyset$ or $[w, +\infty) \cap H(x) = \emptyset$.

The first supposition is very natural, but ii is rather strong and it would be very important to avoid it. If $H(x)$ is an interval then it satisfies ii; but it is worthy allowing more complicated sets of integers, since e.g. results of Gallai concerning vertex-disjoint covering complete-graph-systems [1] can be reduced to such a factorization problem. In fact, every problem concerning complete-graph-systems (e.g. the Hajnal-Corradi conjecture) is equivalent with a factorization problem. The fact of the matter is that the corresponding $H(x)$ does not satisfy ii (except in the case of the mentioned results of Gallai).

Suppose now that $H(x)$ satisfies i and ii. Let x be a vertex. There are four possibilities for the position of the valencies of optimal subgraphs among the integers. If $I_H(x)$ denotes their set, then it may hold:

a. $\max H(x) = \min I_H(x)$; $|I_H(x)| > 1$;

b. $\min H(x) = \max I_H(x)$; $|I_H(x)| > 1$;

c. $I_H(x) \subseteq H(x)$;

d. the remaining cases.

Let A, B, C, D denote the sets of those vertices x for which a, b, c, d hold, respectively. The following properties of these sets can be proved:

$H(x) \cap I_H(x) \neq \emptyset$. If $x \notin C$ then $I_H(x)$ is an interval. If $x \in D$ then $I_H(x) \cap H(x)$ and $I_H(x) \setminus H(x)$ are arithmetical progressions of difference 2.

To state further, more interesting properties we have to introduce some definitions and notations. If our graph is connected and every vertex belongs to D, then it will be said to be H-critical. Let S be the set of all vertices. We denote by $\overline{H}(x)$ the set $\phi(x) - H(x)$, i.e. the set of integers of form $\phi(x) - h$, $h \in H(x)$, and by $m_H(x)$ the maximum of $H(x)$. If $X, Y \subseteq S$, then let $f(X,Y)$ denote the number of edges joining X and Y; form further the set $H^*(z) = H(z) - f(z,Y)$ for $z \in S - X - Y$. Let Z span a connected component of S-X-Y, and denote by H_Z the restriction of H^* on Z. Let $t_1(X,Y)$ denote the number of those connected components Z of S-X-Y for which $\delta(H_Z) \neq 0$; and $t_2(X,Y)$ the number of those Z-s, which are H_Z-critical. Finally, put $u_i(X,Y) = f(X,Y)+t_i(X,Y)-m_H(X)-m_{\overline{H}}(Y)$. Obviously, $u_1 \geqq u_2$.

The main theorem of this lecture is the following:

Theorem. $f(C,D) = 0$. *Further, putting* $H^*(z) = H(z) - f(z,B)$ *for* $z \in D$, *the connected components of the graph spanned by* D *are* H^**-critical, and* $\delta(H) = u_1(A,B) = u_2(A,B)$.

From this theorem it follows easily a characterization of the optimal subgraph analogous to the characterization given in [5], and the following formula:

$$\delta(H) = \max\{u_1(X,Y)\} = \max\{u_2(X,Y)\}$$

where X,Y runs through all disjoint pairs of subsets of S. Now we confine ourselves to the case when $H(x)$ is a set of one element: $H(x) = \{m(x)\}$, $0 \leqq m(x) \leqq \phi(x)$ for every x. The formula for $\delta(H)$ given above seems to be usefull as far as it is of type min=max. But it contains implicitly the solution of the factorization problem in the definition of t_1 and t_2. But this can be avoided easily: if $t(X,Y)$ denotes the number of those connected components Z of the graph spanned by $S-X-Y$ for which

$$m(Z) \not\equiv f(Z,Y) \pmod 2$$

then $t_1(X,Y) \geqq t(X,Y) \geqq t_2(X,Y)$ and hence putting

$$u(X,Y) = f(X,Y) + t(X,Y) - m(X) - [\phi(Y)-m(Y)]$$

we obtain

$$\delta(H) = \max\{u(X,Y)\}.$$

The necessary and sufficient condition for the existance of an m-factor is $\delta(H) = 0$, i.e. $u(X,Y) \leqq 0$ for any $X,Y \subseteq S$, $X \cap Y = \emptyset$. This is the same as Tutte's condition (up to algebraic transformations).

Finally I would like to say some words about simplifications arising in special cases. If our graph is bipartite, then it is rather easy to show that $D = \emptyset$ (in the case $H(x) = \{m(x)\}$). If $H(x) = \{1\}$, then one can substitute $\{-1,1\}$ for $\{1\}$, (this is a set allowed by us), and it is easy to see that nothing changes except that B becomes empty. This observation shows why the conditions for the existance of an 1-factor are simpler. In the case of König's problem (i.e. $H(x) = \{1\}$, the graph is bipartite) it is not worthy substituting $\{-1,1\}$ for $\{1\}$, since the vanishing of D simplifies the results much more than the vanishing of B. This seems to be in the background why are the conditions for the existance of a 1-factor in a bipartite or in an arbitrary graph, of different type.

References

1. Gallai, T., Kritische Graphen II, *Publ. of the Math. Inst. of the Hung. Acad. Sci.*, 8 (1963), 373-395.

2. Gallai, T., Maximale Systeme unabhängiger Kanten, *Publ. Math. Inst. Hung. Acad. Sci.*, 9 (1964), 401-413.

3. Ore, O., Graphs and Subgraphs, *Trans. Amer. Math. Soc.*, 84 (1957), 109-136 and 93 (1959), 185-204.

4. Tutte, The factors of graphs, *Canad. J. of Math.*, 4 (1952), 314-328.

5. Lovász, L., Subgraphs with prescribed valencies, to appear.

GALE DIAGRAMS AND THE UPPER-BOUND CONJECTURE FOR CONVEX POLYTOPES

P. McMullen

Western Washington State College, Bellingham, WA, U.S.A.

In this paper we shall use Gale diagrams to give a new proof of

Theorem. *Among all d-polytopes with $d + 3$ vertices, the simplicial neighbourly polytopes have the maximum number of faces of each dimension.*

We shall follow the terminology of [1] throughout. Without loss of generality, we may restrict our attention to simplicial polytopes P ([1], 5.2). Consider the standard distended Gale diagram $\overline{V}$ of P ([1], 5.4 and 6.3, [2]). If we transpose adjacent diameters of $\overline{V}$, containing the points $\overline{x}$ and $\overline{y}$, then we change the combinatorial type of the corresponding polytope P if and only if $\overline{x}$ and $\overline{y}$ are at opposite ends of L and M. In this case, the increase in the number of k-faces of P is easily calculated. The $d + 1$ points of $\overline{V}$ apart from $\overline{x}$ and $\overline{y}$ fall into two sets, say $\{\overline{z}_1, \ldots, \overline{z}_r\}$ such that

$$o \notin \text{int conv } \{\overline{x}, \overline{y}, \overline{z}_i\}, \quad i = 1, \ldots, r. \tag{1}$$

and $\{\overline{w}_1, \ldots, \overline{w}_s\}$ $(r + s = d + 1)$ such that

$$o \in \text{int conv } \{\overline{x}, \overline{y}, \overline{w}_j\}, \quad j = 1, \ldots, s, \tag{2}$$

as in the first illustration below. We may suppose $r, s, \geq 2$.

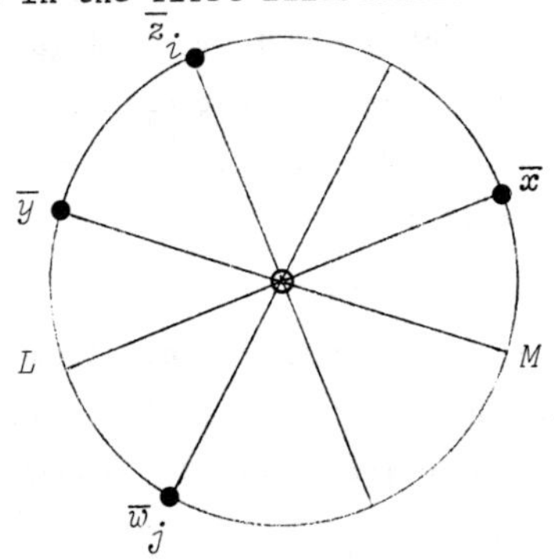

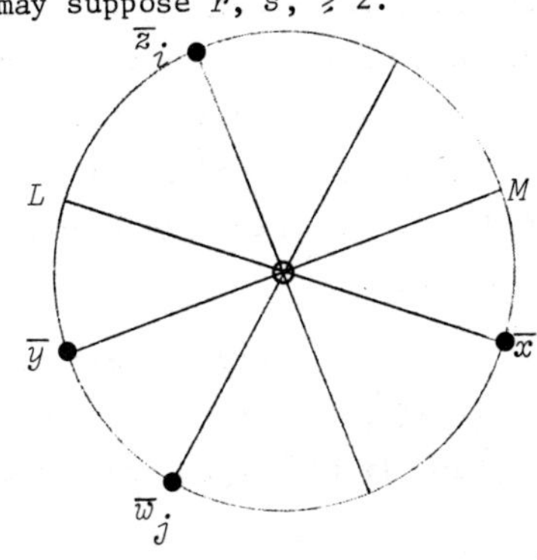

After transposing L and M, the situation is as in the second illustration, and the relations (1) and (2) are reversed. The only k-faces of P which are changed are those which contain neither of the vertices x and y corresponding to $\bar{x}$ and $\bar{y}$. The number of these before the transposition is just the number of ways of choosing $t \geq 0$ of the points $\bar{z}_i$ and $u \geq 1$ of the $\bar{w}_j$, with $t + u = d - k$; that is

$$\sum_{t+u=d+k,\ t\geq 0,\ u\geq 1} \binom{r}{t}\binom{s}{u}$$

After the transposition, we must choose $t \geq 1$ of the $\bar{z}_i$ and $u \geq 0$ of the $\bar{w}_j$, so that the number is

$$\sum_{t+u=d-k,\ t\geq 1,\ u\geq 0} \binom{r}{t}\binom{s}{u}$$

Thus the increase in the number of k-faces of P is

$$\binom{r}{d-k} - \binom{s}{d-k}$$

which is non-negative if $r \geq s$ (for each k), and positive if $r > s$ and $k \geq d - r$ (and so certainly if $k \geq d - 2$).

All that remains to show is that if $\bar{V}$ is not a Gale diagram of a neighbourly polytope ([1], 7.2) then we can transpose a suitably chosen such pair of diameters of $\bar{V}$.

The easiest way to see this appears to be the following, which was suggested by G. C. Shephard ([3], 3.4). We label the diameters of $\bar{V}$ in order $L_1, \ldots, L_{d+3}$, and suppose that L_i has m_i points of $\bar{V}$ on one side and n_i on the other ($m_i + n_i = d + 2$). We write

$$e(\bar{V}) = \sum_{i=1}^{d+3} |m_i - n_i|,$$

which we call the *excess* of $\bar{V}$. Clearly the excess takes its minimal value 0 (d even) or $d + 3$ (d odd) only when P is neighbourly (cf. [1], ex. 7.3.7). If $e(\bar{V})$ is greater than this minimal value, it is not hard to see that a diameter L_i for which $|m_i - n_i|$ is maximal (and thus at least 2) can be transposed with one of the adjacent diameters in the manner described above. This transposition will decrease the excess of $\bar{V}$ by 4, and strictly increase the number of faces of P of some dimension.

So, if we successively perform such transpositions, we simultaneously reduce $e(\bar{V})$ to its minimum, obtaining the Gale diagram of a simplicial neighbourly polytope, and increase the number of faces of each dimension of the corresponding polytope to its maximum. This completes the proof of the theorem.

References

1. Grünbaum, B., Convex Polytopes (John Wiley and Sons, London-New York-Sydney, 1967).

2. McMullen, P., Gale Diagrams of Convex Polytopes (to be published).

3. McMullen, P. and Shephard, G. C., The Upper-bound Conjecture for Convex Polytopes (Lecture notes, University of East Anglia, to be published).

STRONG COVERING OF A BIPARTITE GRAPH

I.P. McWhirter and D.H. Younger

University of Waterloo, Man., Canada

Let G be a finite connected bipartite graph with M a specified subset of the vertex set VG. For $\{V^+, V^-\}$ a bipartition of G, consider each edge α of G to be oriented, with its end in V^+ the positive end $p\alpha$ and its end in V^- the negative end $n\alpha$. For $Y \subseteq VG$, the coboundary δY is *directed* if every edge of δY has its positive end in Y or if every edge of δY has its negative end in Y. A *strong cover* of G is a set of edges that meets each nonnull directed coboundary of G. A strong cover Λ *satisfies matching* if each vertex in M is incident to at most one edge of Λ.

Theorem: *Graph G, with bipartition $\{V^+, V^-\}$, and specified subset $M \subseteq VG$, has a strong cover that satisfies matching with respect to M if and only if for each $X \subseteq M \cap V^+$ or $X \subseteq M \cap V^-$, $c(VG - X) \leq |X|$.*

$c(VG - X)$ is the number of components in the graph $G[VG - X]$, the subgraph of G with vertex set $VG - X$ and edge set those edges of G having both ends in $VG - X$.

The necessity of the stated conditions is easily checked. Sufficiency is established by describing an algorithm. Beginning with a strong cover, say $\Lambda = \Gamma G$ = edge set of G, if Λ does not satisfy matching, then the algorithm finds either

(1) a strong cover Λ' whose deviation from matching is less than that of Λ; or

(2) a set X, a subset of $M \cap V^+$ or $M \cap V^-$, such that $c(VG - X) > |X|$.

The *deviation from matching* of a strong cover Λ is defined as $d(M, \Lambda) = \sum_{v \in M} (|\Lambda\{v\}| - 1)$, where for $Y \subseteq VG$, ΛY is the set of edges in

Λ each incident to some vertex in Y.

The first alternative arises if a decrementing circuit is found, the definition of which is now developed.

A *path* is a nonnull finite sequence $\pi = (v_0, \gamma_1, v_1, \ldots, \gamma_n, v_n)$ of terms, alternately vertices v_i and edges γ_j, such that V_{j-1} and v_j are the ends in G of γ_j. A path is *simple* if its vertex terms are distinct. A path is *circular* if $n \geqslant 1$, the edge terms are distinct, $v_0 = v_n$ and every other pair of vertex terms is distinct.

For path π, let $V\pi$ be the vertex set, $\Gamma\pi$ the edge set. More particularly, $\Gamma_+\pi$ is the set of edges each of which occurs in π as an edge term γ_i for which $v_{j-1} \in V^+$; $\Gamma_-\pi$ is the set of edges γ_j for which $v_{j-1} \in V^-$. A *minus* path is a simple path π_p such that $\Gamma_-\pi_p \subseteq \Lambda$. A *plus* path is one whose inverse is minus. A *minus plus* path is a simple or circular path that is expressible as the product $\pi_m\pi_p$ of a minus path π_m and a plus path π_p, each nondegenerate; in addition, if the terminus v_n of π_m is in M, then $\Lambda\{v_n\}$ meets $\Gamma(\pi_m\pi_p)$. A *decrementing circuit* is a circular minus plus path whose origin is in $M \cap V^+$ or a circular plus minus path whose origin is in $M \cap V^-$.

It is not hard to show that

1. If π_d is a decrementing circuit for strong cover Λ, then $\Lambda' = (\Lambda - \Gamma\lambda_d) \cup \Gamma_-\pi_d$ is a strong cover whose deviation from matching is less than that of Λ.

Thus for any strong cover Λ that does not satisfy matching, the algorithm determines either a decrementing circuit or a set $X \subseteq M$ that fails the conditions.

<u>Algorithm</u>. Since Λ does not satisfy matching, there is an $x \in M$ such that $|\Lambda\{x\}| > 1$. Assume that $x \in V^+$. Let $X_0 = \{x\}$. Suppose $\overline{X}_i = \{X_j: j \in [0, i]\}$ is defined. For $\alpha \in \Lambda X_i$, let $Q(\alpha) = \{w \in VG - \overline{X}_i$: there is from $n\alpha$ to w a minus path that *avoids* $\overline{X}_i$, i.e. that has no internal vertex in $\overline{X}_i\}$. For $\alpha \in \Lambda X_i$, let $R(\alpha) = Q(\alpha) - \{Q(\beta)\colon \beta \in \Lambda X_i - \{\alpha\}\}$. Let $X_{i+1} = \{y \in V - \overline{X}_i\colon \delta\{y\}$ meets $\delta R(\alpha)$ for some $\alpha \in \Lambda X_i\}$.

If for $\alpha \in \Lambda X_i$, $R(\alpha)$ is null, or if $X_{i+1} - M \neq \emptyset$, then G has a decrementing circuit.

On the other hand, if X_{i+1} is null, then $X = \overline{X}_i$. In this case, the algorithm terminates, with $\{G[R(\alpha)]: \alpha \in \Lambda X\}$ the set of components of $G[VG - X]$, whose cardinality exceeds $|X|$.

The remainder of the algorithm is a description of how a decrementing circuit is found in the circumstances just noted. The initial step consists of determining a minus plus path from $p\beta$ to $p\alpha$ that avoids $\bar{X}_i$, where α and β are distinct elements of ΛX_i. If $R(\alpha)$ is null for some $\alpha \in \Lambda X_i$, let $y = n\alpha$. On the other hand, let y be any element in $X_{i+1} - M$, if that is nonnull. In either case, $y \in Q(\beta)$ for some $\beta \in \Lambda X_i - \{\alpha\}$. Let π_β be a minus path from $p\beta$ to y that avoids $VG - Q(\beta)$. Let π_α be a minus path from $p\alpha$ to y that avoids $VG - R(\alpha)$. The product path $\pi_\beta\pi_\alpha^{-1}$ is a minus plus path that avoids $\bar{X}_i$ having both origin $p\beta$ and terminus $p\alpha$ in X_i.

From this minus plus path between two vertices of X_i, a decrementing circuit is found as follows. If $p\alpha = p\beta$, then this path is itself a decrementing circuit. If $p\alpha \neq p\beta$, then let $\alpha', \beta' \in \Lambda X_{i-1}$ be such that $\delta\{p\alpha\}$ meets $\delta R(\alpha')$ and $\delta\{p\beta\}$ meets $\delta R(\beta')$. Let $\pi_{\alpha'}$ be a minus path from $p\alpha'$ to $p\alpha$ that avoids $VG - R(\alpha')$ and $\pi_{\beta'}$ a similar path from $p\beta'$ to $p\beta$. If $\alpha' \neq \beta'$, then $\pi_{\beta'}\pi_\beta\pi_\alpha^{-1}\pi_{\alpha'}^{-1}$ is a minus plus path from $p\beta'$ to $p\alpha'$ that avoids $\bar{X}_{i-1}$. However, this path is not simple if $\alpha' = \beta'$. In this case, $p\alpha \in Q(\gamma')$ for some $\gamma' \in \Lambda X_{i-1} - \{\alpha'\}$. For a minus path from $p\gamma'$ to $p\alpha$ that avoids $\bar{X}_{i-1}$, let π_i be the segment of this path from $p\gamma'$ to its first vertex v_i in $\pi_\beta\pi_\alpha^{-1}$. The product of π_i with either the segment of $\pi_{\beta'}\pi_\beta\pi_\beta^{-1}\pi_{\alpha'}^{-1}$ from v_i to its terminus or with the inverse of the segment from the origin to v_i, is a minus plus path that avoids $\bar{X}_{i-1}$, having origin and terminus in X_{i-1}.

The algorithm proceeds in this way toward the construction of a minus plus path with origin and terminus in $X_0 = \{x\}$. The construction may be terminated by finding a decrementing circuit along the way, but if not, then the minus plus path with origin and terminus x is a decrementing circuit.

The number of elementary steps required to find a strong cover or a set X by recursively applying this algorithm is estimated to be at most ke^2n^2, where $e = |\Gamma G|$, $n = |VG|$, and k is independent of e and n.

A SURVEY OF 3-VALENT 3-POLYTOPES WITH TWO TYPES OF FACES

Joseph Malkevitch

York College of the City University of New York, NY, U.S.A.

Let (k,m,p_k,p_m) $(3 \leqslant k \leqslant 5,\ m \geqslant 7)$ denote any 3-valent, 3-connected and planar graph G consisting of p_k k-gons and p_m m-gons satisfying Euler's relation: $(6-k)p_k = 12 + (m-6)p_m$ (*). Consider the question of when such graphs exist. Note that by Steinitz's theorem (see Grünbaum [1]) if such graphs exist they are 3-polytopal.

If $k = 3$, the 3-connectedness implies that no two triangles can have an edge in common. Hence, if a graph $(3,m,p_3,p_m)$ exists then $m \leqslant 10$, since otherwise when the isolated triangles are shrunk to points, we would obtain a 3-valent graph all of whose faces had six or more sides - a contradiction. For $7 \leqslant m \leqslant 10$, with only a finite number of exceptions, one can construct a $(3,m,p_3,p_m)$ graph using techniques similar to those in Grünbaum-Motzkin [2].

If $k = 4$ and m is a multiple of 4 or $k = 5$ and m is a multiple of 5 then $p_4 + p_m \equiv 2 \bmod 4$ and $p_5 + p_m \equiv 2 \bmod 10$, by Theorem 13.4.4 in Grünbaum [1]. Together with (*), these congruences imply that p_m is even when $k = 4$, $m = 8s$ or $k = 5$, $m = 10s$. Except for the two infinite classes ruled out by this remark, for $k = 4$ or 5 and fixed m, (k,m,p_k,p_m) graphs exist in all but a finite number of cases. The construction of these graphs uses methods similar to those in Grünbaum [1] and Malkevitch [3] and makes prominent use of the face splits shown in Figure 1 (a) and (b) properly applied to the m-gonal prism, or two m-gons joined by $2m$ pentagons, or the dodecahedron.

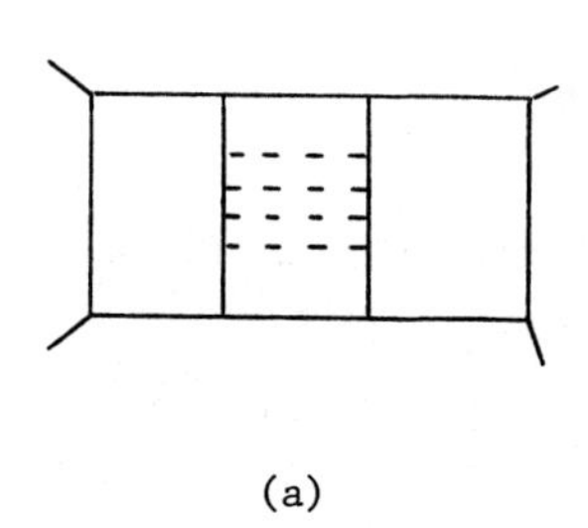

(a)

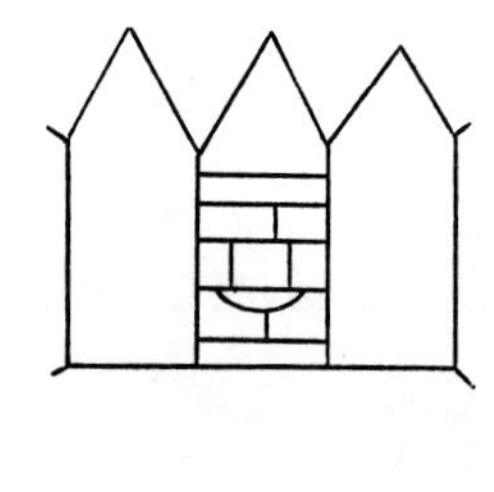

(b)

Figure 1

References

1. Grünbaum, B., Convex Polytopes, Interscience, 1967.

2. Grünbaum, B., and Motzkin, T.S., The number of hexagons and the simplicity of geodesics on certain polyhedra, *Can. J. Math.*, 15 (1963), 744-751.

3. Malkevitch, J., Properties of planar graphs with uniform vertex and face structure, Ph.D. Thesis, University of Wisconsin, 1969.

A CHARACTERIZATION OF TRANSVERSAL INDEPENDENCE SPACES

J.H. Mason*

University of Wisconsin, Madison, WI, U.S.A.

A matroid (E,G_E) will be specified by the set E (finite or infinite) and its set G_E of circuits each of which must be finite. We will need the family $\mathsf{F}(E)$ = the finite unions of circuits of (E,G_E) and by $P_\omega(Y)$ we shall mean the finite subsets of Y.

A deltoid $D = (E,D,Y)$ consists of a pair of sets E and Y and a set $\Delta \leqslant E \times Y$. Under the condition that D is locally right finite, it is well known that D and a matroid (Y,G_Y) induce a matroid (E,G_E) on E, and when $G_Y = \emptyset$, (E,G_E) is called a transversal space.

The proofs of the theorems, while not difficult, are too long to be included here, and so will be presented elsewhere.

Theorem 1. *If (E,G_E) and (Y,G_Y) are matroids, then there exists a locally right finite deltoid $D = (E,D,Y)$ such that (Y,G_Y) and D induce (E,G_E) on E if and only if there exists a map $\sigma : \mathsf{F}(E) \to P_\omega(Y)$ satisfying*

(i) $rk^E(F) = rk^Y(\sigma(F))$ *for any* $F \in \mathsf{F}(E)$

(ii) $\sigma(F_1 \cup F_2) = \sigma(F_1) \cup \sigma(F_2)$ *for any* $F_1, F_2 \in \mathsf{F}(E)$

(iii) $rk^E(A) \leqslant rk^Y [\bigcup_{a\in A} \bigcap_{a\in F\in \mathsf{F}} \text{Span}\,\sigma(F)]$ *for any* $A \in P_\omega(E)$.

Method of proof: Given the deltoid D one produces a larger $D' = (E,\Delta',Y)$ where $\Delta' \geqslant \Delta$ and (Y,G_Y) and D' still induce (E,G_E), and for which, taking $\sigma(A) = \Delta'(A)$ condition (iii) becomes obvious. Given (E,G_E), (Y,G_Y) and the map σ, one puts $\Delta = \{(e,y) : y \in \bigcap_{e\in F\in \mathsf{F}} \text{Span}\sigma(F)\}$ and checks

*This research was supported by the National Research Council of Canada.

that the correct independence space (E,G_E) is induced. It is worth noting that we need only assume that a deltoid is locally left rank finite meaning that $rk^Y(D(e)) < \infty$ for each $e \varepsilon E$, in order that it induce a matroid, but that this does not add any generality.

Corollary. If (E,G_E) is a matroid, then it is transversal if and only if there exists a map $\sigma : F(E) \to P_\omega(Y)$ where Y is a set with cardinality = $rk\ E$ such that

$$\text{(i)}\quad rk^E(F) = |\sigma(F)| \text{ for any } F \varepsilon F(E)$$

$$\text{(ii)}\quad \sigma(F_1) \cup \sigma(F_2) = \sigma(F_1 \cup F_2) \text{ for any } F_1, F_2 \varepsilon F(E)$$

$$\text{(iii)}\quad rk^E\ A \leqslant |\bigcap_{A \leqslant F \varepsilon F} \sigma(F)|\ A \varepsilon P_\omega(E)$$

Definition. If N is an index set, $(A_i : i \varepsilon N)$ a family of sets with $\mathfrak{A} = \left\{ \bigcup_{i \varepsilon I} A_i = A(I) : I \leqslant N \right\}$, and if $\sigma : \mathfrak{A} \to \mathbb{Z}$ then we define the inclusion exclusion function by $IE(\sigma; A_i : i \varepsilon N) = \sum_{I \leqslant N} (-1)^{|I|+1} \sigma(A(I))$.

Thus $IE(\text{card}; A_i : i \varepsilon N) = \text{card} \bigcap_{i \varepsilon N} A_i$.

Theorem 2. *(E,G_E) is transversal if and only if*

$$IE(rk^E; F_1, \ldots, F_n) \geqslant rk^E \bigcap_{i=1} F_i \text{ for any } F_1, \ldots, F_n \varepsilon F(E).$$

Method of proof: The IE condition is essentially the condition that a family of finite subsets of a set Y can be found satisfying properties (i) and (ii). Using condition (i), it is also a direct translation of condition (iii).

Corollary. If (E,G) has rank r, and if there are n distinct circuits $C_1, \ldots, C_n$ each of rank $r-1$ such that $rk(C_i \cup C_j) = r$ for $i \neq j$, then $n > r$ implies that (E,G_E) is not transversal.

Proof. $IE(rk^E; C, \ldots, C_n) = \binom{n}{1} \cdot (r-1) - \binom{n}{2} r + \binom{n}{3} r \ldots + (-1)^{n+1} \binom{n}{n} \cdot r$

$$= r - \binom{n}{1} + r[-1 + \binom{n}{1} - \binom{n}{2} + \ldots + (-1)^{n+1}$$

$$= r - n < 0$$

contradicting the condition in Theorem 2.

If (E,G) is a matroid, then it has a Whitney Dual if and only if the family G^* of subsets of E minimal with respect to intersecing each basis of (E,G) non-trivially, forms a set of circuits of a matroid on E, usually called the co-circuits. It may be shown that (E,G^*) is a matroid if and only if (E,G) is a sum of finite matroids.

The proof will appear elsewhere.

NEW INSTRUCTION TYPES FOR IMPLEMENTING COMBINATORIAL ALGORITHMS ON A COMPUTER

David W. Matula

Washington University, St. Louis, Missouri, MO, U.S.A.

Numerous algorithmic procedures for resolving combinatorial problems should be classified as non-numeric since they are logical in character rather than arithmetic. Consider for example the fact that 0,1 incidence matrices are the data structures typically used to represent binary relations, graphs and families of subsets. Furthermore, permutations may be described by 0,1 matrices, and linear 0,1 vectors can adequately describe particular selections of objects such as combinations.

Current day large scale computers have very sophisticated computational power for effecting arithmetic operations, however, machine hardware instructions of a logical character for bit (0,1) string and bit array processing are more noticeable by their absence. This situation exists despite the fact that boolian bit by bit logic operations parallel the primitive electrical circuitry whereas arithmetic instructions involve complex circuit design problems.

As the demand increases for efficiency in combinatorial computation it is likely that more convenient data structures and instruction types will be provided in computer hardware. The data structures thus obtained should include "square words" of $n \times n$ bits in addition to the n bit linear words currently available. Although a fixed word size will probably prevail, the facility to readily extend instructions to multiple words should be present. In order to channel the creation of instructions for combinatorial computation towards providing the best run-time performance the following hardware-software criteria should be observed and balanced when possible.

Criteria for Fundamental Instructions on a Square Word Computer for Combinatorial Computation

Hardware: 1) The instruction should be subject to convenient design in machine hardware allowing large parallelism but little sequencing time delay.

2) The instruction should be readily extendable to multiple word operation.

Software: The instruction should be a well designed primitive common to many combinatorial algorithms.

With regards to the latter point, the algorithmically inclined combinatorial analysts can provide a valuable service by categorizing known efficient algorithms and seeking the common fundamental procedures that would be candidates for machine instruction primitives.

To this end we shall close by considering two types of machine feasible combinatorial instructions which appear to have wide applicability in combinatorial computation.

1) Combinatorial Incrementors (Combinations and Permutations)

a) Combination Incrementor: On a binary computer the process of adding unity, i.e. $x \leftarrow x + 1$, provides a convenient way of passing through all subsets of a set simply by interpreting the 1-bits as indicators of the set members. However, it is not difficult to design an incrementation process to sequentially pass through only the k-membered subsets, thus forming a combination incrementor instruction.

b) Permutation Incrementor: Having placed n bits in an $n \times n$ bit square word so as to represent a permutation, it would be very useful to have a single instruction to "increment" the bit pattern to the next permutation in a cyclic manner covering all $n!$ permutations. Considerable work has been done on procedures for orderly listings of permutations[1], and these techniques should be scrutinized for the purpose of possible direct hardware implementation of a permutation incrementor instruction.

2) Boolian Matrix Operations

It is readily observed that boolian ANDing and ORing operations between linear words, square words and between a linear and a square word in a matrix product fashion can all be conveniently designed in computer circuitry utilizing high parallelism. Such operations appear quite fundamental to the statement of many combinatorial algorithms and this is illustrated with the example program of Table I for finding the maximum number of edge disjoint paths between two vertices in a graph along with a vertex set and its complement which determine a minimum cut separating the same two vertices. The new instructions utilized in this FORTRAN like program are indicated in Table II, and other terms should be self explanitory.

The importance of such an example as the program of Table I is the simplicity of stating an important combinatorial algorithm in a form closely allied to feasible primitive machine instructions.

References

1. Lehmer, D.H., The Machine Tools of Combinatorics, in Beckenbach, *Applied Combinatorial Mathematics*, Univ. of California Press, 1964.

Table I: A program for a combinatorial algorithm for finding the maximum number of edge disjoint paths between the vertices 1 and N along with the vertices determining a minimum cut set of edges separating 1 and N.

```
      INTEGER I,J,N,MC
      LINEAR S(N),T(N),B1,B2
      SQUARE AJ,A,P
      READ INPUT N,AJ
      S(0) = E;1
      T(0) = E;N
      A = AJ
      P = 0
   5  DΘ 10 I = 1,N
         S(I) = S(I-1)∨ A
  10     IF( S(I)∧T(0) ≠ 0) GΘ TΘ 20
      GΘ TΘ 40
  20  DΘ 30 J = 1,I
         T(J) = INB((T(J-1)∨ A*) ∧ S(I-J))
  30     P = (T(J)∧*T(J-1))∨ P
      P = P∧⌐(P∧P*)
      A = AJ∧⌐P
      GΘ TΘ 5
  40  B(1) = S(0)
      DΘ 50 J = 1,N
  50     B1 = B1 ∨ S(J)
      B2 = ⌐ B1∧E;I,I=1,N
      MC = CNT(E;1 ∨ P)
      READ ΘUTPUT MC, B1, B2
      STΘP
```

Table II. The meaning of the logical functions utilized in the program of table I.

l - linear word of n bits

w - square word of $n \times n$ bits

i - integer (in n bit word)

Instruction Form	
$l \vee l \rightarrow l$	logical OR (bit by bit)
$l \wedge l \rightarrow l$	logical AND (bit by bit)
$l_1 \wedge^* l_2 \rightarrow w$	matrix product of column l_1 and row l_2
$l \vee w \rightarrow l$	boolian matrix product (ORing of rows)
$l \wedge w \rightarrow l$	boolian matrix product (ANDing of rows)
$w^* \rightarrow w$	transpose $n \times n$ bit matrix word
$\neg w \rightarrow w$	interchange 0's and 1's
$w \wedge w \rightarrow w$	logical AND of full matrix words
$w \vee w \rightarrow w$	logical OR of full matrix words
$INB(l) \rightarrow l$	all 1's except leading 1 are set to zero
$CNT(l) \rightarrow i$	bit count of linear word

SOME COMBINATORIAL PROBLEMS RELATED TO ABSTRACT DISTANCES

Robert A. Melter

University of South Carolina, Columbia, SC, U.S.A.

An *abstract distance space* is a triple (S,P,d) where S is a set, P is a set, and d is a function from $S \times S$ to P; d has properties analogous to those of the usual real metric. If $S = P$ the space is said to be autometrized. In [3] the author considered the class of unary algebras S determined by a function f with the following properties:

(1) for x, $y \in S$ there exist non-negative integers m, n such that $f^m(x) = f^n(y)$, (2) f has at most one fixed point in S and (3) if x is not a fixed point then $f^n(x) \neq x$ for all positive integers n. Under the order $\{x \leqslant y$ if there exists a non-negative integer n such that $f^n(x) = y\}$ S becomes a join semlattice and $d(x,y) = x \cup y$ is an autometrization. One example of such a function is the Euler ϕ function where $\phi(n)$ is the number of positive integers which are less than or equal to and prime to n. Here the distance between positive integers r and s is the largest integer K such that $K = \phi^i(r) = \phi^i(s)$ for some positive integers i and j. It would seem to be a difficult problem to provide a "formula" for this distance as a function of r and s. In this respect let us recall that Carmichael's conjecture (still unresolved) states that the equation $\phi(x) = n$ never has a unique solution [2]. In the context of the autometrized unary algebra the question becomes: Given an integer x, does there always exist another integer y and a third integer z at a minimal distance from both x and y. [The distance function of an autometrized unary algebra satisfies, $d(x,y) = d(x,y)$, $d(x,z) \leqslant d(x,y) \cup d(y,z)$, and $d(x,x) = d(y,y)$, iff $x = y$.]

Another important class of abstract distance spaces are those obtained from Boolean valued rings. A Boolean valuation for a ring R is a mapping ϕ of R into a Boolean algebra B such that (1) $\phi(x) = 0$ iff $x = 0$, (2) $\phi(x+y) \leqslant \phi(x) \cup \phi(y)$ and (3) $\phi(xy) \leqslant \phi(x) \cap \phi(y)$. If instead of (3) we have (3′) $\phi(xy) = \phi(x) \cap \phi(y)$, then ϕ is said to be a strong Boolean valuation. It has long been known that a strong Boolean valuation can be defined on a commutative ring R if and only if R is a subdirect sum of rings without divisors of zero. A recent result of Andrunakievic and Rjabuhin [1] allows this result to be extended to the non-commutative case. Although various examples of Boolean valuations are known the following theorem provides a means of defining intrinsic Boolean valuations in arbitrary rings.

Theorem. *Let R be a ring. Represent R as a ring of global sections of a sheaf over a topological space X.* [See [6] for definition and construction.] *Then the mapping $\phi(\sigma)$ = support of σ is a Boolean valuation. Here B is the Boolean algebra of all subsets of X. Furthermore the valuation is non-trivial if and only if R contains a central idempotent different from zero and one. The valuation is essentially unique if the left and right anihilators of each element of R coincide and are generated by central idempotents.*

If ϕ is a Boolean valuation then $d(x,y) = \phi(x,y)$ is an abstract distance, [4].

The following concepts familiar in geometry often provide interesting combinatorial connections: metric bases, equilateral sets, groups of motions and congruence indices. (For definitions, see: [5]). In particular:

1) If R has an identity and is of characteristic zero, then the set of summands of the identity (in the intrinsic geometry) is equilateral of side one. This contrasts with the Boolean geometry of a p-ring in which all equilateral sets are finite. (See [5]).

2) An integer n is the order of the group of weak isometries of an autometrized unary algebra if and only if $n = 2^k \cdot n_1!\, n_2! \ldots n_r!$ see

3) The introduction of analytic geometry in the distance space associated with a p-ring provided an application of theorems due to Raussnitz 1882 [7] about the representation of permutations by polynomials with coefficients in $GF(p)$.

References

1. Andrunakievic, V.A., and Rjabuhin, Ju. M., Rings without nilpotent elements and completely simple ideals. *Soviet Math. Dokl.*, 9 (1968) 565-68.

2. Klee, V., Is there an n for which $\phi(x) = n$ has a unique solution? *Amer. Math. Monthly*, 76 (1969), 288-289.

3. Melter, R.A., Autometrized Unary Algebras, *J. Comb. Theor.*, (1968), 21-29.

4. Melter, R.A., Boolean Valued Rings and Boolean Vector Spaces, *Arch. Math.*, 15 (1964), 354-363.

5. Melter, R.A., Contributions to Boolean Geometry of p-rings, *Pacific J. Math.*, 14 (1964), 995-1017.

6. Pierce, R.S., Modules over Commutative Regular Rings, *Amer. Math. Soc. Memoir* #70, 1967.

7. Raussnitz, G., Zur theorie der congruenzen hoheren grades, *Math und Natur. Berichte aus Ungarn I* (1882), 266-278.

8. White, G., Iterations of Generalized Euler Functions, *Pacific J. Math.*, 12 (1962), 777-783.

INDECOMPOSABLE POLYTOPES

Walter Meyer

University of Wisconsin, Madison, WI, U.S.A.

If A, B are convex subsets of R^n, their vector sum is defined: $A + B = \{a+b: a \in A, b \in B\}$. Similarly $tA = \{ta: a \in A\}$ for $t \geq 0$. We are concerned here with the question of when a polytope P is indecomposable, i.e. when is $P = Q + R$ possible only when Q and R are each of the form $tP + \{x\}$, where t is non-negative and real and x is a point of R^n, for suitable t's and x's. Hitherto all known indecomposable polytopes have satisfied the conditions of Theorem 3 (below), due to Shephard, and have therefore been indecomposable by virtue of their combinatorial structure. We have found necessary and sufficient conditions for a polytope to be indecomposable which suggest that indecomposability is not an entirely combinatorial concept. By way of confirmation, Fig. 1 is a Schlegel diagram with both indecomposable and decomposable realizations. A decomposable realization can be produced by truncating a cube at each vertex so that the truncating plane bisects the incident edges. It is possible to deform this polytope by small displacements of certain vertices and 2-faces to make a new polytope of the same combinatorial type where: *AFJE*, *DEIH*, *CHIG* remain squares but *ABCD* has no pair of parallel edges. Such a polytope is indecomposable.

We now give, without proof, two theorems besides that of Shephard, dealing with conditions on a combinatorial type for all its realizations to be indecomposable. It is not hard to show that Theorem 1 implies Theorem 2 and Theorem 3 in the case where $d = 3$. However, Fig. 2 shows a combinatorial type which satisfies the conditions of Theorems 1 and 2, but not Theorem 3.

Theorem 1. *A 3-polytope P is indecomposable if there exist 4 2-faces, F_1, F_2, F_3, F_4, so that for each vertex v there is a sequence*

$F_1, F_2, F_3, F_4, \ldots, F_k = v$ with these properties: (a) each vertex in the sequence is preceded in the sequence by at least 3 containing 2-faces; (b) each 2-face in the sequence, except F_1, F_2, F_3, F_4, is preceded by at least one vertex which lies on it; (c) each member of the sequence is a vertex or 2-face.

Theorem 2. *A 3-polytope P is indecomposable if its combinatorial type can be reduced to that of a tetrahedron by the Steinitz' Theorem reductions n_0, n_1, n_2 (see Grünbaum, p.238, for details about these reductions).*

Theorem 3. (Shephard). *A polytope P is indecomposable if there exists an edge E such that for every vertex v of P there is an edge-wise connected chain of simplicial faces, the first and last faces of the chain containing E and v respectively.*

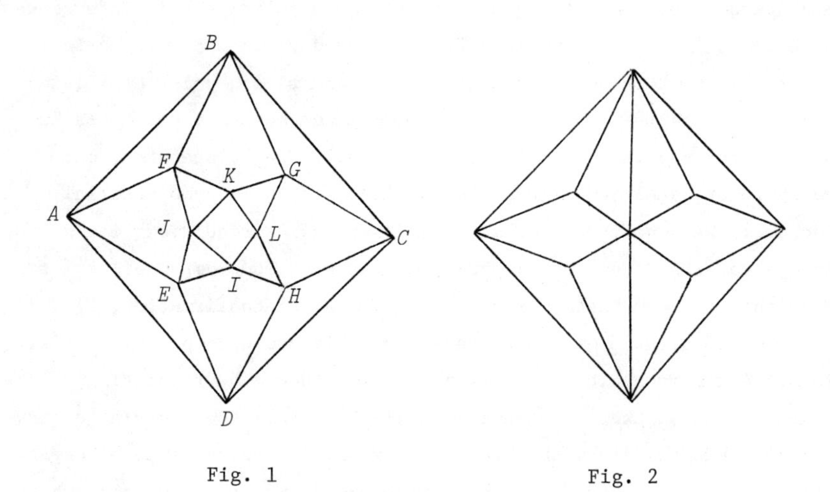

Fig. 1 Fig. 2

References

1. Grünbaum, B., Convex polytopes, Interscience Series, Vol.16, John Wiley and Sons, London, 1967.

2. Shephard, G., Decomposable Convex Polyhedra, *Mathematika*, 10 (1963), 89-95.

EDGE EQUIVALENCES RELATED TO CARTESIAN DECOMPOSITONS OF GRAPHS

D.J. Miller

University of Victoria, B.C., Canada

Sabidussi [4] proved that a connected graph of finite type can be written uniquely as a cartesian product of indecomposable factors. Vizing [5] gave a new proof of this theorem and Imrich [1] extended this result to systems of sets. In [2] we extended Sabisussi's theorem, showing that every connected graph has a unique prime factorization with respect to weak cartesian multiplication. This was established by proving that the set of acyclic equivalence relations (Definitions 1) that contain $\alpha \cup \beta$ (Definitions 1) was a principal filter in the lattice of all equivalences on the edge set of a graph. Moreover it was shown that each equivalence relation in this lattice gives rise to a weak cartesian decomposition and that the least element decomposes the graph into its prime factorization. In [3] we use this least element to determine the automorphism group of a graph that is the weak cartesian product of non-isomorphic prime graphs. The purpose of this paper is to investigate further this sublattice of equivalence relations. In particular we define the acyclic completion and show that if we restrict the completion to the lattice of equivalences containing $\alpha \cup \beta$ then it is in fact a closure operation, hence deducing that the acyclic equivalences containing $\alpha \cup \beta$ are a complete lattice. The proofs are omitted but may be found in section VIII of the author's doctoral thesis written at McMaster University. In the following definitions $E(X)$ denotes the edge set of a graph X.

Definition 1. The following two binary relations α and β on $E(X)$ were introduced by Sabidussi [4]. For $e, e' \in E(X)$ $e \alpha e'$ if and only if e and e' are adjacent, and among the saturated subgraphs of X which contain e and e' there is no 4-circuit. $e \beta e'$ if and only if e and e' are not adjacent, and among the saturated subgraphs of X which contain e and e' there is a 4-circuit.

Let ρ be an equivalence relation on $E(X)$. A circuit C of X is called ρ-*compatible* if and only if C can be written as the union $P_0 \cup \ldots \cup P_n$ of proper paths P_i, $i=0, \ldots, n$ such that $E(P_i) \times E(P_j) \subset \rho$ or $\bar{\rho}$ according as $i = j$ or $i \neq j$. (Here $\bar{\rho}$ denotes the complement of ρ). n is called the ρ-*degree* of C and will be denoted by $\deg_\rho C$. A ρ-compatible circuit C is called *simple* if for every $e \in E(C)$, $\deg_\rho C$ is minimal among the ρ-compatible circuits that contain the edge e. ρ will be called *acyclic* if and only if X does not contain any ρ-compatible circuits.

Definition 2. Let ρ be any equivalence on $E(X)$. We will define the *acyclic completion* of ρ, which we denote by ρ^*. Put $\rho^{(0)} = \rho$. Assume that $\rho^{(n)}$, $n \geqslant 0$ has already been defined. For $e, e' \in E(X)$, we define a binary relation $\tau^{(n)}$ by: $e \, \tau^{(n)} e'$ if and only if there exists a simple $\rho^{(n)}$-compatible circuit C with $e, e' \in E(C)$. Let $\rho^{(n+1)}$ be the smallest equivalence on $E(X)$ containing $\rho^{(n)} \cup \tau^{(n)}$. Finally we take $\rho^* = \bigcup_{n=0}^{\infty} \rho^{(n)}$.

Proposition 1: *Let ρ be any equivalence on $E(X)$ and ρ^* the acyclic completion of ρ. Then*

(i) $\rho \subset \rho^*$,

(ii) ρ^* *is acyclic,*

(iii) $\rho=\rho^*$ *if ρ is acyclic,*

(iv) *if $\alpha \cup \beta \subset \rho \subset \sigma$ then $\rho^* \subset \sigma^*$.*

The above proposition shows that if we restrict the acyclic completion to the complete lattice of all equivalences containing $\alpha \cup \beta$ then it is in fact a closure operation.

In order to prove part (iv) of the above we first showed that if $\rho \supset \alpha \cup \beta$ then ρ^* is the smallest acyclic equivalence containing ρ. The following example shows that there need not exist a smallest acyclic equivalence containing a given equivalence relation. Let X be a circuit of order $n \geqslant 4$, $E(X) = \{e_1, \ldots, e_n\}$ and δ the identity relation on $E(X)$. Let e_i, e_j be two distinct non-adjacent edges of X. Put $\rho_{ij} = \delta \cup \{(e_i, e_j), (e_j, e_i)\}$. ρ_{ij} is a minimal acyclic equivalence on $E(X)$. Hence there are $\frac{1}{2}n(n-3)$ distinct minimal acyclic equivalences on $E(X)$. Here $\delta^* = E(X) \times E(X)$, i.e., δ^* is not a minimal acyclic equivalen

We now give an example of a connected graph Y and an equivalence

σ on $E(Y)$, containing $\alpha \cup \beta$, with the property that for every integer $n \geqslant 2$ there exists a simple σ-compatible circuit C in Y with $\deg_\sigma C = n$. Let X be any graph, E a subset of $E(X)$ and let K_2 denote the complete graph on two vertices say 0 and 1. We define the *interchange* X_E of X *relative to* E as follows: For each $e = [x,y] \in E$, we have that $[(x,0),(x,1),(y,1),(y,0)]$ is a saturated 4-circuit in $X \times K_2$. X_E is obtained from $X \times K_2$ by deleting the edges $[(x,0),(y,0)]$, $[(x,1),(y,1)]$ and adjoining the diagonals $[(x,0),(y,1)]$, $[(x,1),(y,0)]$. If ρ is an equivalence on $E(X)$ then ρ induces an equivalence ρ_E on $E(X_E)$ as follows: For $e = [x,y]$, $e' = [x',y'] \in E(X_E)$, $e \,\rho_E\, e'$ if and only if either (i) $pr_1\, x = pr_1\, y$ and $pr_1\, x' = pr_1\, y'$, or (ii) $[pr_1 x, pr_1\, y]$, $[pr_1\, x', pr_1\, y'] \in E(X)$ and $[pr_1\, x, pr_1\, y] \,\rho\, [pr_1\, x', pr_1\, y']$.

Proposition 2: *Let ρ be an equivalence on $E(X)$ containing $\alpha \cup \beta$, E an equivalence class of $E(X)$ modulo ρ with E containing no triangles. Then the induced equivalence ρ_E on the interchange graph X_E of X relative to E contains $\alpha \cup \beta$.*

Let X be a 4-circuit, $e \in E(X)$, $\rho = E(X) \times E(X)$. Let X_2 be the interchange of X relative to e and ρ_2 be the equivalence on $E(X_2)$ induced by ρ. It is easily verified that ρ_2 contains $\alpha \cup \beta$ on X_2 and that $C_2 = [x_1, x_2, x_3, x_4, x_5]$ is a ρ_2-compatible circuit (Fig. 1). Take $E_2 = \{[x,y] \in X_2\colon\ pr_1\, x = pr_1\, y\}$.

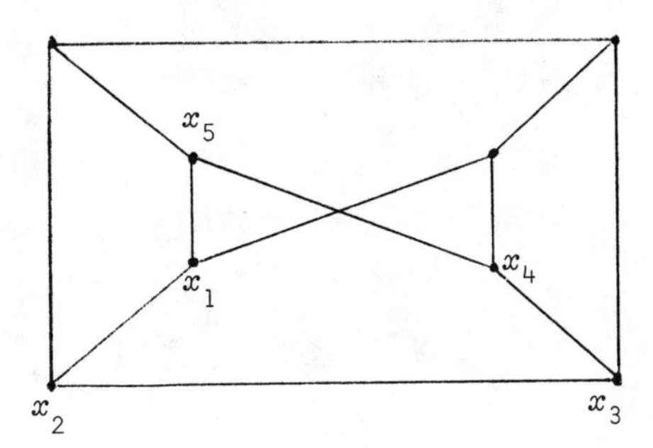

Figure 1.

Let X_3 be the interchange of X_2 relative to E_2 and ρ_3 the equivalence induced on $E(X\ )$ by ρ_2. Since E_2 is an equivalence class of $E(X_2)$ modulo ρ_2 Proposition 2 implies that $\rho_3 \supset \alpha \cup \beta$. $C_3 = [(x_1,0), (x_2,0), (x_3,0), (x_4,0), (x_5,0), (x_1,1)]$ is a ρ_3-compatible circuit with ρ_3-degree $= 3$. Moreover C_3 is simple.

Continuing this process we can construct for each integer n, $n \geq 2$, a connected graph X_n, an equivalence ρ_n on $E(X_n)$, and a simple ρ_n-compatible circuit C_n with ρ_n-degree = n.

Take $Y = \prod_{n=2}^{\infty} (X_n, x_n)$, $\sigma = \bigcap_{n=2}^{\infty} \tilde{\rho}_n$, where $x_n \in V(X_n)$. (For the definition of $\tilde{\rho}_n$ see [1], Definition 6). Then Y is connected, $\sigma \supset \alpha \cup \beta$, and for each integer $n \geq 2$, there exists a simple σ-compatible circuit C in Y with $\deg_\rho C = n$.

References

1. Imrich, W., Kartesisches Produkt von Mengensystemen und Graphen, *Studia Sci. Math. Hungar.*, 2 (1967), 285-290.

2. Miller, D.J., Weak Cartesian Product of Graphs, to appear in *Colloq. Math.*

3. Miller, D.J., The Automorphism Group of a Product of Graphs, to be submitted.

4. Sabidussi, G., Graph Multiplication, *Math. Zeitschr.*, 72 (1960), 446-457.

5. Vizing, V., The Cartesian Product of Graphs. (Russian), *Vycisl. Sistemy*, 9 (1963), 30-43.

THE EXPECTED STRENGTHS OF LOSERS IN KNOCKOUT TOURNAMENTS

J. W. Moon

University of Alberta, Edmonton, Alta.,
Canada

1. Introduction. In a typical knockout tournament some of the players are matched off and the losers withdraw from further play; this process is repeated and the winner is the player who eventually emerges undefeated. Let us suppose no two players have the same strength so there is some (unknown) labelling 1, 2, $\cdots$, n of the players in order of increasing strength; we also assume the stronger player always wins in any match so the only element of randomness arises in the way in which the players are matched off. After the tournament has been played we know the winner must be player n, but in general we do not know with certainty the strengths (ranks) of the remaining players. Our main object here is to determine the expected strength of a player with a given performance pattern in two types of knockout tournaments. We omit many of the details for the sake of brevity; we shall make frequent use of the identity

$$\sum_{j=a}^{m-b} \binom{j}{a}\binom{m-j}{b} = \binom{m+1}{a+b+1}. \tag{1}$$

2. The classical case. Let C_t denote a knockout tournament on $n = 2^t$ players in which all undefeated players are matched off at random in each round and the strongest player emerges undefeated after t rounds.

I. If $n = 2^t$, then there are $n!/2^{n-1}$ tournaments C_t possible.

II. If $S(i,r)$ denotes the probability that player i survives round r, then

$$S(i,r) = \binom{i-1}{2^r-1} \cdot \binom{2^t-1}{2^r-1}^{-1}.$$

III. If λ_i denotes the round in which player i ($< 2^t$) is defeated, then

$$E(\lambda_i) = 1 + \sum_{r=1}^{t-1} S(i,r) \text{ and } E\big(\lambda_i(\lambda_i - 1)\big) = 2 \sum_{r=1}^{t-1} rS(i,r).$$

The first two results follow quite directly from first principles (it follows from II, for example, that the second strongest player survives until the final round with probability $2^{t-1}/(2^t-1)$; see [1] and [4]); the third follows from the observation that the probability that player i is defeated in round $r + 1$ equals $S(i,r) - S(i,r+1)$. H. Morin has prepared tables of the mean and variance of λ_i for $t \leq 8$. It is possible to give closed formulas for $E(\lambda_i)$ and $\sigma^2(\lambda_i)$ for a few large values of i; for example, if i is the second strongest player, $2^t - 1$, then $E(\lambda_i) = (t-1) + t(2^t-1)^{-1}$ and $\sigma^2(\lambda_i) = 2 - t^2 2^t(2^t-1)^{-2}$. If ρ denotes the number of games played by an unspecified player in any given tournament C_t, then it is a routine exercise to show that $E(\rho) = 2-2(1/2)^t$ and $\sigma^2(\rho) = 2 + 2\,(1/2)^t(1-2t-2(1/2)^t)$. The next result follows from result II and identity (1).

IV. If player x survives round r and player y is defeated in round r, then

$$E(x) = (2^t+1) \cdot \frac{2^r}{2^r + 1} \quad \text{and} \quad E(y) = (2^t+1) \cdot \frac{2^r}{2^r + 1} \cdot \frac{2^{r-1}}{2^{r-1} + 1},$$

where the average is taken over all tournaments C_t.

We say a player (of unknown strength) x is of type $(\alpha, \beta, \cdots, \nu)$ if x loses in round $\alpha + 1$ to someone who loses in round $\beta + 1 \cdots$ to someone who loses in round $\nu + 1$ to the strongest player, 2^t. There is a one-to-one correspondence between the $2^t - 1$ losers in a tournament C_t and the 2^{t-1} increasing sequences $(\alpha, \beta, \cdots, \nu)$ of length less than t that can be formed from the integers $0, 1, \cdots, t - 1$. Let $E_t(\alpha,\beta,\cdots,\nu)$ and $T_t(\alpha, \beta, \cdots, \nu)$ denote the expected values of x and $x(x+1)$ given that player x is of type $(\alpha, \beta, \cdots, \nu)$ in a random tournament C_t.

V. $$E_t(\alpha, \beta, \cdots, \nu) = 2^t \cdot \frac{2^\nu}{2^\nu + 1} \cdot \cdots \cdot \frac{2^\alpha}{2^\alpha + 1} \quad \text{and}$$

$$T_t(\alpha, \beta, \cdots, \nu) = 2^t(2^t+1) \cdot \frac{2^\nu}{2^\nu + 2} \cdot \cdots \cdot \frac{2}{2^\alpha + 2}.$$

The main step in proving these results is showing that $P_t(i:\alpha, \cdots, \mu, \nu)$, the probability that player i is of type $(\alpha, \cdots, \mu, \nu)$ in a random tournament C_t, satisfies the relation

$$P_t(i:\alpha, \cdots, \mu, \nu) = (2^{t-\nu}-1)^{-1} \binom{2^t-1}{2^\nu-1}^{-1} \sum_{w=0}^{i-1} \binom{i-1}{w} \cdot \binom{2^t-1-i}{2^\nu-1-w} P_\nu (w+1:\alpha, \cdots, \mu). \tag{2}$$

Consider the subtournament C_ν determined by the (unique) player y of type (ν) and the $2^\nu - 1$ players who lose to y either directly or indirectly. The probability that C_ν contains player i and exactly w players weaker than i is given by the binomial coefficients in the right hand side of (2) (the proof of this is similar to the proof of II). For each fixed value of w, the probability that i is of type $(\alpha, \cdots, \mu, \nu)$ in C_t equals $P_\nu(w + 1:\alpha, \cdots, \mu)$, the probability that i is of type $(\alpha, \cdots, \mu)$ in C_ν, times $(2^{t-\nu}-1)^{-1}$, the probability that y loses to the strongest player in round $\nu + 1$. Relation (2) now follows upon summing over w.

Therefore,

$$\begin{aligned} E_t(\alpha, \beta, \cdots, \mu, \nu) &= \sum_i iP_t (i:\alpha, \cdots, \mu, \nu) \\ &= (2^{t-\nu}-1)^{-1} \binom{2^t-1}{2^\nu-1}^{-1} \sum_w (w+1) P_\nu(w+1;\alpha,\ldots,\mu) \\ &\qquad \cdot \sum_i \binom{i}{w+1} \binom{2^t-1-i}{2^\nu-1-w} \\ &= \frac{2^t}{2^\nu + 1} E_\nu(\alpha, \beta, \cdots, \mu), \end{aligned}$$

upon interchanging the order of summation and applying identity (1). The first formula in *V* now follows by induction and the second is proved in the same way (the results in the case of players of type (ν) follow readily from II and identity (1)). These formulas are equivalent to those obtained by Hartigan [3] from a somewhat different point of view.

3. Another case. Let K_n denote a knockout tournament on n players in which any two players compete in the first round and the winner of

each round is challenged by some undefeated player in the next round until the strongest player emerges undefeated after $n - 1$ rounds. Narayana and Zidek [5] have investigated some properties of this method of conducting a tournament.

VI. If $n \geq 2$, then there are $\frac{1}{2} \cdot n!$ tournaments K_n possible.

VII. If $s(i,r)$ denotes the probability that player i plays in round r and survives, then

$$s(i,r) = \binom{i-1}{r} \cdot \binom{n}{r+1}^{-1}.$$

VIII. If λ_i denotes the round in which player i ($< n$) is defeated, then

$$E(\lambda_i) = \tfrac{1}{2}(n-1) + 1/n + \sum_{r=1}^{n-2} s(i,r) \quad \text{and}$$

$$E(\lambda_i(\lambda_i - 1)) = \tfrac{1}{3}(n-1)\ (n-2) + 2 \sum_{r=1}^{n-2} rs(i,r).$$

IX. If player x plays in round r and survives and player y plays in round r and is defeated, then

$$E(x) = (n+1)\,\frac{r+1}{r+2} \quad \text{and}$$

$$E(y) = \tfrac{1}{2}(n+1)\,\frac{r}{r+1} \cdot \frac{r+3}{r+2}.$$

The proofs of VI and VII are quite straightforward; the proofs of VIII and IX use the fact that the probability that player i is defeated in round r equals $1/n + s(i, r-1) - s(i,r)$.

We say a player x is of type $(\alpha, \beta, \cdots, \nu)$ if x wins at least one match before losing in round α to someone who loses in round $\beta \cdots$ to someone who loses in round ν to the strongest player, n. Let $e_n(\alpha, \beta, \cdots, \nu)$ and $t_n(\alpha, \beta, \cdots, \nu)$ denote the expected values of x and $x(x+1)$ given that player x is of type $(\alpha, \beta, \cdots, \nu)$ in a random tournament K_n.

X. If $1 < \alpha < \beta < \cdots < \nu \leq n - 1$, then the probability that a tournament K_n has a player of type $(\alpha, \beta, \cdots, \nu)$ is $(n \cdot \nu \cdot \cdots \cdot \beta)^{-1}$.

XI. $e_n(\alpha, \beta, \cdots, \nu) = n \cdot \frac{\nu}{\nu + 1} \cdot \cdots \cdot \frac{\alpha}{\alpha + 1}$ and

$$t_n(\alpha, \beta, \cdots, \nu) = n(n + 1) \cdot \frac{\nu}{\nu + 2} \cdot \cdots \cdot \frac{\alpha}{\alpha + 2} .$$

Result X follows by a simple induction argument; the main step in proving XI is showing that $p_n(i:\alpha, \cdots, \mu, \nu)$, the probability that player i is of type $(\alpha, \cdots, \mu, \nu)$ in a random tournament K_n given that some player in K_n is of this type, satisfies the relation

$$p_n(i:\alpha, \cdots, \mu, \nu) = \frac{n}{\nu+1} \binom{n}{\nu+1}^{-1} \sum_{w=0}^{i-1} \binom{i-1}{w} \binom{n-1-i}{\nu-1-w} \cdot p_\nu(w+1:\alpha, \cdots, \mu). \tag{3}$$

The derivation of (3) and the remaining steps in the proof are quite similar to the argument outlined in the proof of V and will be omitted; the following results can also be proved in essentially the same way.

XII. Suppose player x plays for the first time in round r and loses immediately to player y. If $y = n$, then $E(x) = \frac{1}{2}n$ and $E(x(x+1)) = \frac{1}{3}n$; if y is of type $(\alpha, \beta, \cdots, \nu)$, then $E(x) = \frac{1}{2}e_n(\alpha, \beta, \cdots, \nu)$ and $E(x(x+1)) = \frac{1}{3}t_n(\alpha, \beta, \cdots, \nu)$.

We remark in closing that T. V. Narayana and some of his students are comparing these and other knockout tournaments with respect to various criteria (see also [2]).

References

1. Lewis Carroll, Lawn Tennis Tournaments. The Works of Lewis Carroll. Hamlyn, London, 1965, pp. 1059-1066.

2. H. A. David, The Method of Paired Comparisons. Griffin, London, 1963.

3. J. A. Hartigan, Probabilistic completion of a knockout tournament, *Ann. Math. Statist.* 37 (1966) 495-503.

4. F. Mosteller, Fifty Challenging Problems in Probability with Solutions. Addison-Wesley, Reading, 1965, pp. 30-31.

5. T. V. Narayana and J. Zidek, Contribution to the Theory of Tournaments, Part I, The Combinatorics of Knockout Tournaments. University of Alberta Preprint Series, 1968.

THE SECOND MOMENT METHOD IN COMBINATORIAL ANALYSIS

L. Moser

University of Hawaii, Honolulu, HI, U.S.A.

The *second moment method* is applied to a variety of combinatorial problems some of which are described here.

1. The set $\{1,2,\ldots,2n\}$ is divided into two classes, $A = \{a_1, a_2, \ldots, a_n\}$ and $B = \{b_1,\ldots,b_n\}$. Let $M_k(A) = \sum_{a_i - b_j = k} 1$, and

   $f(n) = \min_A \max_k M_k(A)$. We find lower bounds for $f(n)$.

2. Let $f(n)$ be the least number of integers $a_1, a_2, \ldots, a_{f(n)}$ such that the integers $1,2,\ldots,n$ can all be written as the sum of four a_i's. We find lower bounds for $f(n)$.

3. Let θ_i be real and $\sum_{i=1}^{n} \theta_i^2 = 1$. We show that, given any integer $k > 1$, there exist integers χ_i, $i = 1,2,\ldots,n$, not all zero, such that $|x_i| < k$ and

$$|\theta_1 x_1 + \ldots + \theta_n x_n| \leqslant \sqrt{\frac{k^2 - 1}{k^{2n} - 1}} .$$

4. Given m equations in $n > m$ unknowns:

$$a_{11}\, x_1 + \ldots + a_{1n}\, x_n = 0$$
$$\vdots$$
$$a_{m1}\, x_1 + \ldots + a_{mn}\, x_n = 0,$$

with integer coefficients a_{ij} bounded, in absolute value, by A. Then there exists a non-trivial integral solution $x_1, x_2, \ldots, x_n$ with

$$\max_{i=1,\ldots,n} |x_i| < (c \sqrt{n} A)^{\frac{m}{n-m}} .$$

As special cases of the method used above we have:

(i) Given n coins, each weighing 0 or 1. Question. How many subsets msut be weighed in order to determine the weights of all the coins?
Answer. At least $\dfrac{n \log 4}{2 + \log n}$.

(ii) Given k positive integers $a_1 < a_2 < \ldots < a_k$, with the $2^k - 1$ non-empty subsets having distinct sums, we show that

$$\sum_{i=1}^{k} a_i^2 \geq \frac{4^k - 1}{3} .$$

CONSTRUCTION OF A BASIS OF ELEMENTARY CIRCUITS AND COCIRCUITS IN A DIRECTED GRAPH

John D. Murchland

Kernforschungszentrum Korlsruhe
and University of Karlsruhe, Germany

Introduction.

The notation of Berge [1] and Berge and Ghouila-Houri [2] is employed here. Finite, connected, directed graphs are considered, with n_v vertices and n_a arcs. The directed entities path, circuit, arborescence and cocircuit correspond to the undirected chain, cycle, tree and cocycle.

These entities may be represented by vectors of size n_v which specify the number of times each arc is taken in the sense of its direction. An old result is that any cycle may be uniquely expressed as a linear combination of the cycles in a basis of (elementary and simple) cycles. Such a basis must have n_c members, where

$$n_c = n_a - n_v + 1.$$

Simarilarly, any cocycle may be expressed as a linear combination of n_{cc} (elementary) cocycles, where

$$n_{cc} = n_v - 1.$$

Constructively, a spanning tree gives a basic cycle when each arc outside it is added to it, and a basic cocycle for each edge removed from it.

In [1] and [2] Berge proves that a strongly connected graph necessarily has a basis of n_c elementary circuits, and a graph without circuits a basis of n_{cc} elementary cocircuits. It will be seen that the actual constructions given here lack the simplicity and duality of the undirected case. They each need of the order of $n_a n_v$ operations at the lowest level.

Basis of elementary circuits.

Berge's proof in [1] or [2] shows that a cycle may be expressed as a linear combination of circuits. The following process is more direct than this implied purification and selection.

1. Choose any vertex r in the graph as root.

2. Construct a spanning arborescence A_1 *from* the root r (that is, a directed tree providing a path from r to every other vertex).

3. Construct a spanning arborescence A_2 *into* the root r.

4. For each arc $[vw)$ which does not belong to either A_1 or A_2 construct the circuit

$$[r \ldots vw \ldots r),$$

where the path from r to v is taken from A_1, and that from w to r from A_2. From this possibly non-elementary circuit extract an elementary circuit containing $[vw)$. Call this an 'external' circuit with defining arc $[vw)$.

5. Repeat the step 4 operation for each arc of A_2 which does not occur in A_1. Call these 'internal' circuits.

Because the graph is strongly connected, the arborescences required by the algorithm can always be found. The external circuits are independe[nt] since their defining arc occurs in no other circuit. The defining arc in each internal circuit is, for each, the furthest arc from the root in A_2 (which does not occur in A_1). This establishes their independence also. Finally, n_c elementary circuits are obtained. This shows the correctness of the algorithm (and, since the number of independent circuits could not exceed the number of independent cycles, gives another proof of Berge's result).

In the expression of an arbitrary elementary circuit, the coefficien[t] of each external circuit must be zero or one. Since the defining arc of an internal circuit may occur in several other circuits, its coefficient may have negative values. However, the basis found by the algorithm cannot have more than n_v-1 internal circuits and may have only 1 if A_1 and A_2 can be chosen from a Hamiltonian circuit. (Clearly it must be at least one if the graph has a non-basic elementary circuit).

A basis for all paths from one vertex to another, or to all others, may be found by adding suitable return arcs, carrying out the constructio[n] and deleting them. The choices made in the construction will control the number of circuits and paths found, and which are internal.

Basis of elementary cocircuits.

In a graph without circuits, call a vertex without predecessors an origin, and one without successors a destination.

Berge's proof in [2] that a basis of cocircuits will number n_{cc} is by induction. A destination v, taken by itself, will determine a cocircuit. Berge shows that the $n_{cc} - 1$ cocircuits which form a basis in the graph in which v has been identified with one of its predecessors will persist when v is expanded again. The predecessor must be chosen so that no circuit is created.

This implies an algorithm, but the cocircuits formed will not be elementary unless each vertex when it is shrunk is not an articulation point. Since a graph without circuits must have at least one origin or destination (indeed, two) which is not an articulation point [Harary, Norman and Cartwright, 3] a favourable choice may always be made.

1. Number the vertices of the graph so that every arc runs from a lower to a higher-numbered vertex.

2. If only one vertex remains, go to 6.

3. Find an origin or destination vertex v which is not an articulation point. If v is a destination go to 5.

4. Find the lowest-numbered successor, w, to v. Shrink v into w by changing every arc $[vx)$ emergent from v into $[wx)$ and deleting those which become $[ww)$. Add the arc $[vw)$ to the list of arcs for the cocircuit tree. Go to 2.

5. Find the highest-numbered predecessor, u, to v; change every arc $[tv)$ to $[tu)$ deleting those which become $[uu)$; add $[uv)$ to the cocircuit list.

6. The arcs placed in the cocircuit list will now number n_{cc} and will form a tree. The n_{cc} cocycles generated by the vertices of the components of the tree when each arc in turn is removed will be the desired basis of independent elementary cocircuits.

The important parts of a proof for this algorithm are to show that the graph can be numbered as stated, that the shrinkage of steps 4 and 5 will not produce any circuits, and that the information contained in the cocircuit list of arcs is sufficient and permits the elementary cocircuits to be extracted from it in the way stated.

Note that some of the arcs which occur in the cocircuit tree may not be present in the original graph.

References

1. Berge, C., Théorie des Graphes and ses Applications, Dunod, Paris, 1958, English translation by A. Doig, Methuen, London, 1962, 27-30 and 155.

2. Berge, C., and Ghouila-Houri, A., Programmes jeux et Réseaux de Transport, Dunod, Paris, 1962, English translation by M. Merrington and C. Ramanujacharyulu, Methuen, London, 1965, 123-132.

3. Harary, F., Norman, R.Z., and Cartwright, D., Structural Models: an Introduction to the Theory of Directed Graphs, John Wiley, New York, 1965.

EQUICARDINAL MATROIDS AND FINITE GEOMETRIES

U.S.R. Murty

University of Waterloo, Ont., Canada

1. Introduction. There are many equivalent ways of defining a "matroid", e.g., see Whitney [5]. We give a definition of a "matroid" in terms of the concept of a "hyperplane". Let E be a finite set. Let H be a collection of subsets of E. Let the following postulates be satisfied.

I. No member of H is strictly contained in another.

II. If H_1, H_2 are members of H and x is a member of E such that $x \notin H_1 \cup H_2$, then there exists a member H_3 of H such that

$$x \in H_3 \supseteq H_1 \cap H_2.$$

Then the ordered pair $\langle E,H \rangle$ is called a *matroid*. The members of E and H are respectively called the *cells* and *hyperplanes* of the matroid. The complements of hyperplanes are called the *co-circuits* of the matroid.

One way of obtaining examples of matroids is as follows: Let S be a projective (affine) space of dimension at least two. Let E be a finite set of points in S. For $A \subseteq E$, define *rank of* A, $r(A)$, as the maximal number of elements in a subset B of A with the property that no k of the elements of B are in a $(k-2)$-dimensional subspace of S, for $3 \leq k \leq |B|$. Let H be the collection of maximal subsets of E of rank $r(E) - 1$. Then the pair $\langle E,H \rangle$ is a matroid. If S is a finite projective (affine) geometry and E is the set of all points of S, then we say that such a matroid is of *type I (type II)*. Now one can ask the following question: What restrictions must be imposed on a matroid in order that it may be isomorphic to a matroid of type I (type II). An almost complete solution to this problem appears in disguise in two papers on block designs, [1] and [2]. A more general problem is to find all matroids in which any two hyperplanes have the same cardinality. In [3], the author characterized all "binary matroids" which have this property. A matroid is called a *binary matroid* if each non-null mod 2 sum of co-circuits of the matroid is a union of disjoint co-circuits of the matroid. A matroid of type I (type II) constructed from a geometry over $GF(2)$ is binary, and we say that it is of *binary-type I (type II)*.

In the next section we mention some of the known results. In the last section we pose a very general problem.

2. Theorems. A matroid is called an *equicardinal matroid* if all the hyperplanes of the matroid are of the same cardinality. A matroid is said to be *connected* if any two cells of the matroid are contained together in some co-circuit. For the purposes of establishing a complete list of equicardinal matroids it is sufficient to look for connected equicardinal matroids. A matroid is called a *k-matroid* if it is connected and its hyperplanes have k cells. A matroid is said to be a *proper matroid* if given any two cells p and q, there is a hyperplane containing p and not containing q.

Let E be a finite set, $|E| \geq k + 1$, $k \geq 1$. Let H be the collection of k-subsets of E. Then the pair $\langle E, H \rangle$ is an equicardinal matroid, and we say that it is of *type 0*. When $|E| = k + 2$, then this matroid will be binary and we say that it is of *binary-type 0*. We are now ready to state our theorems.

Theorem 1 (Murty). *A proper binary k-matroid is, up to isomorphism, a matroid of binary type 0, binary type I or binary type II.*

Remark. A q-expansion of a matroid consists in replacing each cell of the matroid by a q-set of cells, preserving incidence with hyperplanes, sets replacing two different cells being disjoint. All binary equicardinal matroids can be obtained from proper ones by the operation of q-expansion.

Theorem 2 (Dembowski-Wagner). *A proper k-matroid is isomorphic to a matroid of type I if and only if the hyperplanes of the matroid constitute a (v, k, λ)-design on the cells of the matroid, with $v \geq k + 2$.*

Theorem 3 (Kantor). *A proper k-matroid is isomorphic to a matroid of type II if and only if it has the following properties*

(i) $|E| \geq k + 2$

(ii) *there is an integer $u < k$ such that two intersecting hyperplanes have u cells in common.*

(iii) *for some non-incident cell-hyperplane pair there is precisely one hyperplane not meeting the given hyperplane and containing the given cell.*

In fact Kantor [2] proved a theorem which includes both theorems 2 and 3. He also describes a proper 6-matroid which is neither of type 0, type I or type II, due to Witt.

3. A Very General Problem. The *rank* of a matroid is the least number of cells in a subset of the set of cells of the matroid that meets every co-circuit of the matroid. We define an (n,r,k)-*matroid* as a k-matroid of rank k which has n cells. Then a very general problem is to determine whether an (n,r,k)-matroid exists for given integers n, r and k. A theorem of the author [4] implies that every $(n,r,n-3)$-matroid is of type 0, in which case r will be equal to $n-2$. There is always an $(n,k+1,k)$-matroid of type 0, for $n-2 \geq k$. Very few general results are known.

References

1. Dembowski, P., and Wagner, A., Some Characterizations of Finite Projective Spaces, *Arch. Math.*, 11 (1960), 465-469.

2. Kantor, W.M., Characterizations of Finite Projective and Affine Spaces, *Can. J. Math.*, 21 (1969), 64-75.

3. Murty, U.S.R., Equicardinal Matroids, submitted for publication (1968).

4. Murty, U.S.R., Matroids with Sylvester Property, to appear in *Aequationes Mathematicae*.

5. Whitney, H., On the abstract properties of linear dependence, *Amer. J. Math.*, 57 (1935), 507-533.

A SURVEY OF THE THEORY OF WELL-QUASI-ORDERED SETS

C.St.J.A. Nash-Williams

University of Waterloo, Waterloo, Ont., Canada

We shall use the following terminology and notation. P denotes the set of all positive integers. A sequence *on* a set A is a sequence whose terms belong to A. $F(A)$ denotes the set of all finite sequences on A. $I(A)$ denotes the set of all infinite sequences on A. (In this paper "infinite sequence" always means an infinite sequence $a_1,a_2,\ldots$ of the kind used in analysis, with a term a_i corresponding to each positive integer i.) $T(A)$ denotes the class of all transfinite sequences on A. If α is an ordinal number, $T_\alpha(A)$ is the set of all transfinite sequences on A of length α. The set of all subsets of A will be denoted by PA. A *quasi-ordered* (*qo*) set is a set Q on which a reflexive and transitive relation $\leqslant$ is defined: if $q,q' \varepsilon Q$ and $q \leqslant q'$ we say that q *anticipates* q'. (The difference between *qo* sets and partially ordered sets is that, in a *qo* set, we do not insist that if $q \leqslant q' \leqslant q$ then q must equal q': this difference is relatively unimportant since any *qo* set can be reduced to a partially ordered set by identifying elements in the same equivalence class under an equivalence relation $\sim$, where $q \sim q'$ means that $q \leqslant q' \leqslant q$.) Throughout this lecture, Q denotes a *qo* set. An infinite sequence $q_1,q_2,\ldots$ belonging to $I(Q)$ is *good* if there exist $i,j \varepsilon$ P such that $i < j$ and $q_i \leqslant q_j$, and is *bad* otherwise. A *qo* set Q is *well-quasi-ordered* (*wqo*) if all sequences which belong to $I(Q)$ are good. There are several equivalent formulations of this definition. The version just given is the one I have found easiest to work with as a rule, but perhaps the following gives a better intuitive grasp of the underlying idea. Call two elements of Q *comparable* if one of them anticipates the other, and let us say that Q is *of finite width* if there is no infinite set of mutually incomparable elements of Q. Let us further say that Q satisfies the *descending chain condition* if there is no sequence $q_1, q_2, \ldots$ belonging to $I(Q)$ such that $q_1 > q_2 > \ldots$, where the notation $q > q'$ means that q' anticipates q and q does not anticipate q'. Then it is easily provable that Q is *wqo* if and only if it is of finite width and satisfies the descending chain condition.

It is easily seen that every well-ordered set is *wqo*. To construct slightly less trivial examples of *wqo* sets we can use a theorem of Higman [2]. For any *qo* set Q, let us quasi-order $F(Q)$ by the rule that a finite sequence $q_1, \ldots, q_m$ belonging to $F(Q)$ *anticipates* a finite sequence $q'_1, \ldots, q'_n$ belonging to $F(Q)$ if the latter sequence has a subsequence $q'_{i(1)}, q'_{i(2)}, \ldots, q'_{i(m)}$ $(i(1) < i(2) < \ldots < i(m))$ such that q_t anticipates $q'_{i(t)}$ (in Q) for $t = 1, \ldots, m$. Thus if Q is the well-ordered set P then in $F(Q)$ the sequence 5,4,8,6 anticipates the sequence 1,1,7,3,4,9,2,4,7,6,3 as may be seen by comparing the first sequence with the subsequence 7,4,9,7 (or the subsequence 7,4,9,6) of the second sequence. Higman's theorem states that *if Q is wqo then $F(Q)$ is wqo*. The proof, even for the special case of F(P), seems to be a not entirely easy exercise if one has no previous knowledge of the subject: it seems to be a feature of well-quasi-ordering theory that things may be tolerably easy if one knows just how to tackle them but there is often little to help one in guessing the right approach. The following is a method of starting a proof of Higman's theorem, for further details of which the reader can be referred to [5]: this proof is a little different from the original one of Higman [2]. Suppose that Q is *wqo* and $F(Q)$ is not. Let B_1 be the set of all $s \in F(Q)$ such that s is the first term of some bad infinite sequence on $F(Q)$. Select an element s_1 of B_1 with as few terms as possible. Let B_2 be the set of all $s \in F(Q)$ such that s_1, s are, in that order, the first two terms of some bad infinite sequence on $F(Q)$. Select an $s_2 \in B_2$ with as few terms as possible. Let B_3 be the set of all $s \in F(Q)$ such that s_1, s_2, s are, in that order, the first three terms of some bad infinite sequence on $F(Q)$. Select an $s_3 \in B_3$ with as few terms as possible; and so forth. It can be shown that this leads to a contradiction.

If we quasi-order $I(Q)$, or $T(Q)$, in a manner analogous to the above quasi-ordering of $F(Q)$, it is not necessarily true that $I(Q)$ or $T(Q)$ is *wqo* when Q is *wqo*. Nevertheless, Higman's theorem does generalize to transfinite sequences in the following sense. Call a transfinite sequence on Q *restricted* if it has only finitely many distinct terms, e.g., if q and q' belong to Q, the element $q,q',q,q',q,q',\ldots$ of $I(Q) = T_\omega(Q)$ is restricted. Finite sequences are automatically restricted. Rado [9] conjectured that the set of all restricted transfinite sequences on Q is *wqo* if Q is *wqo* and proved this for transfinite sequences of length less than ω^3. Subsequently Kruskal, in unpublished work, and Erdős and Rado, in [1], proved the conjecture for restricted transfinite sequences of length less than ω^ω, and finally the author, in [6], proved it in full. The proof is somewhat complicated and tricky.

Although it is not in general true that $T(Q)$ must be *wqo* when Q is *wqo*, yet this does hold for many *wqo* sets Q, and indeed it seems to require rather artificial constructions to produce *wqo* sets Q for which $T(Q)$ is not *wqo*. Consideration of this kind of thing suggests that some well-quasi-ordered sets are not really as "well" quasi-ordered as would be desirable for some purposes, and that there is a need to identify a class of really well-behaved well-quasi-ordered sets. An idea of Rado [9], supplemented by further investigation, led to the following fairly formidable definition. Let us call a finite sequence $a_1,a_2,\ldots,a_n$ belonging to F(P) or an infinite sequence $b_1,b_2,\ldots$ belonging to I(P) *ascending* if $a_1 < a_2 < \ldots < a_n$, or $b_1 < b_2 < b_3 < \ldots$, respectively. Let AF(P), AI(P) denote respectively the sets of ascending sequences belonging to F(P), I(P). A *left-segment* of an infinite sequence $b_1,b_2,\ldots$ is a finite sequence of the form $b_1,b_2,\ldots,b_n$. Call a subset B of AF(P) a *block* if every element of AI(P) has a left-segment belonging to B. (If we imagine ourselves trying to write down successively the terms of a sequence in AI(P) but being compelled to stop if at any stage the terms already written down form a finite sequence belonging to some specified subset of AF(P), then that subset of AF(P) is a block if it blocks *every* member of AI(P) at some stage.) For $s,t \in AF$(P), let us write $s \lhd t$ if there exists a sequence $a_1,a_2,\ldots,a_n$ belonging to AF(P) such that t is the sequence $a_2,a_3,\ldots,a_n$ and s is $a_1,a_2,\ldots,a_m$ for some $m < n$. Let us say that Q is *better-quasi-ordered* (*bqo*) if, for every pair B,f such that B is a block and f is a function from B into Q, there exist $s,t \in B$ such that $s \lhd t$ and $f(s) \leqslant f(t)$. It is easily proved that all well-ordered sets are *bqo* and that all *bqo* sets are *wqo*; and in fact *bqo* sets are the particularly well-behaved *wqo* sets which we sought to identify. In [8], by a lengthy argument, I proved that if Q is *bqo* then $T(Q)$ is *bqo* (and therefore *a fortiori wqo*). This settled a conjecture of Milner that $T(Q)$ is *wqo* whenever Q is a set of ordinal numbers (quasi-ordered in the usual way), which Milner [4] had proved for transfinite sequences of length less than ω^3. I am reasonably confident that it can also be proved that if $T(Q)$ is *wqo* then Q is *bqo*; but hitherto I have never found time to develop the details of such a proof.

My own interest in well-quasi-ordering arose initially from its connections with graph theory. Let G denote the class of all graphs, T denote the class of all trees, FT denote the class of all finite trees, G_n denote the class of all graphs in which all vertices have valency $\leq n$, and FT_n denote $(FT) \cap G_n$. A *subdivision* of a graph G is a graph obtained from G by inserting into each edge λ of G some non-negative finite number $n(\lambda)$ of vertices of valency 2 so as to subdivide λ into $n(\lambda)+1$ edges, as illustrated in Figure 1. Any class of graphs may be quasi-ordered by the

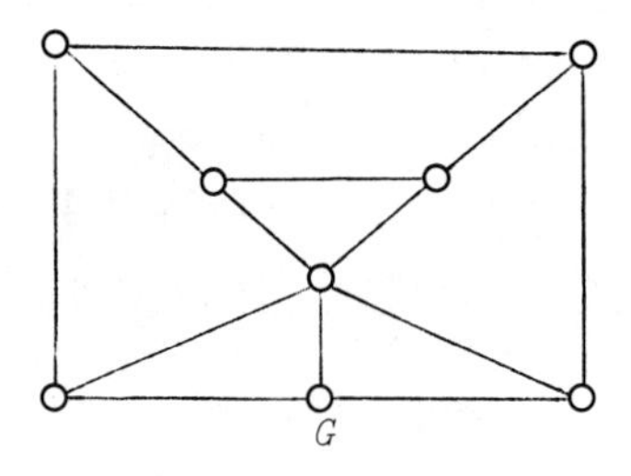

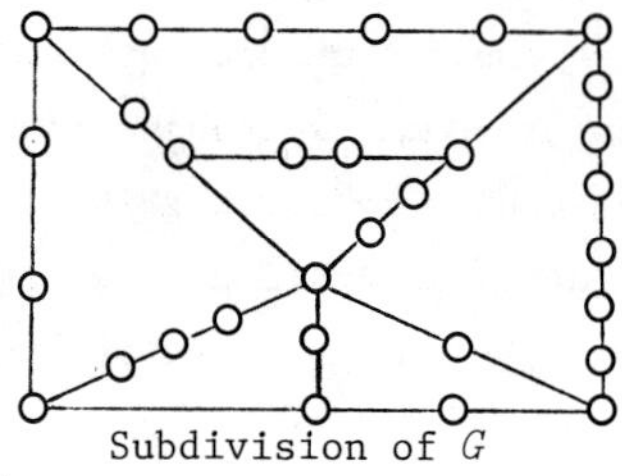

Figure 1

rule that a graph G *anticipates* a graph G' if and only if some subdivision of G is isomorphic to a subgraph of G' (which implies that G is homeomorphic to a subgraph of G' but is in fact slightly stronger than this). With this definition, it is easily seen that G is not *wqo*, but it was proved by Higman [1] that FT_n is *wqo* for any $n \in P$. Kruskal [3] proved that FT is *wqo* and a subsequent but independent proof of this theorem was obtained by Tarkowski and briefly announced in [10]. A substantially shorter proof that FT is *wqo* was given by the author in [5], using the same type of argument as was described above in connection with the proof that $F(Q)$ is *wqo* if Q is *wqo*. Finally, by a very complicated argument in [7], I was able to prove that (as conjectured by Kruskal in [3]) T is *wqo*. The use of properties of *bqo* sets played a major role in the proof of this theorem, as also in the proof of Milner's conjecture concerning transfinite sequences, and indeed the stronger result that T is *bqo* was obtained.

It might help to give just a little of the flavour of this very lengthy proof if I describe one of the lemmas used in it. For the purpose of the proof, it was found convenient to use *rooted* trees, i.e. trees in which a particular vertex is selected as "root", and to make a slight sharpening in the definition of our quasi-ordering to take account of the rooting. With this done, let $\mathcal{R}$ denote the quasi-ordered class of all rooted trees. If ξ is a vertex of a rooted tree R we define the *branches of R above* ξ to be those components of $R-\xi$ (the graph obtained from R by deleting ξ and its incident edges) which do not include the root of R. Each of these branches is considered as a rooted tree, the root of the branch being that vertex of it which is adjacent to ξ in R. The general idea of our lemma is to try to reduce the consideration of a rooted tree R to the consideration of smaller rooted trees in some sense: this might be the basis of something like an argument by induction, although of course, since we may be considering infinite trees, it would have to be in the nature of transfinite and not ordinary induction. One might consider that branches of R above its vertices could be the smaller rooted trees required for this purpose, but one here runs into the difficulty that some of these may be as big as R itself, if R is an infinite tree. Our lemma is designed to provide a way round this difficulty. Call a branch B of R above ξ *large* if R anticipates B (in the quasi-ordering of $\mathcal{R}$) and *small* otherwise. Let $l(\xi)$ denote the *number* of large branches of R above ξ, and $S(\xi)$ denote the *set* of small branches of R above ξ. Let $V(R)$ denote the set of vertices of R and $\Theta(R)$ denote the element $\{(l(\xi),S(\xi)) : \xi \in V(R)\}$ of $P(\mathrm{C} \times P\mathcal{R})$, where C is the class of all cardinal numbers. Then our lemma states that R_1 anticipates R_2 in $\mathcal{R}$ if $\Theta(R_1)$ anticipates $\Theta(R_2)$ in a certain quasi-ordering of $P(\mathrm{C} \times P\mathcal{R})$ which is derived in a natural way from the quasi-ordering on $\mathcal{R}$. Roughly speaking, this tells us that, in determining whether R_1 anticipates R_2, it is only the small branches of R_1 and R_2 whose structure need be considered in detail, since the only information we need about large branches is how many of them there are above a vertex of either tree.

I am informed by Professor R. Laver that he has recently proved that the class of all *totally* ordered denumerable sets is *wqo* under the rule that an ordered set O_1 *anticipates* an ordered set O_2 if O_2 has a subset which is order-isomorphic to O_1.

One of the most interesting outstanding problems in well-quasi-ordering theory is a conjecture of Vázsonyi that G_3 is *wqo*. Mr. T.A. Jenkyns is at present studying some special cases of this apparently very difficult problem, and it will be interesting to see what emerges from these investigations. It is, of course, within the bounds of possibility that the conjecture might be disproved. If it is true, however, I have suggested in [7] that there is a possibility that one could pleasantly unify it with the above theorem concerning T by considering a certain new quasi-ordering on G which is in general somewhat "weaker" than the one which we have been considering but happens to be equivalent to it on both T and G_3. It seems conceivable that G might be *wqo* under this "weaker" quasi-ordering, and this would imply that both T and G_3 are *wqo* in the sense previously discussed.

Perhaps, if we could obtain simpler or more insightful proofs of the more difficult theorems described in this lecture, and a better grasp of the theory underlying them, this might be a good first step towards sharpening our weapons for tackling further problems. For instance, is there a simpler or more convenient way of characterising *bqo* sets? I hav at present a few faint thoughts concerning this last question, but am not really sure that they amount to anything.

References

1. Erdős, P. and Rado, R., A theorem on partial well-ordering of sets of vectors, *J. London Math. Soc.* 34 (1959), 222-224.

2. Higman, G., Ordering by divisibility in abstract algebras, *Proc. London Math. Soc.* (3) 2 (1952), 326-336.

3. Kruskal, J.B., Well-quasi-ordering, the tree theorem and Vázsonyi's conjecture. *Trans. Amer. Math. Soc.* 95 (1960), 210-225.

4. Milner, E.C., Well-quasi-ordering of transfinite sequences of ordina numbers, *J. London Math. Soc.* 43 (1968), 291-296.

5. Nash-Williams, C.St.J.A., On well-quasi-ordering finite trees, *Proc. Cambridge Philos. Soc.* 59 (1963), 833-835.

6. Nash-Williams, C.St.J.A., On well-quasi-ordering transfinite sequen *Proc. Cambridge Philos. Soc.* 61 (1965), 33-39.

7. Nash-Williams, C.St.J.A., On well-quasi-ordering infinite trees, *Proc. Cambridge Philos. Soc.* 61 (1965), 697-720.

8. Nash-Williams, C.St.J.A., On better-quasi-ordering transfinite sequences, *Proc. Cambridge Philos. Soc.* 64 (1968), 273-290.

9. Rado R., Partial well-ordering of sets of vectors, *Mathematika* 1

(1954), 89-95.

10. Tarkowski, S., On the comparability of dendrites, *Bull. Acad. Polon. Sci. Sér. Sci. Math. Astronom. Phys.* 8 (1960), 39-41.

k-SOCIETIES WITH GIVEN SEMIGROUP

Jaroslav Nešetřil and Pavol Hell

McMaster University, Hamilton, Ont., Canada.

A *k-society* $\mathcal{G}$ (k-uniform set system), where $k \geq 2$ is a natural number, is a couple (X,R) where $R \subset P(X)$ such that $A \in R \Rightarrow |A| = k$. This is a natural generalization of the notion of an undirected graph without loops. The compatible mapping between two k-societies is defined by the same manner as between graphs:

If $\mathcal{G} = (X,R)$, $\mathcal{H} = (Y,S)$ are k-societies, $f : X \to Y$ then f is compatible iff $f(A) \in S$ whenever $A \in R$.

We denote $c(\mathcal{G},\mathcal{H})$ the set of all compatible mappings $\mathcal{G} \to \mathcal{H}$. It is well known that for every monoid S^1 there is an undirected graph without loops G (hence 2-society) such that $c(G) \approx S^1$. We want to extend this result for k-societies. Let us denote by $\mathcal{R}_k$ the category of all k-societies and all their compatible mappings. $\mathcal{R}_2$ is the category of all undirected graphs without loops. Let us denote by $\mathcal{R}$ the category of all (directed) graphs. It suffices to find the full embedding $\phi : \mathcal{R} \to \mathcal{R}_k$, moreover we prove that $\mathcal{R}_k$ is a so called binding category. Our method is based on the idea that (with respect to compatible mappings) some graphs behave like k-societies. Let $k \geq 2$ be fixed from now on. Denote by $\mathcal{R}_2^k$ the category of all undirected graphs G satisfying the following conditions:

(i) G has no loops.

(ii) each edge of G belongs to some k-complete subgraph of G.

(A k-complete graph is a complete graph of cardinality k). For every graph $G \in \mathcal{R}_2^k$, $G = (X,R)$ we define the k-society $G^0 = (X,R^0)$ by $A \in R^0$ if A is the carrier of some k-complete subgraph of G.

Lemma. Let G, H be graphs with (*). Then $c(G,H) = c(G^0,H^0)$. (The proofs of this note can be found in [3]).

Consequently, we have a full embedding $\phi_1 : R_2^k \to R_k$ defined by $\phi_1(G) = G^0$, $\phi_1(f) = f$.

By [2], there is a full embedding $\phi_3 : R \to R_2'$, where R_2' is a full subcategory of R_2 of all graphs without isolated vertices. Thus it remains to find a full embedding $\phi_2 : R_2' \to R_2^k$. By [2], it suffices to find, for every $k \geq 2$, a strongly rigid graph (I_k,a,b) with (*). An example of such a graph is $I_n^m = ((X,R),\ 0,\ 2n)$ where $X = \{0,1,\dots,4n\}$ and

$$(i,j) \in R \Longleftrightarrow |i-j| \leq n \quad \text{or} \quad i = 0, j \geq 4n - m.$$

If $0 < m < n$ then it is possible to show that I_n^m is a strongly rigid graph. If $m \geq k$, then I_n^m satisfies (*).

Proposition: There is a full embedding $R \to R_k$; consequently given any monoid S^1 there is a k-society g with $c(g) \simeq S^1$. By the above constructions we have immediately:

Corollary: Given any $k \geq 2$ and any infinite set X, then there is a rigid k-society $g = (X,R)$.

It is known (see [1]) that there is no rigid 2-society (X,R) for $|X| < 8$ and there is such a 2-society for every $|X| \geq 8$. It can be easily proved that there is no rigid k-society (X,R) for $|X| \leq k+1$. By the following, this bound is the best possible. Let $G = (X,R)$ be a k-society, $|X| = k+2$. Define the graph $\overline{G} = (X,\overline{R})$ by $(a,b) \in \overline{R} \Longleftrightarrow X - \{a,b\} \in R$.

Proposition: A k-society $G = (X,R)$, $|X| = k+2$ is rigid iff there is no non-identical automorphism of the graph $\overline{G}$.

Let us remark that given any set X, $|X| > 5$, there is a graph without non identical automorphism. We can also construct rigid k-societies whenever $|X| > k+2$.

References

1. Hedrlin, Z. and Pultr, On rigid undirected graphs, 1966, *Can. J. Math.*, 18 (1966), 1237-1242.
2. Hedrlin, Z. and Pultr, A., Symmetric relation (undirected graphs) with given semigroup, *Monatsh. Math.*, 69 (1965), 318-322.
3. Hell, P. and Nešetřil, J., Graphs and k-societies (to appear).

ON THE NUMBER OF k-TUPLES IN MAXIMAL SYSTEMS $m(k,l,n)$

Scott Niven

The University of Calgary, Alta., Canada

Let $m(k,l,n)$ be a maximal system of k-tuples in a set E, $|E| = n$, such that each l-tuple in E is contained in at most one k-tuple. The cardinality of such a system is denoted by $\bar{\bar{m}}(k,l,n)$. Also let $S(k,l,n)$ denote a tactical system of k-tuples in a set E, $|E| = n$, such that every l-tuple is contained in *exactly* one k-tuple. The cardinality of such a system is denoted by $\bar{\bar{S}}(k,l,n)$.

J. Schönheim [1] showed that

$$(1)\quad \bar{\bar{m}}(k,l,n) \leqslant [\tfrac{n}{k}[\tfrac{n-1}{k-1}\ [\ldots[\tfrac{n-l+1}{k-l+1}]\ldots]]]$$

where brackets denote the greatest integer function.* We have shown that when n is not too large relative to k and l, we may improve this bound usually. Such improved bounds and certain exact results are given by the following theorems. The arguments used to prove the theorems involve taking the dual designs of given designs.

Theorem 1. *Suppose $km = pn + q$ where $0 \leqslant q < n$ by the division algorithm. Also suppose*

$$(n-q)\ \binom{p}{2} + q\binom{p+1}{2} > l\binom{m}{2}.$$

Then $\bar{\bar{m}}(k,\ l+1,\ n) < m$.

From this theorem we can get the following general bound which is often not quite as sharp as that given by Theorem 1, but can be used more readily than Theorem 1.

---

*We recently learned that (1), and Theorems 1 and 2 were also obtained by Selmer M. Johnson in a paper on error-correcting codes [2].

Theorem 2. *If* $k^2 > ln$, *then* $\overset{=}{m}(k, l+1, n) \leqslant [\frac{n(k-l)}{k^2-ln}]$.

We can get an exact result in certain cases.

Theorem 3. *Suppose* $\lambda\frac{n-1}{k-1}$ *and* $\lambda\frac{n(n-1)}{k(k-1)}$ *are integers (where* λ, n, k *are all integers), and that* $n > k$. *A balanced incomplete block design with parameters* $v = n$, $k = k$, $\lambda = \lambda$, *(and hence* $r = \lambda\frac{n-1}{k-1}$, $b = \lambda\frac{n(n-1)}{k(k-1)}$) *exists if and only if* $\overset{=}{m}(\lambda\frac{n-1}{k-1}, \lambda+1, \lambda\frac{n(n-1)}{k(k-1)}) = n$.

Proof. Suppose first $\overset{=}{m}(\lambda\frac{n-1}{k-1}, \lambda+1, \lambda\frac{n(n-1)}{k(k-1)}) = n$. Let $E = \{1,2,\ldots\lambda\frac{n(n-1)}{k(k-1)}\}$. Then there exist $A_1, A_2, \ldots, A_n$ which are $\lambda\frac{n-1}{k-1}$-tuples in E such that each $\lambda+1$-tuple of E is contained in at most one A_i. Form the dual design $B_1, B_2, \ldots, B_b$. Thus $B_i = \{j \mid i \in A_j\} \subset \{1,2,\ldots,n\}$ for each i, $1 \leqslant i \leqslant b$. Suppose some pair $\{x,y\}$ occurs in $\lambda+1$ of the B_i, say in $B_{i_1}, B_{i_2}, \ldots B_{i_{\lambda+1}}$. By definition of the dual design, this means $\{i_1, i_2, \ldots, i_{\lambda+1}\} \subset A_x$ and $\{i_1, i_2, \ldots, i_{\lambda+1}\} \subset A_y$, a contradiction. Hence any pair $\{x,y\} \subset \{1,2,\ldots,n\}$ occurs at most λ times in the B_i $(i = 1,2,\ldots,b)$.

Counting pairs we obtain

$$(2) \quad \binom{|B_1|}{2} + \binom{|B_2|}{2} + \ldots + \binom{|B_b|}{2} \leqslant \lambda\binom{n}{2}.$$

Next note that easily

$$\sum_{i=1}^{b} |B_i| = \sum_{i=1}^{n} |A_i| = \lambda\frac{n(n-1)}{k-1}, \text{ so that } \sum_{i=1}^{b} |B_i| \text{ is a fixed value.}$$

Under this condition it is known that the left-hand side of (2) is an *absolute* minimum only when the $|B_i|$ are all equal, or as nearly equal as possible. In this case the $|B_i|$ are as nearly equal as possible when

$$|B_i| = k = \frac{\sum_{i=1}^{b} |B_i|}{b} \text{ for all } i,\ 1 \leqslant i \leqslant b.$$

Under these circumstances the left hand side of (2) equals $b\binom{k}{2} = \lambda\frac{n(n-1)}{k(k-1)}\frac{k(k-1)}{2} = \lambda\binom{n}{2}$, which is the right hand side of (2). Hence $|B_i| = k$ for *each* $i = 1, 2, \ldots, b$, or else (2) is not satisfied. Also, if any pair $\{x,y\} \subset \{1, 2, \ldots n\}$ occurred fewer than λ times in the B_i, again (2) would not be satisfied since the right-hand side of (2) would be decreased. Hence we have k-tuples $B_1, B_2, \ldots, B_{\lambda\frac{n(n-1)}{k(k-1)}}$ in a set

$\{1, 2, \ldots n\}$ such that each pair of $\{1, 2, \ldots n\}$ occurs exactly λ times in the B_i. Thus the desired block design exists.

On the other hand, suppose the balanced incomplete block design exists. Hence, if $E = \{1, 2, \ldots n\}$, let $B_1, B_2, \ldots B_b$, k-tuples of E, be the blocks of the design. Take the dual of this design, giving sets $A_1, A_2, \ldots A_n$ contained in $\{1, 2, \ldots b\}$, where $|A_i| = r = \lambda\frac{(n-1)}{k-1}$. By arguments like those already given, if $|A_i \cap A_j| \geq \lambda + 1$ for some $i \neq j$, we would have the pair $\{i,j\}$ occurring in at least $\lambda+1$ of the B_k, a contradiction. Hence $|A_i \cap A_j| \leq \lambda$ for all $i \neq j$, so that we have proved

$$\bar{\bar{m}}(\lambda\frac{n-1}{k-1}, \lambda+1, \lambda\frac{n(n-1)}{k(k-1)}) \geq n.$$

But also Theorem 2 applies since the hypotheses are satisfied, namely $(\lambda\frac{n-1}{k-1})^2 > \lambda\lambda\frac{n(n-1)}{k(k-1)}$ or $\frac{n-1}{k-1} > \frac{n}{k}$. This theorem gives, upon simplification,

$$\bar{\bar{m}}(\lambda\frac{n-1}{k-1}, \lambda+1, \lambda\frac{n(n-1)}{k(k-1)}) \leq [n] = n.$$

Theorem 3 gives us the exact value of $\bar{\bar{m}}(k,l,n)$ in many instances, since the existence of many balanced incomplete block designs is known. For example, Steiner systems $S(k,2,n)$ are known to exist for $k = 3$ (Ref. [3,4]), $k = 4$ or 5 (Ref. [5]) under the conditions $n \equiv 1 \bmod k(k-1)$ or $n \equiv k \bmod k(k-1)$. We have easily

Corollary. If $S(k,2,n)$ exists $(n > k)$, then

$$\bar{\bar{m}}(\lambda\frac{n-1}{k-1}, \lambda+1, \lambda\frac{n(n-1)}{k(k-1)}) = n.$$

We take an example to compare this to the bound given in (1). If we set $\lambda = 1$, (1) would give us

$$\bar{\bar{m}}(\frac{n-1}{k-1}, 2, \frac{n(n-1)}{k(k-1)}) \leq \left[\frac{\frac{n(n-1)}{k(k-1)}}{\frac{n-1}{k-1}}\left[\frac{\frac{n(n-1)}{k(k-1)}-1}{\frac{n-1}{k-1}-1}\right]\right] \sim \frac{n^2}{k^2},$$

so that if k is fixed the ratio of the exact bound n given by the corollary to the bound given by (1) goes to 0 as $n \to \infty$.

Let $N(k,l,m)$ denote the smallest cardinal with the following property: given a set E, $|E| = N(k,l,m)$, there exists a system of m k-tuples in E such that each l-tuple of E is contained in at most one k-tuple of the system. We have shown

Theorem 4. *If* $(m-1)(l-1) \leq k$, *then*

$$N(k,l,m) = (l-1)\binom{m}{2} + [k - (m-1)(l-1)]m.$$

This theorem allows exact evaluation of $\bar{\bar{m}}(k,l,n)$ when n is small. For example, for $l = 2$, (and $k \geq 2$), we have $0 = N(k,2,0) < N(k,2,1) < N(k,2,2) < \dots < N(k,2,k) = N(k,2,k+1) = \frac{k(k+1)}{2}$. Clearly then, given any $n < \frac{k(k+1)}{2}$, there exists m, $0 \leq m \leq k-1$, so that $N(k,2,m) \leq n < N(k,2,m+1)$, which clearly implies $\bar{\bar{m}}(k,2,n) = m$. Also it is easily shown that $\bar{\bar{m}}(k,2,\frac{k(k+1)}{2}) = k+1$, so that we have the exact value of $\bar{\bar{m}}(k,2,n)$ for $n \leq \frac{k(k+1)}{2}$.

References

1. Schönheim, J., On maximal systems of k-tuples, *Stud. Scient. Math. Hung.* I(1966), 363-368.

2. Johnson, S. M., A new upper bound for error-correcting codes, *IRE Transactions of the Professional Group on Information Theory*, IT-8, 3(1962).

3. Moore, E. H., Tactical Memoranda, *Amer. J. Math.* 18(1896) 264-303.

4. Reiss, M., Über eine Steinerische combinatorische Aufgabe, *J. für Reine u. Angewandte Mathematik* 56(1859) 326-344.

5. Hanani, H., The existence and construction of balanced incomplete block designs, *Annals of Math. Statistics* 6(1961) 362-386.

ON A FUNCTION DEFINED ON A PARTITION OF INTEGERS

Robert E. Odeh

University of Victoria, BC., Canada

Introduction

Let $\Omega = \{1,2,\ldots,km\}$. Consider a partition of Ω into k ordered subsets $S_1,\ldots,S_k$, each of size m. Denote by $a_{i1}, a_{i2},\ldots, a_{im}$ the m elements of S_i. For $i > 1$, define $\phi(a_{ij})$ to be the number of elements in $\bigcup_{r=1}^{i-1} S_r$ which are smaller than a_{ij}. Let $T_i=\sum_{j=1}^{m} \phi(a_{ij})$ and let $T=\sum_{i=2}^{k} T_i$.

Of the possible $(km)!/(m!)^k$ ordered partitions of Ω, let $W(t;k,m)$ denote the number of such partitions which yield a value of $T=t$, $0 \leq t \leq k(k-1)m^2/2$.

In this paper a recursive algorithm is derived for the evaluation of $W(t;k,m)$. The algorithm has proven in practice to be particularly useful for evaluating $W(t;k,m)$ on a computer.

Preliminary Notions

By an (x,y) subset we mean a set of x integers whose sum is y. Let $N_n(x,y)$ be the number of (x,y) subsets of $\{1,\ldots,n\}$ for $x = 1,\ldots,n$.

It can be shown that the following recurrence formula will hold (A proof is given in Odeh and Cockayne [1]).

(1) $N_{n+1}(x,y) = N_n(x,y) + N_n(x-1,\ y - (n+1))$, with the boundary conditions

$$N_n(x,y) > 0, \text{ if } x(x+1)/2 \leq y \leq x(2n - x + 1)/2$$

(2) $N_n(0,0) = 1$,

and

$$N_n(x,y) = 0, \text{ otherwise.}$$

Note also that

$$(3) \quad N_n(1,y) = 1 \text{ if } 1 \leq y \leq n,$$

and by symmetry

$$(4) \quad N_n(x,y) = N_n(n-x, \frac{n(n+1)}{2} - y),$$

$$(5) \quad N_n(x,y) = N_n(x, (n+1)x - y).$$

From the relations (1) - (5) $N_n(x,y)$ can be evaluated recursively.

The formula for $W(t;k,m)$

For $k = 2$; $W(t;2,m)$ is given by

$$(6) \quad W(t;2,m) = N_{2m}(m, t+m(m+1)/2),$$

where $0 \leq t \leq m^2$.

For $k > 2$ and $0 \leq t \leq k(k-1)m^2/2$ we have

$$(7) \quad W(t;k,m) = \sum_s W(s;k-1,m) \cdot N_{km}(m, t-s+m(m+1)/2),$$

where the sum is over the values of s in the range

$$\max[0, t-m^2(k-1)] \leq s \leq \min[t, (k-1)(k-2)m^2/2].$$

To prove (6) and (7) we first let

(8) $T = \sum_{i=2}^{k} T_i = T^* + T_k$, where $T^* = \sum_{i=2}^{k-1} T_i$. We may assume, without loss of generality, the elements in S_k are ordered so that $a_{k1} < a_{k2} < \ldots < a_{km}$. Then a_{kj} is larger than $a_{kj} - 1$ elements from Ω, of which $(j-1)$ are contained in S_k. It follows that

$$(9) \quad \phi(a_{kj}) = a_{kj} - 1 - (j-1) = a_{kj} - j.$$

Therefore

$$(10) \quad \begin{aligned} T_k &= \sum_{j=1}^{m} \phi(a_{kj}) = \sum_{j=1}^{m} (a_{kj} - j) \\ &= \sum_{j=1}^{m} a_{kj} - m(m+1)/2. \end{aligned}$$

For $k = 2$, $T = T_2$. From (10) the number of ways that $T = t$ equals the number of ways that $\sum_{j=1}^{m} a_{kj} = t + m(m+1)/2$. It follows from the definition of $N_n(x\ y)$ that

$$W(t;2,m) = N_{2m}(m,t + m(m+1)/2).$$

To prove (7) we may proceed as follows. Given a partition $S_1,\ldots,S_k$ we order the elements in $\bigcup_{r=1}^{k-1} S_r$ from 1 to $(k-1)m$. Then for $i < k$ replace the elements in S_i by the ordered values. Note that the value of T^* is unchanged by this mapping. Therefore, if S_k is held fixed there are $[(k-1)m]!/(m!)^{k-1}$ possible ordered partitions of $\Omega - S_k$. Thus the number of ways that $T^* = s$ is given by $W(s;k-1,m)$. We note also that this result is independent of the choice of S_k.

Now, since $T = T^* + T_k$, $T = t$ if $T^* = s$ and $T_k = t-s$. From the definition of $N_n(x,y)$ the number of ways in which $T_k = t-s$ is equal to

$$N_{km}(m,\ t-s + m(m+1)/2).$$

For fixed s the total number of ways in which $T = t$ is $W(s;k-1,m)\cdot N_{km}(m,t-s + m(m+1)/2)$. Summing over all possible values of s which make both $W(\cdot)$ and $N(\cdot)$ positive yields (7).

From symmetry it can also be shown that

$$W(t;k,m) = W(k(k-1)m^2/2 - t;k,m).$$

The above arguments can easily be extended to the case where the partition is into subsets of unequal size.

Acknowledgement

This work was supported by the Canadian National Research Council under grant No. NRC A5203.

References

1. Odeh, R.E. and Cockayne, E.J., "Balancing equal weights on the integer line," *Journal of Combinatorial Theory*, 6 (1969).

INTERSECTION GRAPHS OF FAMILIES OF CONVEX SETS WITH DISTINGUISHED POINTS

William F. Ogden
Case Western Reserve University, Cleveland, OH, U.S.A.

Fred S. Roberts
The RAND Corporation, Santa Monica, CA, U.S.A.

Let K_n be the family of convex sets in R^n. Let $c(n)$ be the collection of all (finite)* graphs (V,E) so that there is an assignment to each x in V of a set $C(x)$ in K_n so that for $x \neq y$, $(x,y) \in E$ iff $C(x) \cap C(y) \neq \emptyset$. These are the intersection graphs of families of convex sets in R^n.

It has been shown by Wegner [6], using an argument due to B. Grünbaum, that every graph is in $c(3)$. Wegner has shown that there are graphs not in $c(2)$, but no nice necessary and sufficient conditions are known for membership in $c(2)$. $c(1)$ is the class of interval graphs studied by Fulkerson and Gross [1], Gilmore and Hoffman [2], and Lekkerkerker and Boland [3]. Let us say a graph in $c(1)$ is a *unit interval graph* if the $C(x)$ can be taken as closed unit intervals. Unit interval graphs have been characterized by Wegner [6] and by Roberts [4].

Define a second collection of intersection graphs $c'(n)$ as follows. $c'(n)$ is the collection of all graphs (V,E) so that there is an assignment to each x in V of a set $C(x)$ in K_n and a distinguished point $f(x)$ in $C(x)$ so that for $x \neq y$, $(x,y) \in E$ iff $f(x) \in C(y)$. (This definition is motivated by the notion of threshold or just-noticeable-difference in psychophysics.) A graph in $c'(1)$ is an *indifference graph* if each $C(x)$ can be taken as a closed unit interval and $f(x)$ as its midpoint. Indifference graphs have been characterized by Roberts [4].

<u>Theorem 1</u>: $c'(1)$ = *the class of indifference graphs* = *the class of unit interval graphs.*

---

*All graphs in this paper are finite.

Proof: The first equality is proved in Roberts [5] and the second equality is straightforward.

Theorem 2: *Every graph is in* $c'(2)$.

Proof: Given a graph (V,E), list its vertex set V as $a_1, a_2, \ldots, a_n$. Choose any circle in R^2 and place the points $f(a_1), f(a_2), \ldots, f(a_n)$ in a uniform manner about the circle in the given order. Then, for each i, join in consecutive order by straight line segments the point $f(a_i)$ and those points $f(a_j)$ corresponding to those a_j adjacent to a_i in (V,E). The resulting figure is a convex neighborhood $C(a_i)$ of $f(a_i)$. For example, if a_1 is adjacent to a_3, a_5, a_6, then $C(a_1)$ is shown below:

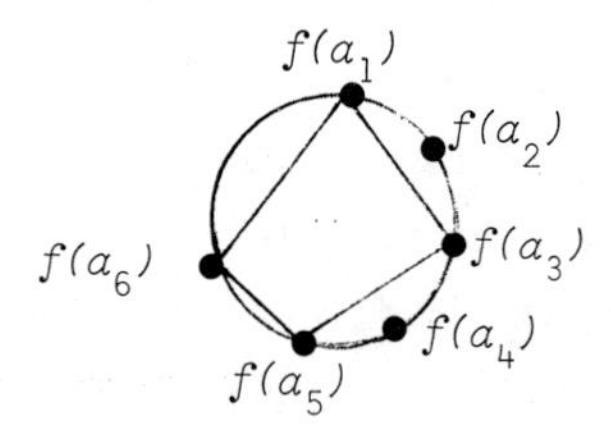

The resulting neighborhoods are easily seen to have the desired properties.

Acknowledgement

The authors are indebted to Victor Klee for pointing out the results of Wegner.

References

1. Fulkerson, D.R., and Gross, O.A., Incidence Matrices and Interval Graphs, *Pacific J. Math.*, 15 (1965), 835-855.

2. Gilmore, P.C., and Hoffman, A.J., A Characterization of Comparability Graphs and of Interval Graphs, *Canad. J. Math.*, 16 (1964), 539-548.

3. Lekkerkerker, C.G., and Boland, J.Ch., Representation of a Finite Graph by a Set of Intervals on the Real Line, *Fund. Math.*, 51 (1962), 45-64.

4. Roberts, F.S., Indifference Graphs, in *Proof Techniques in Graph Theory (Proc. 2nd Ann Arbor Graph Theory Conf.)*, F. Harary (ed.), New York, Academic Press, 1969.

5. Roberts, F.S., On the Compatibility between a Graph and a Simple Order, The RAND Corporation RM-5778 (Submitted to the *J. of Comb. Theory.)*.

6. Wegner, G., Eigenschaften der Nerven Homologisch-eigenfacher Familien in R^n, Doctoral dissertation, Göttingen, 1967.

SYSTEMS OF DISTINCT REPRESENTATIVES

Phillip A. Ostrand

University of California, Santa Barbara, CA, U.S.A.

We take as a starting point the transfinite symmetric form of the theorem of Philip Hall on systems of distinct representatives. For our purposes this is best expressed in terms of binary relations. Let X,Y be sets and $T \subset X \times Y$. Let $A \subset X$ and $B \subset Y$. Adopting functional notation, let $T(A) = \{y \in Y | (x,y) \in T$ for some $x \in A\}$ and define $T^{-1}(B)$ similarly. We say that T spans A if $|T(I)| \geq |I|$ for every finite $I \subset A$ (where $|\ |$ denotes cardinality) and T spans B if $|T^{-1}(J)| \geq |J|$ for all finite $J \subset B$. We say that T is scattered if T is a one-to-one correspondence between subsets of X and Y. Note that if T is scattered then T spans A and B if and only if $A \subset T^{-1}(Y)$ and $B \subset T(X)$. With this terminology we have:

Theorem. *Let $T \subset X \times Y$, $A \subset X$ and $B \subset Y$ such that $T(x)$ and $T^{-1}(y)$ are finite for each $x \in A$ and $y \in B$. There is a scattered set $S \subset T$ which spans A and B if and only if T spans A and B.*

For X and Y finite this theorem expressed in terms of incidence matrices was proved by Mendelsohn and Dulmage [1]. One proves the transfinite case by showing, via Zorns Lemma, that there is a minimal $S \subset T$ which spans A and B, and that the minimality of S implies that S is scattered. Note that for $A = X$ and $B = \Phi$ we have the normal transfinite form of Philip Hall's theorem, in which the sets being represented are the $T(x)$, $x \in X$.

As an application of this theorem we derive the well-known Banach Mapping Theorem.

Theorem. *If $\phi : X \to Y$ and $\theta : Y \to X$ are injective mappings then there are partitions $X = X^1 \cup X^2$ and $Y = Y^1 \cup Y^2$ such that $\phi(X^1) = Y^2$ and $\theta(Y^1) = X^2$.*

Proof. Let $T = \{(x,y) \mid y = \phi(x) \text{ or } x = \theta(y)\}$. Since ϕ and θ are injective we conclude that $|T(x)| \leq 2$ and $|T^{-1}(y)| \leq 2$ for each $x \in X$ and $y \in Y$, and that $|T(I)| \geq |\phi(I)| = |I|$, $|T^{-1}(J)| \geq |\theta(J)| = |J|$ for each finite $I \subset X$ and $J \subset Y$. It then follows that there is a scattered $S \subset T$ which spans X and Y. Partition $X = X^1 \cup X^2$ and $Y = Y^1 \cup Y^2$ where $X^1 = \{x \in X \mid (x,\phi(x)) \in S\}$ and $Y^2 = \phi(X^1)$. Then for $(x,y) \in S$, $x \in X_1$ if and only if $y = \phi(x)$ if and only if $y \in Y^2$, so $x \in X^2$ if and only if $y \in Y^1$ in which case $x = \theta(y)$. Thus we have $\theta(Y^1) = X^2$. The extended form of Banach's Theorem due to Perfect and Pym [2] may be similarly derived.

Banach's theorem generalizes in the following manner.

Theorem. *Let* $\phi_i : X_i \to X_{i+1}$ *be an injective mapping for* $1 \leq i \leq 2n$ *(where* X_{2n+1} *means* X_1*). There are partitions* $X_i = X_i^1 \cup X_i^2$ *such that* $\phi_i(X_i^1) = X_{i+1}^2$ *for* $1 \leq i \leq 2n$.

This suggests the possibility of generalizing the theorem on SDR's to relations on more than two arguments. Let $T \subset \prod_{i=1}^{n} X_i$. For $1 \leq i \leq n$ let T_i be the projection of T onto X_i. In particular for $x \in \prod_{i=1}^{n} X_i$, x_i is the ith component of x. T is scattered if for each $1 \leq i \leq n$ and each $a \in X_i$ there is at most one $x \in T$ such that $x_i = a$. For $A \subset X_i$ we say T spans A if for every finite $I \subset A$ there is a scattered $S \subset T$ such that $S_i = I$. For $n = 1$, Philip Hall's theorem shows that this apparently stronger definition of spanning is in fact equivalent to the earlier one. As before a scattered S spans $A \subset X_i$ if and only if $A \subset S_i$. We may now state:

Conjecture. Let $T \subset \prod_{i=1}^{n} X_i$ and $A_i \subset X_i$, such that for $1 \leq i \leq n$ and for each $a \in A_i$, $\{x \in T \mid x_i = a\}$ is finite. Then there is a scattered $S \subset T$ which spans each A_i if and only if T spans each A_i.

This conjecture is attractive because it reduces to the previous theorem in case $n = 1$ and the generalized Banach theorem follows from it by an argument completely analogous to the preceding one.

Unfortunately the conjecture is false. A counterexample is $T = \{(1,2,1),(2,3,2),(2,1,1),(3,2,2)\}$ where $X_1 = X_2 = X_3 = \{1,2,3\}$ and $A_1 = A_2 = A_3 = \{1,2\}$. T spans A_1, A_2 and A_3 but any subset of T which spans them must necessarily contain $(1,2,1)$ and $(2,1,1)$ and thus cannot be scattered.

Thus it remains an open question whether there are any reasonable conditions one can impose on $T \subset \prod_{i=1}^{n} X_i$ that will insure the existence of a scattered subset $S \subset T$ such that $A_i \subset S_i$ for all i.

References

1. Mendelsohn, N.S. and Dulmage, A.L., Some generalizations of the problem of distinct representatives, *Can. J. Math.* 10 (1958), 230-241.

2. Perfect, H. and Pym, J.S., An extension of Banach's mapping theorem with applications to problems concerning common representatives. *Proc. Cambridge Phil. Soc.* 62 (1966), 187-192.

ASPECTS OF TRANSVERSAL THEORY

Hazel Perfect

The University of Sheffield, England

1. The theorems of P. Hall and M. Hall. A recent proof due to R. Rado; a defect form, the Hoffman-Kuhn criterion for a transversal with marginal elements, the Ford-Fulkerson criterion for a common transversal (*CT*) of two families deduced from Hall's theorem by the method of 'elementary constructions'.

2. Common representatives and extensions of Banach's mapping theorem. Translation of P. Hall's theorem to give conditions for a system of common representatives (*SCR*) of two finite families; an extension of Banach's mapping theorem, and its use to generalize theorems on *SCR*s to apply to infinite families.

3. Applications of Menger's graph theorem in transversal theory. Theorems on common transversals (*CT*s) and *CT*s with marginal elements of two families deduced from Menger's theorem; a method of L. Mirsky for obtaining more general results on *SCR*s with repetitions.

4. Independence structures and submodular functions. The submodularity of the rank function of an independence structure and its use to prove Rado's theorem on independent transversals; induced structures; a theorem of Edmonds and Rota on independence structures induced by submodular functions.

5. Transversal independence. Set-theoretic examples of independence structures, in particular the collection of partial transversals of a family of sets; their use, in conjunction with Rado's theorem, to obtain further results in transversal theory.

*Synopsis of Instructional Series of Lectures.

LATTICOID PRODUCT, SUM AND PRODUCT OF GRAPHS

Claude François Picard

Centre National de la Recherche Scientifique, Paris, France

Arcs will mean directed arcs; graphs will mean directed graphs, without multiple arcs but possibly containing loops (x,x).

1. The *Kronecker product* of two graphs $G = (X,\Gamma)$ and $H = (Y,\Delta)$ is the graph $GH = (XY,\Gamma\Delta)$, where $(x_i,x_k) \in \Gamma$, $(y_j,y_l) \in \Delta \iff (x_iy_j,x_ky_l) \in \Gamma\Delta$.

Prop. 1. Iff H is a single vertex with a loop, then GH is isomorphic to G.

If the condition is satisfied, we put $H = I_y$; I will denote more generally the set of all possible loops. If $I_Y = (Y,I)$ has all (y,y) and no other arc, then GI_Y has $|Y|$ subgraphs which are all isomorphic to G. It is similar for I_XH; we call these graphs *extensions* of G,H. The *sum* of graphs (or cartesian product) is the graph $G + H = GI_Y \cup I_XH$, denoted by $(XY,\Gamma+\Delta)$ where $(x_iy_j,x_ky_l) \in \Gamma + \Delta$ iff $[x_i = x_k$ and $(y_i,y_l) \in \Delta]$ or $[(x_i,x_k) \in \Gamma$ and $y_j = y_l]$. $G + H$ is isomorphic to $H + G$.

GH and $G + H$ are well-known and called by many different names in the extensive literature (Albrecht, Berge, Čulik, McAndrews, Welsch, A.T. White, ...).

Prop. 2. The arcs of GH are included in the transitive closure of $G + H$; GH and $G + H$ have no common arc iff G and H have no loop.

Prop. 3. $G + H$ has a circuit iff there is one in G or in H.

Prop. 4. Let p_G be the number of connected components of G; $p_{G+H} = p_Gp_H$. We choose this property due to Albrecht to show how to use the extensions of G and H and give the scheme of our new proof -

GI_Y has $p_G\ |Y|$ components, for all $x \in X$.

$GI_Y \cup I_xH$ has $p_G\ |Y| - (|Y| - p_H)$ components,

$$I_X H = \bigcup_x I_x H \Rightarrow G + H \text{ has } p_G\ |Y| - (|Y| - p_H) p_G \text{ components.}$$

2. We call *latticoid* a finite graph without circuit, for which there exists a (unique) origin of paths leading to all other vertices.

G, H and K will denote latticoids.

Prop. 5. A latticoid is a rooted-tree iff it does not contain any vertex which is terminal vertex of more than one arc.

A tree (finite, connected graph without cycle) is a rooted-tree iff the same condition is true. A finite graph without circuits has at least one vertex, called a root, which is terminal extremity of no arc. A latticoid can be constructed from such a graph G by adding a vertex and arcs from a to the roots of G.

Let a, b be the unique roots of G, H; let $E = \{\xi_1,\ldots,\xi_T\} \subset X$, $E' = \{\eta_1,\ldots,\eta_T\} \subset Y$ be the sets of terminal vertices. The graphs GI_b and $I_E H$ have the vertices $\xi_1 b,\ldots,\xi_T b$ in common and no other elements.

We call *latticoid product* of G and H the graph

$$G \llcorner H = G\, I_b \cup I_E H,$$

which is clearly a latticoid with ab as root and with EE' as set of terminal vertices. $H \llcorner G$ has also $|EE'|$ terminal vertices but is not isomorphic to $G \llcorner H$. There exists a bijection between the paths from ab to $\xi_i\, \eta_j$ in $G \llcorner H$ and the paths from ba to $\eta_j\, \xi_i$ in $H \llcorner G$.

Prop. 6. If G and H are two latticoids, then $G + H$ is also a latticoid and all vertices and arcs of $G \llcorner H$ belong to $G + H$: $G \llcorner H$ is a subgraph of $G + H$.

Prop. 7. $(G + H) + K = G + (H + K)$ and $(G \llcorner H) \llcorner K = G \llcorner (H \llcorner K)$.

Prop. 8. If G and H are degenerated into two paths, then $G \llcorner H$ is isomorphic to the graph obtained by concatenation of G and H.

We call *Dedekind latticoid* a latticoid for which all paths joining two vertices have the same number of arcs.

Prop. 9. If G and H are two Dedekind latticoids, then $G + H$ and $G \llcorner H$ are also Dedekind latticoids; if G and H are two rooted-trees, then $G + H$ and $G \llcorner H$ are also rooted-trees.

ON A FAMILY OF SYMMETRY CODES OVER *GF*(3) AND RELATED NEW FIVE-DESIGNS

Vera Pless

Air Force Cambridge Research Laboratories, L.G. Hanscom Field, Bedford, MA, U.S.A.

We will define a new family of error-correcting codes over the field of 3 elements, and will exhibit several new 5-designs which are related to the first five codes in the family.

First we define an (n,k) error correcting code over $GF(3)$. We consider an n-dimensional vector space over $GF(3)$ with a concrete orthonormal basis. An (n,k) code is simply a k-dimensional subspace of such a space. What is of interest when one considers a code, rather than a subspace, is the weight of a vector, that is, the number of non-zero components it has with respect to the fixed basis. The minimum weight, d, of a code is the weight of the non-zero vector in it of lowest weight.

We will now define the codes in the family. Let q be a power of an odd prime such that $q \equiv -1(3)$. Each code $C(q)$ is given in terms of a basis. The vectors in this basis are the rows of the matrix (I, S_q) where I is the $(q+1) \times (q+1)$ identity matrix and S_q is the $(q+1) \times (q+1)$ matrix below.

$$S_q = \begin{array}{c} \\ \infty \\ 0 \\ \vdots \\ i \\ \vdots \\ (q-1) \end{array} \begin{array}{c} \begin{array}{cccccc} \infty & 0\ldots & & j & \ldots & (q-1) \end{array} \\ \left[\begin{array}{cccc} 0 & 1 & 1 & 1 \\ \chi(-1) & & \chi(j) & \chi(q-1) \\ & & & \\ \chi(-1) & & \chi(j-i) & \\ & & & \\ \chi(-1) & & & \end{array}\right] \end{array}$$

where χ(a square) = 1, χ(a non-square) = -1, and $\chi(0) = 0$. These basis vectors generate a $(2q+2,q+1)$ code in the family. The following facts about this family can be demonstrated.

1. The codes are self-orthogonal; hence all weights are divisible by 3.

2. For $q \equiv 1(4)$, $(-S_q,I)$ is also a basis of $C(q)$.
   For $q \equiv 3(4)$, (S_q,I) is also a basis of $C(q)$.

These properties are useful for computing minimum weights for these codes.

We recall the definition of a $\lambda; 5 - r - n$ design. We have a set of n elements. The design D is a collection of subsets of the n elements such that every subset in D contains r elements, and every 5-subset of the n elements is contained in the same number λ of the subsets of D. Consider the vectors of a fixed weight in a code. If the coordinate indices of the non-zero coordinates of such vectors form a 5-design, then the vectors are said to yield a 5-design. It is possible to say that vectors of a certain weight yield 5-designs under certain conditions by means of the Assmus-Mattson Theorem [1]. Two 5-designs are said to be equivalent if there is a permutation of the n integers so that the subsets of one design go onto the subsets of the other design.

We will discuss the first five codes in the family and indicate which 5-designs are present.

<u>Case 1</u>. $q = 5$. This is a (12,6) code, and it can be shown that $d = 6$. Hence [2] this is equivalent to the well known Golay code, and its minimum weight vectors yield the equally well known 1; 5 - 6 - 12 design.

<u>Case 2</u>. $q = 11$. This is a (24,12) code and it can be shown that $d = 9$. By the Assmus-Mattson Theorem [1], the vectors of weights 9, 12 and 15 yield 5-designs. We list the λ's.

6; 5 - 9 - 24

576; 5 - 12 - 24

8,580; 5 - 15 - 24

It can be shown that the 6; 5 - 9 - 24 design is not equivalent to the 6; 5 - 9 - 24 design associated to the quadratic residue (24,12) code over $GF(3)$, but we do not know whether these designs are equivalent to known 5-designs.

Case 3. $q = 17$. This is a (36,18) code. The minimum weight $d = 12$ was found in part by computer. Here we definitely get new 5-designs via the Assmus-Mattson Theorem; no 5-designs were known for their parameters before. We list them and their λ's.

45; 5 - 12 - 36
5,577; 5 - 15 - 36
209,685; 5 - 18 - 36
2,438,973; 5 - 21 - 36

Case 4. $q = 23$. This is a (48,24) code. The minimum weight $d = 15$ was found in part by computer. We list the 5-designs and their λ's. The λ's for these designs are the same as the λ's [in 1] for designs associated with the quadratic residue (48,24) code over $GF(3)$.

364; 5 - 15 - 48
50,456; 5 - 18 - 48
2,957,388; 5 - 21 - 48
71,307,600; 5 - 24 - 48
749,999,640; 5 - 27 - 48

It can be shown that the first design above is not equivalent to the known one.

Case 5. $q = 29$. This is a (60,30) code. The minimum weight $d = 18$ was found in part by computer. From this it follows as before that the vectors of weights 18, 21, 24, 27, 30 and 33 yield 5-designs. These are new 5-designs. There are no other known 5-designs with these parameters, but their λ's are as yet unknown.

Notice that for the five codes above $d = \frac{q+7}{2}$.

The group of a code is the group of monomial transformations which send the code onto itself. Let $G(q)$ denote the group of $C(q)$.

Theorem. *Let R be such that R modulo $\{I,-I\}$ is isomorphic to $PSL_2(q)$.*

If $q \equiv 1(4)$, $G(q)$ contains a subgroup isomorphic to RZ_4 where R and Z_4 commute elementwise and $R \cap Z_4 = \{I,-I\}$.

If $q \equiv 3(4)$, $G(q)$ contains a subgroup isomorphic to $Z_2 \times R$.

The proofs of these results will appear in the Journal of Combinatorial Theory.

References

1. Assmus, Jr., E.F. and Mattson, H.F., New Five Designs, *J. Comb. Theory*, 6 (1969), 122-151.

2. Pless, V., On the Uniqueness of the Golay Codes, *J. Comb. Theory*, 5 (1968), 215-228.

TENSOR PRODUCTS ON THE CATEGORY OF GRAPHS

A. Pultr*

McMaster University, Hamilton, Ont., Canada
and
Caroline University, Prague, Czechoslovakia

A tensor product on a concrete category (K,U) (K is a category and U a fixed forgetful functor on it) is a covariant functor $\otimes : K \times K \to K$ such that

- **T1.** There is a functor Hom: $K \times K \to K$ such that $\langle X \otimes Y, Z\rangle \approx \langle X, \mathrm{Hom}(Y,Z)\rangle$ and $U \cdot \mathrm{Hom}(X,Y) \approx \langle X,Y\rangle$ naturally in X,Y,Z.
- T2. There is an object X_0 in K such that $X \otimes X_0 \approx X$ naturally in X.
- T3. $X \otimes Y \approx Y \otimes X$ naturally in X,Y.
- T4. $(X \otimes Y) \otimes Z \approx X \otimes (Y \otimes Z)$ naturally in X,Y,Z.

A functor $\otimes : K \times K \to K$ satisfying T1, T2, T3 is called a WKT-product. A tensor product (a WKT-product) such that $U(X \otimes Y) = UX \times UY$ is said to be regular, otherwise it is said to be singular.

Let R designate the category of graphs (sets with binary relations) and graph homomorphisms (the relation preserving mappings), U the natural forgetful functor $R \to$ Set. For a graph X denote by $r(X)$ its binary relation $r(X) \subset UX \times UX$.

Denote by A the graph $(2, \{(0,1)\})$ (2 is the set consisting of 0 and 1) and by Q the graph $(2 \times 2, r)$ where $((i,j), (k,\ell)) \in r$ iff either $i = k$, $j = 0$, $\ell = 1$ or $i = 0$, $k = 1$, $j = \ell$. Let

$$\eta : Q \to R$$

be a homomorphism such that (1) $U\eta$ is onto (2) if $\eta(i,j) = \eta(k,\ell)$ then

---

*Support from the Canadian National Research Council and the McMaster University is gratefully acknowledged.

$\eta(j,i) = \eta(\ell,k)$ (3) if $(\eta(i,j),\eta(k,\ell)) \in r(R)$ then $(\eta(j,i),\eta(\ell,k)) \in r(R)$.

Construct $\otimes_\eta : R \times R \to R$ as follows:

$$U(X \otimes_\eta Y) = UX \times UY/_\sim$$

where $\sim$ is an equivalence generated by the relation e, where $(x_i,y_j)e(x_k,y_\ell)$ iff there exist homorphisms $\phi : A \to X$ and $\psi : A \to Y$ with $\phi(m) = x_m$, $\psi(m) = y_n$ and $\eta(i,j) = \eta(k,\ell)$,

$r(X \otimes Y) \ni (\overline{(x_i,y_j)}, \overline{(x_k,y_\ell)})$ (the bars denote classes of equivalence) iff either $x_i = x_k$ and $(y_j,y_\ell) \in r(Y)$ or $y_j = y_\ell$ and $(x_i,x_k) \in r(X)$ or there exist homomorphisms $\phi: A \to X$ and $\psi: A \to Y$ with $\phi(m) = x_m$, $\psi(n) = y_n$ and $(\eta(i,j), \eta(k,\ell)) \in r(R)$.

If $\phi_i : X_i \to Y_i$ $(i = 0,1)$ are homomorphisms, put $\phi_0 \otimes \phi_1 \overline{(x,y)} = \overline{(\phi(x),\phi(y))}$. The following theorems hold:

Theorem. *Every* $\otimes_\eta$ *is a* WKT-*product. Every* WKT-*product* $\otimes : R \times R \to R$ *is naturally equivalent to some* $\otimes_\eta$.

Theorem. *Every singular* WKT-*product is a tensor product.*

Theorem. $\otimes_\eta$ *is regular iff* η *is one-to-one.*

The regular tensor products $\otimes_\eta$ are obtained from homomorphisms η described as follows (in the figures the upper left (lower left, upper right, lower right respectively) points indicate the images of (0,0) ((0,1), (1,0), (1,1) respectively) under η, the arrows indicate the relations concerning *distinct* points):

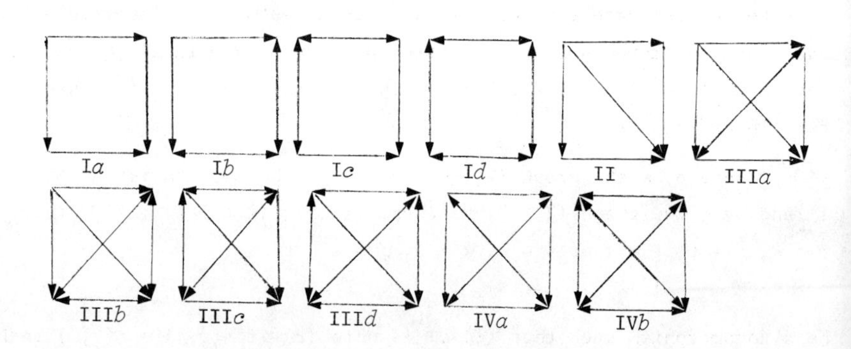

In the cases Ia and II we may add any system of loops satisfying (3) above, in the cases IIIa and IVa we cannot add any, in the cases Ib and IIIb (Ic and IIIc respectively) we may add any systems of loops satisfying (3) and not containing the left upper (right lower respectively) loop. In the remaining cases the systems containing the only loop at the left upper point or only the loop at the right lower point are prohibited.

Corollary. There are exactly 149 singular and 52 regular tensor products on $\mathcal{R}$.

The category $\mathcal{R}_s$ of symmetric graphs was also studied. We obtained 3 singular and 4 regular tensor products there.

ON THE EXISTENCE OF RESOLVABLE BALANCED INCOMPLETE BLOCK DESIGNS*

D. K. Ray-Chaudhuri and Richard M. Wilson

The Ohio State University, Columbus, OH, U.S.A.

1. Introduction.

Let X be a finite set, the elements of which will be called points or treatments. Let $\mathcal{L} = (B_i / i \in I)$ be a list (or family) of subsets of X. These subsets which occur in X are called lines or blocks. Any given subset may occur in $\mathcal{L}$ more than once. The pair $(X, \mathcal{L})$ is called a design. According to graph theorists (Claude Berge, Calgary Combinatorics Conference, June, 1969), a design is a hypergraph where the points are the vertices and the lines are the hyperedges.

Let v and λ be positive integers and K be a set of positive integers.

A design $(X, \mathcal{L})$ is said to be a (v, K, λ)--pairwise balanced design (PBD) iff

(i) $|X| = v$,

(ii) $|B_i| \in K$ for every $i \in I$, and

(iii) $x, y \in X$, $x \neq y$, the number of indices i for which $\{x,y\} \subset B_i$ is precisely λ. For any finite set X, $|X|$ denotes the number of elements.

The concept of PBD was introduced independently by Hanani [6] and Bose and Srikhande [4]. If K consists of a single positive integer, then a (v, K, λ) - PBD is called a (v, k, λ)- balanced incomplete block design (BIBD). A $(v, K, 1)$ - PBD is a rank 3 combinatorial geometry in the sense of Crapo and Rota [9].

The following is the simplest nontrivial BIBD:

*This research was supported in part by NSF Research Grant GP 9375.

$v = 7 \quad k = 3,\ \lambda = 1, \qquad X = \{1,2,3,4,5,6,7\}$

$B_1 = (1,2,3), \qquad B_4 = (2,4,6), \qquad B_7 = (3,4,7).$

$B_2 = (1,4,5), \qquad B_5 = (2,5,7),$

$B_3 = (1,6,7), \qquad B_6 = (3,5,6)$, and

This BIBD is of course the finite projective plane of order 2.

A class of blocks $\mathcal{B}_1$ is said to be a parallel class iff $\forall x \in X$, $\exists$ exactly one block B such that $x \in B \in \mathcal{B}_1$. A (v, k, λ) - BIBD is said to be resolvable iff the list of blocks $\mathcal{L}$ can be partitioned into parallel classes $\mathcal{B}_1, \mathcal{B}_2, \ldots, \mathcal{B}_r$ where r is a positive integer. The following is a resolvable BIBD with $v = 9$, $k = 3$, $\lambda = 1$, $r = 4$, $X = \{1,2,3,4,5,6,7,8,9\}$ and

$\mathcal{B}_1$	$\mathcal{B}_2$	$\mathcal{B}_3$	$\mathcal{B}_4$
(1,2,3)	(1,4,7)	(1,5,9)	(3,5,7)
(4,5,6)	(2,5,8)	(3,4,8)	(1,6,8)
(7,8,9)	(3,6,9)	(2,6,7)	(2,4,9).

Each parallel class consists of three blocks.

2. Statement of the Main Theorem and Some Historical Remarks

Necessary conditions for the existence of a (v, k, λ) - BIBD are

(1) $\qquad \lambda(v-1) \equiv 0 \pmod{(k-1)}$ and

(2) $\qquad \lambda v(v-1) \equiv 0 \pmod{k\,(k-1)}$.

BIBD's with parameters $k = 3$ and $\lambda = 1$ are called Steiner triple systems. The necessary conditions on v for the existence of a $(v, 3, 1)$ - BIBD reduce to $v \equiv 1$ or $3 \pmod 6$. The question of whether this is also sufficient was raised by Steiner [21] in 1853. The answer is yes and a proof was given by Reiss [61] in 1859. However, as Marshall Hall, Jr. points out [5, p. 237] the problem had been raised and solved by Kirkman [13] in 1847. Steiner was anticipated by Kirkman by at least 6 years. Indeed Steiner triple systems should be named "Kirkman-Steiner systems." In the twentieth century first significant contributions to the problem of existence of BIBD were made by Bose [2] and Hanani [6]. Bose introduced the "difference method" which enabled one to construct several infinite families of (v, k, λ) - BIBD's. Bose and Srikhande [4] and Hanani [6] introduced the "composition methods" which build up designs on larger number of treatments starting from designs on smaller number of treatments. Hanani [6] proved that the necessary conditions (1) and (2) are also sufficient for $k = 3$ and 4. For $k = 5$, he showed that the condition is sufficient except for $(v, k, \lambda) = (15, 5, 2)$ in which case the design does not exist. Necessary conditions on the existence of a (v, k, λ)- resolvable BIBD are (1), (2) and $v \equiv 0 \pmod{k}$ (3).

For $\lambda = 1$, these three conditions reduce to $v \equiv k \pmod{k(k-1)}$ (4). This condition is not sufficient in general. From the theorem of Bruck and Ryser [10], it is possible to find infinite number of integers k for which the resolvable $(k^2, k, 1)$ - BIBD does not exist. For instance for $k = 6$ and 14, the $(k^2, k, 1)$ - resolvable BIBD does not exist. The main theorem of this paper is that the necessary condition (4) is also "asymptotically sufficient." Let

$$B^*(k) = \{v/(v, k, 1) - \text{resolvable BIBD exists}\} \quad \text{and}$$

$$R_k^* = \{r/\ r(k-1) + 1 \in B^*(k)\}.$$

Theorem 4. (Main theorem.) Let k be a given positive integer. Then there exists a constant $c(k)$ such that if $v \geq c(k)$ and $v \equiv k \pmod{k(k-1)}$, then $v \in B^*(k)$. An equivalent statement is

Theorem 4' Let k be a given positive integer. Then there exists a constant $c(k)$ such that if $r \geq c(k)$ and $r \equiv 1 \pmod{k}$, then $r \in R_k^*$.

Proof of theorem 4 is unfortunately very long and will be published elsewhere. Resolvable BIBD's with $k = 3$ and $\lambda = 1$ are also known as Kirkman Designs or Kirkman's school girl arrangements. In 1847 in an article "On a problem in Combinations" published in Cambridge and Dublin Mathematical Journal, Rev. T. J. Kirkman [13] introduced the school girl problem. A teacher would like to take 15 school girls out for a walk, the girls being arranged in 5 rows of three. The teacher would like to insure equal chances of friendship between any two girls. Hence it is desirable to have different row arrangements for the 7 days of the week such that any pair of girls walk in the same row exactly one day of the week. Clearly such an arrangement for 15 girls is equivalent to a (15, 3, 1) - resolvable BIBD, the 15 girls correspond to the 15 points, every row corresponds to a block of size 3 and the 7 days correspond to the 7 parallel classes. In the general case one wants to arrange v girls in $\frac{v}{3}$ rows of three. The problem is to find different row arrangements for $\frac{v-1}{2}$ consecutive days such that every pair of girls walk in the same row exactly one day out of the $\frac{v-1}{2}$ days. Of course such an arrangement is equivalent to a $(v, 3, 1)$ - resolvable BIBD and hence a necessary condition is $v \equiv 3 \pmod 6$. Kirkman's school girl problem generated great interest in the late 19th century and early 20th century. Between 1847 and 1947, more than 50 articles had been written on the problem. Celebrated Mathematicians like Burnside, Cayley, Moore, Sylvester and others contributed to the problem. It was proved that for several infinite families of integers n, Kirkman arrangements for $6n+3$ girls exist. However, for an arbitrary integer n, no solution was known. Marshall P. Hall, Jr. [5, 242] observes, "Solutions of the school girl problem are known for a number of specific values of $v = 6t+3$, but no solution for the general case is known to the writer" in his book, "Combinatorial Theory," Blaisdell Publishing Company, 1967. In 1968, in a paper, "Solution of Kirkman's School girl Problem" presented at the Combinatories Symposium of the American Mathematical Society at Los Angeles, California, the present authors proved that for any positive integer n, Kirkman arrangements for $(6n+3)$ - girls exist. Stated differently, $(v, 3, 1)$ resolvable BIBD exists if $v \equiv 3 \pmod 6$. The paper, "Solution of Kirkman's School girl Problem" will be published in the Proceedings of the Symposia in "Pure" Mathematics, Combinatorics, Vol. 19, of the American Mathematical Society. Oscar Eckenstein gives a bibliography of the school girl problem in Messenger of Mathematics, 41 (1912). For the sake of historical interest, this list of paper is included at the end of the paper. The main theorem of this paper is a generalization of our result on the Kirkman's school girl problem.

3. Concepts and Theorems Used in the Proof

Let $(X, \mathcal{B})$ be a PBD with $\lambda = 1$. Such a PBD is said to be a (k,t) - completed resolvable design if and only if (i) $|X| = kt+1$ and (ii) $\mathcal{B}$ consists of blocks of size $(k+1)$ and a single block of size t unless $t = k+1$. The block of size t is called the block at infinity. The relation of (k,t)- completed resolvable designs to $((k-1)t+1, k, 1)$ - resolvable BIBD's is the same as that of finite projective planes to finite affine planes.

An (m, n, d, λ) orthogonal array is a quadruple (S_1, S_2, S_3, A) where S_1, S_2, and S_3 are finite sets with cardinalities m, λn^d and n respectively and A is a mapping from $S_1 \times S_2$ to S_3 such that for distinct elements $a_1, a_2, \ldots a_d$ of S_1 and arbitrary elements $b_1, b_2, \ldots, b_d$ of S_3 the following system of equations for x have exactly λ solutions:

$$A(a_1, x) = b_1,$$
$$\vdots \quad \vdots$$
$$A(a_d, x) = b_d \quad \text{and}$$
$$X \in S_2.$$

Usually S_1 is regarded as the set of rows of the array, S_2 as the set of columns and S_3 as the set of symbols and in the i *th* row and j *th* column one writes down the symbol $A(i,j)$. The concept of orthogonal array was introduced by Rao [8]. It is easily proved that a set of $(m-2)$ mutually orthogonal latin squares of order n exist if and only if an $(m, n, 2, 1)$ - orthogonal array exists. Let $N(n)$ denote the maximum number of mutually orthogonal latin squares of order n. Generalizing a result of Parker [7], Bose and Srikhande [3] proved that if $k \leq N(m)+1$, then for $1 \leq x < m$, $N(k\,m+x) \geq Min\,(N(m), N(x), N(k)-1, N(k+1)-1)$.

Using this result, Chowla, Erdös and Straus [11], proved that $\lim_{n\to\infty} N(n) = \infty$. Theorem of Chowla, Erdös and Straus play an important role in the proof of our main theorem.

Let (S_1, S_2, S_3, A) be an (m, n, d, λ) - orthogonal array. A subset S_2' of columns is said to be a parallel class of columns if for all $a \in S_1$ and $b \in S_3$, the equation $A(a,x) = b$, $x \in S_2'$ has exactly one solution for x. The orthogonal array is said to be resolvable if and only if the set of columns S_2 can be partitioned into parallel classes of columns. It is easily proved that an $(m, n, 2, 1)$ orthogonal array exists if and only if an $(m-1, n, 2, 1)$ - resolvable orthogonal array exists. The following is a resolvable orthogonal array with $m = n = 3$, $d = 2$, $\lambda = 1$ and $S_3 = \{a, b, c\}$:

$S_2{}^1$	$S_2{}^2$	$S_2{}^3$
$a\ b\ c$	$a\ b\ c$	$a\ b\ c$
$a\ b\ c$	$b\ c\ a$	$c\ a\ b$
$a\ b\ c$	$c\ a\ b$	$b\ c\ a$

Resolvable orthogonal arrays are very important for the "composition" of resolvable BIBD's.

Let $B(K) = \{v \mid (v, K, 1)$ - P B D exists$\}$. The mapping $K \to B(K)$ is a closure operation on the subsets of positive integers, i.e.

(i) $B(K) \supseteq K$

(ii) $B(B(K)) = B(K)$ and

(iii) $K_1 \supseteq K_2 \Rightarrow B(K_1) \supseteq B(K_2)$.

A subset K is said to be closed if and only if $B(K) = K$. Let $R_k = \{r | r(k-1)+1 \in B(k)\}$. Hanani [6] proved that R_k is a closed set i.e. $B(R_k) = R_k$. Hanani presented the result in a different form. The existence problem of designs is essentially that of characterizing the closed sets. Let $\beta(K)$ denote the unique positive integer which generates the ideal generated by the set $\{k(k-1) | k \in K\}$.

Since the ring of integers is a principal ideal domain, $\beta(K)$ exists. If K is finite, then $\beta(K) = g.c.d. \{k(k-1) | k \in K\}$. A fiber f of a closed set K is a residue class f mod $\beta(K)$ such that $\exists k \in K$ and $k \equiv f \pmod{\beta(K)}$. A fiber f is said to be complete if and only if $\exists$ a constant M such that $\{v | v \geq M, v \equiv f \pmod{\beta(K)}\} \subseteq K$. If all fibers are complete, the closed set K is said to be eventually periodic with period $\beta(K)$. Richard M. Wilson in his Ph.D. dissertation proved the following theorem:

Every closed set K is eventually periodic with period $\beta(K)$. The proof of this theorem is very long. This theorem and its various consequences will be published in a paper of Richard M. Wilson in The Journal of Combinatorial Theory.

Wilson's theorem is very important in the proof of our main theorem.

4. Statement of the Auxiliary Results

In this section, we describe the theorems that we need to prove to establish our main theorem.

Lemma 1. A (k,t) - completed resolvable design exists if and only if $t \in R_k^*$.

Theorem 1. *R_k^* is a closed set, i.e. $B(R_k^*) = R_k^*$.*

Theorem 2. *Let k be a given positive integer. There exists a constant $c(k)$ such that if $q \geq c(k)$, $q \equiv 1 \pmod{k(k-1)}$ and q is a prime power, then $kq \in B^*[k]$ or equivalently $k^2t + 1 \in R_k$ where $q = k(k-1)t + 1$.*

Theorem 3. $\beta(R_k^*) = \begin{cases} k & \text{if } k \text{ is even,} \\ 2k & \text{if } k \text{ is odd.} \end{cases}$

References

1. W. W. Rouse Ball, (Revised by H. S. M. Coxeter), Mathematical Recreations and Essays, New York, The Macmillan Company, (1947), pp. 267-298.

2. R. C. Bose, On the construction of balanced incomplete block design, *Ann. Eugenics.*, 9 (1939), 353-399.

3. R. C. Bose and S. S. Srikhande, On the construction of sets of mutually orthogonal latin squares and the falsity of a conjecture of Euler, *Trans. Amer. Math. Soc.*, 95 (1960), 191-209.

4. R. C. Bose and S. S. Srikhande, On the composition of balanced incomplete block designs, *Canad. J. Math.*, 12 (1960), 177-188.

5. Marshall Hall, Jr., Combinatorial Theory, Blaisdell Publishing Company, Waltham, Mass., (1967).

6. H. Hanani, The existence and construction of balanced incomplete block designs, *Ann. Math. Stat.*, 32 (1961), 361-386.

7. E. T. Parker, Orthogonal Latin Squares, *Proc. Nat. Accad. Sci., U.S.A.*, 45 (1959), 859-862.

8. C. R. Rao, Factorial experiments derivable from combinatorial arrangements of arrays, *Suppl. J. Roy. Statist. Soc.*, 9 (1947), 128-139.

9. H. H. Crapo and G. C. Rota, On the foundation of combinatorial theory: Combinatorial Geometry, Proceedings of the Symposia in "Pure" Mathematics, Combinatories, Vol. 19, American Math. Soc.

10. R. H. Bruck and H. J. Ryser, The non existence of certain finite projective planes, *Canad. Jour. Math.*, 1 (1949), 88-93.

11. S. Chowla, P. Erdös and E. G. Straus, On the maximal number of pair wise orthogonal latin squares of a given order. *Canad. Jour. Math.*, 12 (1960), 204-208.

12. Oscar Eckenstein, Bibliography of Kirkman's School girl problem, *Messenger of Mathematics*, 41-42 (1912) 33-36.

13. T. P. Kirkman, On a problem in combinations, *Cambridge and Dublin Mathematical Journal*, 2 (1847), 191-204.

14. T. P. Kirkman, Query, *Lady's and Gentleman's Diary* (1850), p. 48.

15. A. Cayley, On the triadic arrangements of seven and fifteen things, *London, Edinburgh, and Dublin Philosophical Magazine and Journal of Science*, 37 (1850), 50-53.

16. T. P. Kirkman, Note on an unanswered prize question, *Cambridge and Dublin Mathematical Journal*, 5 (1850), 255-262.

17. T. P. Kirkman, On the triads made with fifteen things, *London, Edinburgh, and Dublin Philosophical Magazine and Journal of Science*, 37 (1850), 169-171.

18. Solutions to Query VI, *Lady's and Gentleman's Diary* (1851), p. 48.

19. R. R. Anstice, On a problem in combinations, *Cambridge and Dublin Mathematical Journal*, 7 (1852), 279-292.

20. R. R. Anstice, On a problem in combinations (continued from vol. vii., p. 292), *Cambridge and Dublin Mathematical Journal*, 8 (1853), 149-154.

21. J. Steiner, Combinatorische Aufgabe, *Crelle's Journal für die reine und angewandte Mathematik*, 14 (1853), 181-182.

22. T. P. Kirkman, Theorems on combinations, *Cambridge and Dublin Mathematical Journal*, 8 (1853), 38-45.

23. T. P. Kirkman, On the perfect r partitions of $r^2 - r + 1$, *Transactions of the Historic Society of Lancashire and Cheshire*, 9 (1856-1857), 127-142.

24. B. Pierce, Cyclic solutions of the school-girl puzzle, *Astronomical Journal (U.S.A.)*, 6 (1860), 169-174.

25. J. J. Sylvester, Note on the historical origin of the unsymmetrical six-valued function of six letters, *London, Edinburgh, and Dublin Philosophical Magazine and Journal of Science*, 21 (1861), ser. 4, 369-377.

26. W. S. B. Woolhouse, On the Rev. T. P. Kirkman's problem respecting certain triadic arrangements of fifteen symbols, *London, Edinburgh, and Dublin Philosophical Magazine and Journal of Science*, 21 (1861), 510-515.

27. J. J. Sylvester, On a problem in tactic which serves to disclose the existence of a four-valued function of three sets of three letters each, *London, Edinburgh, and Dublin Philosophical Magazine and Journal of Science*, 21 (1861), ser. 4, 515-520.

28. W. S. B. Woolhouse, On triadic combinations of 15 symbols, *Lady's and Gentleman's Diary* (1862), pp. 84-88.

29. Paper No. 16 is reprinted in the *Assurance Magazine*, 10 (1862), pt. v., No. 49, pp. 275-281.

30. T. P. Kirkman, On the puzzle of the fifteen young ladies, *London, Edinburgh, and Dublin Philosophical Magazine and Journal of Science*, 23 (1862), ser. 4, 198-204.

31. A. Cayley, On a tactical theorem relating to the triads of fifteen things, *London, Edinburgh, and Dublin Philosophical Magazine and Journal of Science*, 25 (1863), ser. 4, 59-61.

32. W. S. B. Woolhouse, On triadic combinations, *Lady's and Gentleman's Diary* (1863), pp. 79-90.

33. J. Power, On the problem of the fifteen schoolgirls, *Quarterly Journal of Pure and Applied Mathematics*, 8 (1867), 236-251.

34. S. Bills, Solution of problem proposed by W. Lea, *Educational Times Reprints*, 8 (1867), 32-33.

35. W. Lea, Solution of problem proposed by himself, *Educational Times Reprints*, 9 (1868), 35-36.

36. T. P. Kirkman, Solution of three problems proposed by W. Lea, *Educational Times Reprints*, 9 (1869), 97-99.

37. A. Frost, General solution and extension of the problem of the fifteen schoolgirls, *Quarterly Journal of Pure and Applied Mathematics*, 11 (1871), 26-37.

38. W. Lea, Solution of problem proposed by himself, *Educational Times Reprints*, 22 (1874), 74-76.

39. J. J. Sylvester, Proposed problem, *Educational Times*, November 1, 1875, p. 193.

40. Appendix note, *Proceedings of the London Mathematical Society*, 7 (1875), 235-237.

41. E. Carpmael, Some solutions of Kirkman's 15-school-girl problem, *Proceedings of the London Mathematical Society*, 12 (1881), 148-156.

42. A fifteen puzzle, *Knowledge*, 1 (1881), 80.

43. A. Bray, The fifteen schoolgirls, *Knowledge*, 2 (1882), 80-81.

44. E. Marsden, The school-girls' problem, *Knowledge*, 3 (1883), 183.

45. A. Bray, Twenty-one school-girl puzzle, *Knowledge*, 3 (1883), 268.

46. J. J. Sylvester, Note on a nine schoolgirls problem, *Messenger of Mathematics*, 22 (1893), 159-160.

47. Correction to the note on the nine schoolgirls problem, *Messenger of Mathematics*, 22 (1893), 192.

48. A. C. Dixon, Note on Kirkman's problem, *Messenger of Mathematics*, 23 (1893), 88-89.

49. W. Burnside, On an application of the theory of groups to Kirkman's problem, *Messenger of Mathematics*, 23 (1894), 137-143.

50. E. H. Moore, Tactical memoranda, *American Journal of Mathematics*, 18 (1896), 264-303.

51. E. W. Davis, A geometric picture of the fifteen school-girl problem, *Annals of Mathematics (U.S.A.)*, 2 (1896-1897), 156-157.

52. A. F. H. Mertelsmann, Das Problem der 15 Pensionatsdamen, *Zeitschrift für Mathematik und Physik*, 43 (1898), 329-334.

53. W. Ahrens, Review of Schubert's Mathematische Mussestunden, *Zeitschrift für mathematischen und naturwissenschaftlichen Unterricht*, 31 (1900), 386-388.

54. H. E. Dudeney, Solution, *Educational Times Reprints*, 14 (1908), 97-99.

55. H. E. Dudeney, Solution (continued), *Educational Times Reprints*, 15 (1909), 17-19.

56. O. Eckenstein, Note, *Educational Times Reprints*, 16 (1909), 76-77.

57. H. E. Dudeney, Solution (continued), *Educational Times Reprints*, 17 (1910), 35-38.

58. O. Eckenstein, Solutions, *Educational Times Reprints*, 17 (1910), 38-39.

59. O. Eckenstein, Note, *Educational Times Reprints*, 17 (1910), 49-53.

60. W. W. Rouse Ball, Proposed problem, *Educational Times* (February 1, 1911), p. 82.

61. M. Reiss, Über eine Steinersche Combinatorische Aufgabe welche im 45sten Bande dieses Journals, Seite 181, gestellt worden ist, *J. reine u. angew. Math.*, 56 (1859) 326-344.

62. D. K. Ray-Chaudhuri and Richard M. Wilson, Solution of Kirkman's school-girl problem, Proceedings of the Symposia in "Pure" Mathematics, Combinatorics, Vol. 19, American Math. Soc.

HOW TO GROW TREES

(Some Problems of Combinatorial Arboriculture)

Ronald C. Read

University of West Indies, Kingston, Jamaica

Enumeration problems relating to trees have been studied for a long time, and many formulae giving the numbers of various kinds are known. In this paper we consider the problem of constructing, rather than just counting, all the trees of a given kind.

Labelled trees. Little need be said about labelled trees. The construction of all spanning trees of a given labelled graph is of importance in electrical circuit theory, and algorithms for doing this are well-known (see [2], [3]). Applied to the complete graph on n nodes these algorithms will give all labelled trees on n nodes.

Rooted trees. If we delete the root of a rooted tree we obtain a number, say k, of rooted trees. Let k_i of these have i nodes ($i = 1,2,3,...$). Reversing this process we construct all rooted trees as follows. Choose an integer k, and an integer k_i such that $\Sigma\, k_i = k$ and $\Sigma\, ik_i = n-1$, where n is the number of nodes in the trees being constructed. Then, in all possible ways, choose k_i rooted trees on i nodes ($i = 1,2,3,...$) allowing repetitions, and join their roots to a new node. Do this for all choices of k and the k_i.

This method follows exactly the method by which Cayley first enumerated rooted trees [1]. It presupposes that the rooted trees on fewer than n nodes have already been constructed.

Unrooted trees. One fairly practical method of constructing all unrooted trees on n nodes is the following. We suppose that we have a list of unrooted trees on n-1 nodes. We take each of these and add an extra node, joining it in turn to each of the existing nodes. Thus from each tree in the input list we obtain n-1 trees. All trees on n

nodes will be constructed in this way, but will occur many times. We therefore need a method for checking whether a tree just produced is one that we have had before.

This is best done by means of a *coding algorithm*, that is, an algorithm which associates with each tree a code (a linear string of symbols) in such a way that two trees are isomorphic if, and only if, they have the same code. Efficient algorithms are known for coding trees (see [4]). As each tree is produced, its code is computed, and compared with the codes of the distinct trees so far found. If the code is not among them it is added to the list; otherwise it is rejected.

If this is performed on a computer, a list of the codes of the trees found so far must be stored, and a look-up routine provided to hunt for a given code in the list. Here we meet a difficulty. This list very quickly reaches a stage where it contains the codes of most of the trees we are constructing. Thereafter, almost all the codes produced are rejected, and the bulk of the computer time is spent in picking up the few remaining stragglers. However sophisticated the look-up routine may be, the process of finding whether a code occurs in a long list is bound to be comparatively time-consuming. This is an important point, for it means that in searching for a better method of growing trees we can afford to program even a highly complex routine, taking non-negligible amounts of machine time, provided we can guarantee that each application of it will produce a *new* tree.

To what extent is this possible? Let us first ask whether it is feasible to produce all trees on n nodes without knowing anything about trees on fewer nodes. All trees have either a centre or a bicentre. Those that have a centre can be thought of as rooted trees, the centre being the root. Attached to the root is a certain number, k, of rooted trees, as above. What is special here is that there is an integer h such that none of these k trees has height $> h$, while at least two have height exactly h. Let us temporarily forget this condition, thus returning briefly to the subject of rooted trees.

We can arrange the nodes of a typical tree in "levels", according to their distances from the root, as in Figure 1.

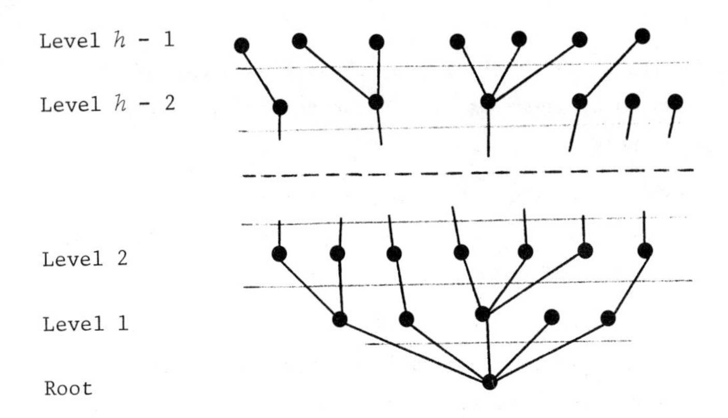

Figure 1.

At level i we have a set L_i of nodes, which will be of various kinds according to what lies above them, i.e., according to the subtrees of which they are roots. The edges between levels i and i-1 define a mapping of L_i into L_{i-1}. It is possible (though by no means easy) to write a computer routine which will construct, in turn, all possible such mappings, taking into account the different kinds of nodes in L_i, and distinguishing the nodes of L_{i-1} (which were indistinguishable until the mapping took place) according to what is mapped onto them.

Suppose that we have chosen a value for h and the numbers of nodes to appear on each level. We set up a mapping routine, as above, for each level, and take all possible combinations of mappings, using a "back-tracking" procedure. We start with the first mapping at each level, and run through all possible mappings $L_2 \to L_1$. When these are exhausted we backtrack, take the next mapping $L_3 \to L_2$, and again run through the mappings $L_2 \to L_1$ (which will in general be different from before, because of the changes of the kinds of nodes in L_2). We carry on until we reach the last mapping on all levels. Each combination of mappings gives a distinct tree. We repeat this for all possible h, and all distributions of nodes among the levels.

Such a program would generate all *rooted* trees. Only a slight modification is required to produce only those trees for which the root is the centre. We keep track of which nodes are roots of subtrees extending upwards to the top level, and at each level we avoid mapping *all* such nodes on to the same node in the level below. Thus at each level, and hence at level 1, there are at least two subtrees of height h.

For bicentral trees the procedure is much the same except that there will be two nodes at the root level, and the subtrees rooted at them must both be of height h.

These are complex procedures, but note that for $n = 20$ (which is about as far as anyone is likely to want to go) either there are few levels, or else the number of nodes per level is small.

A mroe practical procedure, but one which requires lists of smaller trees, is the following. Assume that we have lists of rooted trees on up to n-2 nodes, sorted according to height, and let us first consider the construction of bicentral trees on n nodes. We choose a height h, and in all possible ways choose two rooted trees, each of height h, having n nodes between them. By joining their roots we obtain a bicentral tree on n nodes. Repeat this for all choices of rooted trees, and all values of h.

For central trees the situation is a little more complicated. We choose h, and an integer k (the valency of the root) and proceed as in the construction of rooted trees already described. We make sure, however, that at least two of the k rooted trees chosen have height h, and that none of them has height more than h.

Although this method requires lists of rooted trees on up to n-2 nodes (which, if complete, would contain far more trees than those we are counting) it fortunately does not require the *complete* lists. For as the height h increases the number of nodes m that can be required in the rooted trees decreases. It is easy to show that $m \leqslant n - h - 1$. Thus, for example, we do not need the full list of rooted trees on n-2 nodes, but only those of height 1 (there is only one such). Thus we need construct only those rooted trees which we know are going to be used in 1the construction of unrooted trees, and in this way the amount of input data for the construction program remains manageable.

References

1. Cayley, A., On the theory of the analytical forms called trees, *Phil. Mag.* Vol. XIII (1857), 172-176.

2. Green, D.G. and Hakimi, S.L., Generation and Realization of trees and k-trees. *IEEE Trans.* 1964 CT-11, 247-255.

3. Mayeda, W. and Seshu, S., Generation of trees without duplication. *IEEE Trans.* 1965 CT-12, 181-185.

4. Parris, R., and Read, R.C., Graph isomorphism and the Coding Problem. *Technical Report,* USAFOSR Project 1026-66, 1966.

COMPONENTS OF INTERSECTED UNIONS OF CONVEX SETS

John R. Reay

Western Washington State College, Bellingham, WA, U.S.A.

If each set F_i in a family $\mathcal{F} = \{F_i\}_{i=1}^{n}$ in E^d is the union of k convex sets, let $C(d,n,k)$ denote the largest possible number of components of $\cap\mathcal{F}$. Let $D(d,n,k)$ [respectively, $S(d,n,k)$] denote the same maximum where in addition each $F_i \in \mathcal{F}$ is restricted to be the union of k pair-wise disjoint convex sets [respectively, the complement in E^d of $k - 1$ parallel hyperplanes]. The following lemma gathers certain obvious relations.

Lemma. $S(d,n,k) \leq D(d,n,k) \leq C(d,n,k) \leq k^n$ for all d, n and k. $S(1,n,k) = C(1,n,k) = n(k-1)+1$. If $n \leq d$ then $S(d,n,k) = C(d,n,k) = k^n$.

Theorem. *$S(d,n,k)$ is determined by the generating function*

$$f(s,t) = \Sigma_{d,n}\, S(d,n,k)s^d t^n = 1/[(s-1)(st(k-1) + t - 1)].$$

Proof. It is clear that $S(d,n,k) = k^n$ if $n \leq d$, and $S(1,n,k) = n(k-1) + 1$ on the real line. Also $S(d,n,k) = S(d,n-1,k) + (k-1)\, S(d-1,n-1,k)$, since starting with a family $\mathcal{F}$ of $n - 1$ sets, we may assume, by induction, that $\cap\mathcal{F}$ has $S(d,n-1,k)$ components. The removal of a new independent hyperplane H from the configuration can produce at most $S(d-1,n-1,k)$ new components, since $(\cap\mathcal{F}) \cap H$ can have at most $S(d-1,n-1,k)$ components. Similarly for the $(k-2)$ hyperplanes parallel to H. As in the special case $k = 2$, where this reduces to a problem considered by Steiner [2], Polya [1, page 43] and others, this recursive function and the boundary conditions determine the function $S(d,n,k)$.

Let $f(s,t)$ be defined by $f(s,t) = \Sigma S(d,n,k)s^d t^n$ where the summation runs over all integers d and n, and $S(d,n,k) = 0$ if $d < 0$ or if $n < 0$.

(We interpret $S(0,n,k) = S(d,0,k) = 1$, so the recursion formula is valid if $d \geq 1$, $n \geq 1$.) Then when $|s| < 1$ and $|t| < 1$ the recursion formula yields

$$t\, f(s,t)+st(k-1)\, f(s,t) =$$

$$\Sigma_{d\geq 0,n\geq 1}\, S(d,n-1,k)s^d t^n + \Sigma_{d\geq 1,n\geq 1}\, (k-1)\, S(d-1,n-1,k)s^d t^n =$$

$$\Sigma_{d\geq 1,n\geq 1}\, S(d,n,k)s^d t^n + \Sigma_{n\geq 1}\, S(0,n-1,k)t^n =$$

$$f(s,t) - \Sigma_{d\geq 0}\, S(d,0,k)s^d = f(s,t) - \frac{1}{1-s}\,.$$

Hence $f(s,t) = 1/[(s-1)(st(k-1)+t-1)]$.

Corollary. In the special case $k = 2$, function S has the following closed form:

$$S(d,n,2) = \Sigma_{i=0}^{d} \binom{n}{i}$$

Proof. This follows directly (induction on n) from the recursion formula in the Theorem.

Conjecture. $S(d,n,k) = D(d,n,k) = C(d,n,k)$ for all d, n and k.

Radon's theorem may be used to establish the conjecture in a number of cases, e.g., whenever $d = 1$. We illustrate its use by showing the conjecture true in the easiest case not covered by the Lemma. By the Theorem $S(2,3,2) = 7 \leq D(2,3,2) \leq 8 = k^n$. Assume $F = \{F_i\}_{i=1}^{3}$ is a configuration of sets in E^2 which produces $D(2,3,2)$. Choose a point $x = x(c_1,c_2,c_3)$ in each component, where x lies in the c_i-th convex set of F_i. Radon's theorem asserts that certain x can not exist, so $C(d,n,k)$ is at most n^k minus the number of such points. In the special case at hand, the existence of $x(111)$, $x(122)$, $x(212)$, $x(221)$ and Radon's theorem imply that some F_i would be the union of two non-disjoint convex sets or else $x(222)$ does not exist. Hence $D(2,3,2) = 7 = S(2,3,2)$.

References

1. Polya, G., Induction and Analogy in Mathematics, *Princeton Press*, 1954.

2. Steiner, J., Einige Gesetze über die Theilung der Ebene und des Raumes, *J. Reine Agnew. Math.*, 1 (1826) 349-364.

TWO APPLICATIONS OF TURAN'S THEOREM TO ASYMMETRIC DIGRAPHS

K.B. Reid

Louisiana State University, Baton Rouge, LA, U.S.A.

In 1941 Turan [3] proved the following theorem for ordinary graphs:

Turan's Theorem. Let $n \geq k \geq 3$ and $n = (k-1)t+r$, where k, n, r, t are integers and $1 \leq r \leq k-1$. Let

$$d_k(n) = \frac{k-2}{2(k-1)}(n^2-r^2) + 1/2\ r(r-1).$$

Further, let $\Delta(n,k)$ denote an ordinary graph with n vertices and the following structure: the vertices are partitioned into r classes of $t+1$ elements and $k-r-1$ classes of t elements each, the classes being mutually disjoint, and two vertices are joined by an edge if and only if they do not belong to the same class; $\Delta(n,k)$ contains $d_k(n)$ edges.

Then every ordinary graph with n vertices and more than $d_k(n)$ edges contains at least one complete graph on k vertices as a subgraph, and so does every graph with n vertices and exactly $d_k(n)$ edges which is not isomorphic to $\Delta(n,k)$.

We now give two applications of Turan's Theorem to asymmetric digraphs (incomplete tournaments). By $D(p,q)$ we mean an asymmetric digraph with p vertices and q lines. For omitted definitions see [1].

Definition 1. Let k and p be integers with $p \geq k \geq 3$. Let $e(p,k)$ be the smallest integer q such that every $D(p,q)$ contains a path of length $k-1$.

Theorem 1. $e(p,k) = d_k(p) + 1$.

Proof. Given a $D(p,q)$ with $q = d_k(p) + 1$, $D(p,q)$ contains a tournament of order k by Turan's Theorem applied to $D(p,q)$ considered as an undirected graph. But it is well known that every tournament of order k contains a path of length k-1. Thus, $e(p,k) \leq d_k(p) + 1$. On the other hand, consider $\Delta(p,k)$ of Turan's Theorem. Label the k-1 classes $A_1, A_2, \ldots, A_{k-1}$. Orient an edge from A_i to $A_j (1 \leq i, j \leq k-1)$ if and only if $i < j$. Since a transitive tournament of order k-1 contains no path of length k-1, $\Delta(p,k)$ with this orientation contains no path of length k-1. Thus $e(p,k) \geq d_k(p) + 1$ and Theorem 1 follows.

Definition 2. Let m be a positive integer. $F(m)$ is the smallest integer n such that every tournament of order n contains a transitive tournament of order m (e.g., $F(2) = 2$, $F(3) = 4$, $F(4) = 8$).

Definition 3. Let m and p be integers with $m \geq 3$ and $p \geq F(m)$. $t(p,m)$ is the smallest integer q such that every $D(p,q)$ contains a transitive tournament of order m.

The determination of $t(p,m)$ is reduced to the determination of $F(m)$ as is shown in:

Theorem 2. $t(p,m) = d_{F(m)}(p) + 1$.

Proof. Given a $D(p,q)$ with $q = d_{F(m)}(p) + 1$, $D(p,q)$ contains a tournament of order $F(m)$ by Turan's Theorem applied to $D(p,q)$ considered as an undirected graph. But a tournament of order $F(m)$ contains a transitive tournament of order m. Thus $t(p,m) \leq d_{F(m)}(p) + 1$. On the other hand consider $\Delta(p, F(m))$ of Turan's Theorem. Label the $F(m) - 1$ classes $A_1, A_2, \ldots, A_{F(m)-1}$. By the definition of $F(m)$, there is a tournament, T, of order $F(m) - 1$ which contains no transitive tournament of order m. Suppose the vertex set of T is $\{v_1, v_2, \ldots, v_{F(m)-1}\}$. Orient an edge in $\Delta(p,F(m))$ from A_i to A_j if and only if v_i is directed to v_j in T. $\Delta(p, F(m))$ with this orientation contains no transitive tournament of order m. Thus, $t(p,m) \geq d_{F(m)}(p) + 1$ and Theorem 2 follows.

If $f(n)$ is the largest integer m such that every tournament of order n contains a transitive tournament of order m, then $f(F(m)) = m$. Thus, a knowledge of $f(n)$ for every positive integer n will yield $F(m)$ for ever positive integer m. Consequently, determination of $t(p,m)$ is reduced to the yet unsolved problem of determining the largest transitive sub-tournament contained in a tournament. The latter problem is treated in [2].

References

1. Moon, J.W., Topics on Tournaments. New York, Holt, Rinehart and Winston, 1968.

2. Reid, K.B. and Parker, E.T., Disproof of a conjecture of Erdös and Moser on tournaments. *J. Comb. Theory*, to appear.

3. Turan, P., Eine extremal aufgabe aus der Graphentheorie. *Mat. Fiz. Lapok* 48 (1941), 436-452.

ON THE ENUMERATION OF TREES

Alfred Rényi

Hungarian Academy of Sciences, Budapest, Hungary

1. Introduction.

The aim of this paper is to give another proof for the famous formula of Cayley, according to which, denoting by T_n the total number of trees having n labelled vertices, one has

$$T_n = n^{n-2}. \tag{1}$$

Cayley has stated formula (1) in 1889 (see [1]) without giving a full proof. As a matter of fact, Cayley has formulated in his paper the following more general formula:

$$\sum_T x_1^{v_1(T)-1} x_2^{v_2(T)-1} \ldots x_n^{v_n(T)-1} = (x_1 + x_2 + \ldots + x_n)^{n-2} \tag{2}$$

where the summation is to be extended over all trees T having n vertices labelled by the numbers $1,2,\ldots,n$ and $v_k(T)$ denotes the degree (valency) of the point labelled by k in the tree T. Clearly for $x_1 = x_2 = \ldots = x_n = 1$ formula (2) reduces to (1). Cayley has verified (2) for $n = 6$ by exhibiting all possible trees; after this he remarked that "it will be at once seen that the proof given for this particular case is applicable for any value whatever of n". We shall show in what follows that (2) can in fact be proved by induction very easily: it is not impossible that what Cayley had in mind was essentially the proof given below, or some variant thereof.

Several proofs for Cayley's formula (1) have been given. In his excellent paper [2] J.W. Moon outlines nine proofs of (1), which he attributes to Prüfer, Clarke, Rényi, Dziobek, Katz, Göbel, Moon, Kirchoff and Pólya. Prüfer's proof can be formulated so that it gives the more general formula (2); the proof which Moon attributes to

Kirchhoff leads to a formula which is even more general than (2); we shall return to this question in §3 of this paper. The other seven proofs do not lead, without essential modification, to a proof of (2). It should be emphasized that (2) gives much more information on the family of trees having n labelled vertices than (1): for example it follows immediately from (2) that the number $T_n(v_1,\ldots,v_n)$ of trees with n labelled vertices in which the k-th point has degree $v_k (k = 1,2,\ldots,n)$ is given by

$$T_n(v_1,v_2,\ldots,v_n) = \frac{(n-2)!}{(v_1-1)!(v_2-1)!\ldots(v_n-1)!} \tag{3}$$

if $\sum_{k=1}^{n} v_k = 2n - 2$. Let us mention that Prüfer's proof leads immediately to (3), from which (2) follows.

In what follows we shall call the sum on the right hand side of (2) *Cayley's polynomial of order n*, and denote it by

$$C_n(x_1,x_2,\ldots,x_n) = \sum_T x_1^{v_1(T)-1} \cdot x_2^{v_2(T)-1} \ldots x_n^{v_n(T)-1} . \tag{4}$$

With this notation (2) can be written in the following form:

$$C_n(x_1,x_2,\ldots,x_n) = (x_1 + x_2 + \ldots + x_n)^{n-2}. \tag{5}$$

2. Proof of Cayley's formula.

Our aim is to prove formula (5) where the polynomial $C_n(x_1,x_2,\ldots,x_n)$ is defined by (4). Evidently Cayley's polynomial $C_n(x_1,x_2,\ldots,x_n)$ is a *symmetric* polynomial of its variables, and it is *homogeneous of degree* n-2; the second statement follows from the remark that the degree of each term of Cayley's polynomial is $\left(\sum_{k=1}^{n} v_k\right) - n$, and in every graph the sum of the degrees of the points equals twice the number of edges; as a tree with n points has n-1 edges it follows that

$$\left(\sum_{k=1}^{n} v_k\right) - n = 2(n-1) - n = n - 2.$$

Evidently (5) holds for $n = 2$, because there is only one tree with two vertices and in this both points have degree 1. Let us suppose that (5) holds for n-1 instead of n: we shall prove that this implies that it holds for n too.

If T is a tree with n labelled vertices, let us call the product

$\prod_{k=1}^{n} x_k^{v_k(T)-1}$ in which $v_k(T)$ denotes the degree of the point labelled by the number k, the *profile* of the tree T. Clearly in the profile of a tree T the variable x_k occurs with a positive exponent if and only if the point labelled by the number k is not an endpoint of T; thus we have

$$C_n(x_1,\dots,x_{n-1},0) = \sum{}' x_1^{v_1(T)-1} \cdot x_2^{v_2(T)-1} \dots x_{n-1}^{v_{n-1}(T)-1} \qquad (6)$$

where the dash over the sign Σ indicates that the summation has to be extended over those and only those trees T in which the point labelled by the number n is an endpoint.

If T is such a tree that the point labelled by the number n is an endpoint of T, then removing from T this point together with the unique edge connecting it with another point of T there remains a tree T' having $n-1$ points labelled by the numbers 1, 2, ..., $n-1$. The profile of such a tree T is equal to the profile of the tree T' multiplied by x_k provided that the point labelled by n was connected in T with the point labelled by k. It follows that

$$C_n(x_1,x_2,\dots,x_{n-1},0) = (x_1 + x_2 + \dots + x_{n-1})C_{n-1}(x_1,x_2,\dots,x_{n-1}) \qquad (7)$$

and as by the induction hypothesis

$$C_{n-1}(x_1,x_2,\dots,x_{n-1}) = (x_1 + x_2 + \dots + x_{n-1})^{n-3} \qquad (8)$$

we get

$$C_n(x_1,x_2,\dots,x_{n-1},0) = (x_1 + x_2 + \dots + x_{n-1})^{n-2}. \qquad (9)$$

Let us call the terms $x_1^{k_1} x_2^{k_2} \dots x_n^{k_n}$ and $x_1^{h_1} x_2^{h_2} \dots x_n^{h_n}$ *similar*, if the sequence $(h_1,h_2,\dots h_n)$ is a rearrangement of the sequence $(k_1,k_2,\dots,k_n)$. As $C_n(x_1,x_2,\dots,x_n)$ is a symmetric polynomial, similar terms of $C_n(x_1,x_2,\dots,x_n)$ have equal coefficients. From (9) we get the coefficients of all terms in which x_n is missing; however as $C_n(x_1,\dots,x_n)$ is homogeneous of degree $n-2$, in each term of this polynomial at least two variables are missing; thus to every term there can be found a similar term in which x_n is missing. Thus from (9) we can get the coefficients of all terms of the Cayley-polynomial $C_n(x_1,x_2, \dots,x_n)$. Expressed in another way, $C_n(x_1,x_2,\dots,x_n)$ is the unique polynomial which is symmetric and homogeneous of degree $n-2$ for which (9) holds. As however the polynomial $(x_1 + x_2 + \dots + x_n)^{n-2}$ reduces to

$(x_1 + x_2 + \ldots + x_{n-1})^{n-2}$ if $x_n = 0$, it follows that $(x_1 + x_2 + \ldots + x_n)^{n-2}$ is equal to the Cayley-polynomial of order n, i.e. that (5) holds. This completes the proof of (5).

3. Some further remarks.

In the last step of the proof of (5) given in §2 we have used the following fact: *if $P(x_1,x_2,\ldots,x_n)$ is a symmetric polynomial of its n variables, and it is homogeneous of degree $< n$ then $P(x_1,x_2,\ldots,x_n)$ is uniquely determined by $P(x_1,x_2,\ldots,x_{n-1},0)$.*

Of course $P(x_1,x_2,\ldots,x_n)$ can be expressed by a formula, the right hand side of which can be evaluated provided that $P(x_1,\ldots,x_{n-1},0)$ is given. To express this formula in a compact form, let us denote by S_k the operator which applied to any polynomial $P(x_1,x_2,\ldots,x_n)$ $(n > k)$ substitutes 0 for x_k, i.e.

$$S_k P(x_1,\ldots,x_n) = P(x_1,\ldots,x_{k-1},0,x_{k+1},\ldots,x_n). \tag{10}$$

Let us consider the operator

$$\Delta_n = \prod_{k=1}^{n} (I-S_k) \tag{11}$$

where I is the identity-operator.

It follows by the inclusion-exclusion principle, that if P is any polynomial of n variables, then $\Delta_n P$ is equal to the sum of those terms of P in which every one of the variables $x_1,x_2,\ldots,x_n$ occurs with a positive exponent. In particular if P is such a polynomial which does not contain any term in which all the n variables occur effectively, then we have

$$\Delta_n P = 0. \tag{12}$$

Thus in particular if P is homogeneous of degree $< n$, then (12) holds.

Let us consider now the following generalization of Cayley's polynomials. Let G be any graph having n labelled vertices and let us put

$$C_n(G;x_1,\ldots,x_n) = \sum_T x_1^{v_1(T)-1} \cdot x_2^{v_2(T)-1} \ldots x_n^{v_n(T)-1} \tag{13}$$

where the summation has to be extended over all *spanning trees* T of the graph G, i.e. over all trees having the same n vertices as G and which are subgraphs of G, i.e. consist only of edges belonging to the graph G. We shall call $C_n(G;x_1,\ldots,x_n)$ *the Cayley-polynomial of the graph* G. Evidently the Cayley-polynomial $C_n(x_1,\ldots,x_n)$ considered in the previous sections is the Cayley-polynomial of the complete graph of order n.

Of course the Cayley-polynomial of a graph G is in general not symmetric: however it is homogeneous of degree $n-2$, and thus the formula (12) can be applied. This leads to a proof of the following formula: if G is an arbitrary graph having n labelled points, then we have

$$x_1x_2\ldots x_n\, C_n(G;x_1,x_2,\ldots,x_n) = |A| \tag{14}$$

where $|A|$ denotes the determinant of the $(n-1)$ by $(n-1)$ matrix $A = (a_{jk})$ defined as follows:

$$a_{kk} = x_k \cdot (\Sigma' x_j) \qquad (1 \leqslant k \leqslant n-1) \tag{15}$$

where the summation in Σ' has to be extended over those values of j for which the j-th point is connected in G with the k-th point and if $j \neq k$, $1 \leqslant j \leqslant n-1$, $1 \leqslant k \leqslant n-1$

$$a_{jk} = \begin{cases} -x_jx_k & \text{if the } j\text{-th and } k\text{-th point are connected in } G \\ 0 & \text{otherwise} \end{cases} \tag{16}$$

Formula (14) is a particular case of a more general formula called in electrical engineering *Maxwell's rule*, which is closely related to *Kirchhoff's rule* (see [3] p. 131; for a simple proof see [4]). (14) is obtained from Maxwell's rule if the impedance e_{jk} of the edge connecting the j-th and k-th point is taken to be equal to x_jx_k.

My thanks are due to Professor P.R. Bryant and Professor H. Sachs for the above mentioned references concerning Maxwell's rule.

References

1. Cayley, A., A theorem on trees, *Quarterly Journal of Math.*, 23 (1889), 376-378.

2. Moon, J.W., Various Proofs of Cayley's Formula for Counting Trees; in: Seminar on Graph Theory, edited by F. Harary, 70-78, Holt, Rinehart and Winston, 1967.

3. Bryant, P.R., Graph Theory Applied to Electrical Networks; in: Graph Theory and Theoretical Physics, edited by F. Harary, Academic Press 1967, 111-137.

4. Hutschenreuther, H., Einfacher Beweis des "Matrix-Gerüstsatzes" der Netzwerktheorie, *Wissenschaftliche Zeitschrift der Technischen Hochschule Ilmenau*, 13 (1967), 403-404.

GENUS OF GRAPHS

Gerhard Ringel

Freie Universitat, Berlin, Germany

The genus $\gamma(G)$ of a graph G is defined as the minimum genus of an orientable 2-manifold in which G can be embedded without crossings of arcs. For instance the genus of the complete graph K_4 with 4 vertices is 0. Figure 1, in which opposite sides of the rectangle are to be identified, shows an embedding of the complete bipartite graph $K_{4,4}$ with 4 + 4 vertices into a torus; this shows that $\gamma(K_{4,4}) \leqslant 1$.

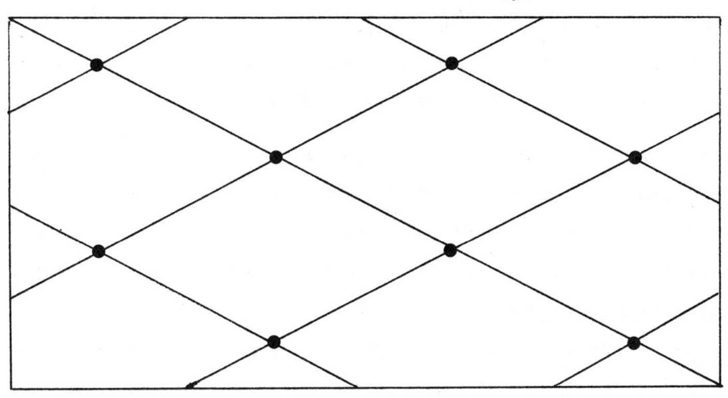

Figure 1

Let Q be the collection of all quadrilaterals (closed paths of length four) in the graph $K_{4,4}$. Q contains 36 quadrilaterals. In the embedding of $K_{4,4}$ into the torus there are only 8 quadrilaterals used as faces. These 8 quadrilaterals form a subset Q_0 of Q. The subset Q_0 has the following three properties.

E1) Each arc of the graph is incident with exactly two quadrilaterals of Q_0.

E2) The subset of all quadrilaterals of Q_0 which are incident with one fixed vertex form a cycle according to their adjacencies.

E3) One can orient the quadrilaterals of Q_0 so that the induced orientations on each arc are one in each direction.

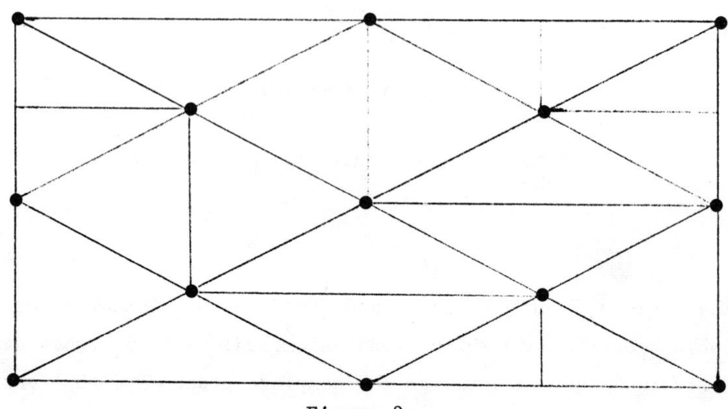

Figure 2.

Figure 2 shows the embedding of the graph $K_{2,2,2,2}$ into the torus. The graph $K_{2,2,2,2}$ contains 32 triangles (closed paths of length 3). In the embedding just 16 triangles are used as faces. The subset T_0 of these 16 triangles has the properties E1), E2), E3) provided one reads triangles instead of quadrilaterals. These ideas we shall use later.

There now follows a list of some results of recent years. Let Q_n be the 1-skeleton of an n-dimensional cube. The notation $K_n - K_2$ (respectively $K_n - K_3$) means that one arc (respectively 3 arcs forming a triangle) are omitted from the complete graph K_n with n vertices. The complete bipartite graph with $p + q$ vertices is denoted by $K_{p,q}$, and $K_{p,q,r}$ means the complete tripartite graph with $p + q + r$ vertices. The notation $\{a\}$ means the integer with $a \leqslant \{a\} < a + 1$.

$$\gamma(Q_n) = 1 + (n-4)2^{n-3} \quad \text{for } n \geqslant 2 \quad \text{(Ringel [1] 1955)}$$

$$\gamma(K_n) = \left\{\frac{(n-3)(n-4)}{12}\right\} \quad \text{for } n \geqslant 3 \quad \text{(Ringel, Youngs [3] 1968)}$$

$$\gamma(K_n - K_2) = \left\{\frac{(n-3)(n-4)-2}{12}\right\} \quad \text{for } n \not\equiv 2 \pmod{24}$$

$$\gamma(K_n - K_3) = \left\{\frac{(n-3)(n-4)-6}{12}\right\} \quad \text{for } n \not\equiv 2 \pmod{3}$$

The last two equations are conjectured for all $n \geqslant 3$.

$$\gamma(K_{p,q}) = \left\{\frac{(p-2)(q-2)}{4}\right\} \quad \text{for } p,q \geqslant 2 \quad \text{(Ringel [2] 1965).}$$

It is easy to obtain from the formula for $\gamma(K_{p,q})$ an inequality for $\gamma(K_{p,q,r})$, namely $\gamma(K_{p,q,r}) \geqslant \left\{\frac{(p-2)(q+r-2)}{4}\right\}$ for $p \geqslant q \geqslant r$ and $p,q+r \geqslant 2$. Ringel and Youngs [5] 1969 conjectured that equality holds and proved this in the following special case.

$$\gamma(K_{m,n,n}) = \left\{\frac{n-2}{2}\right\} \quad \text{for } n \geqslant 2.$$

Another special result was found by A.T. White [6] in 1969:

$$\gamma(K_{2n,n,n}) = (n-1)^2 \quad \text{for } n \geqslant 1.$$

The conjecture

$$\gamma(K_{n,n,n,n}) = (n-1)^2$$

has been proved only for $n = 1,2$ and 4.

The proofs of all these equations have two parts. In the first part one proves the inequality $\geqslant$ by using the Euler fomula and the fact that the shortest closed path in the graph is a triangle or (for bipartite graphs) a quadrilateral.

In the second part one shows the reverse inequality $\leqslant$ by establishing the existence of a certain embedding of the given graph. One can do this by drawing a picture in small cases. In higher cases combinatorial schemes are normally used. The construction of these schemes is often very difficult. We now present a class of graphs for which this second part is relatively easy.

In 1969 T. Alain considered the following graph Π_n having $n!$ vertices. The vertices are represented as the elements of the symmetric group acting on n numbers. Let Δ be the subset consisting of the n-1 elements (12), (23), (34),...,$(n-1,n)$. Two vertices A, B are adjacent by an arc in Π_n if and only if there exists a transposition $\sigma \in \Delta$ with $\sigma A = B$. This arc should be marked by σ. The graph Π_n is bipartite because each even permutation goes into an odd one by using $\sigma \in \Delta$ and vice versa. The graph Π_4 is shown in figure 3.

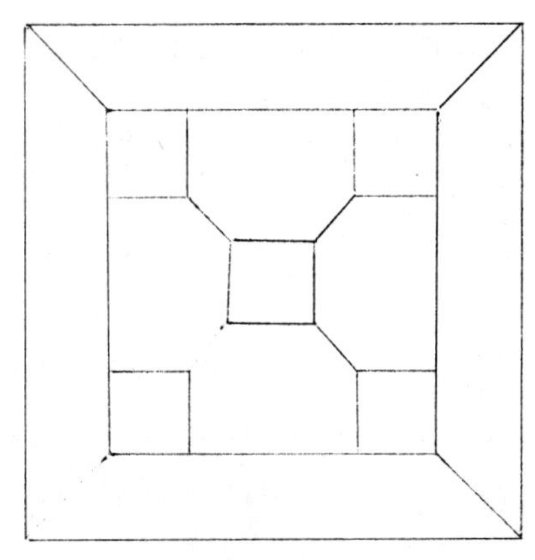

Figure 3.

T. Alain found for the genus of Π_n the formula

$$\gamma(\Pi_n) = 1 + \frac{n!(n-5)}{8} \text{ for } n \geqslant 6.$$

The main step in the proof is to find an embedding of Π_n into an orientable 2-manifold with quadrilateral-faces only. This can be done by constructing a subset Q_0 of the set of all quadrilaterals in Π_n having the three properties E1), E2), E3).

Let us explain this in the case $n = 7$. Put the elements of Δ in cyclic order

$$(12), (45), (23), (56), (34), (67)$$

where any two neighbors have no common number and put all quadrilaterals of the form

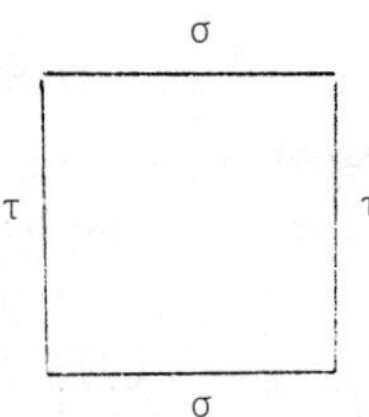

into Q_0 if and only if τ and σ are neighbors in the chosen cyclic order.

It is easy to see that Q_0 satisfies E1), E2), E3). That means that Q_0 gives us the desired embedding.

It might be an interesting suggestion to make the following modifie definition for the genus of multigraphs (possibly with loops and multipl edges). Let M be a multigraph. $\gamma(M)$ is defined as the minimum genus of an orientable 2-manifold such that M can be embedded into it without producing a 1-gon or a 2-gon as faces in the embedding.

The notation $K_n + K_3$ means the graph K_n with three double arcs forming a triangle. For this multigraph it is known that

$$\gamma(K_n+K_3) = \left\{\frac{(n-3)(n-4)+6}{12}\right\} \quad \text{for } n \not\equiv 2 \pmod 3.$$

The question $\gamma(K_m+K_n) = ?$ might be an interesting unsolved problem.

At the Conference on Graphtheory and Computing in Kingston (Jamaica) in January 1969 A.K. Dewdney made the following suggestion.

Denote the orientable 2-manifold of genus p by S_p. Identify two distinct points in S_p. The new "surface" you get is denoted by $S_{p,1}$, and is called a sphere with p handles and 1 identification. One can ask which graphs are embeddable into $S_{p,1}$. It is known that K_g is embeddable into $S_{2,1}$ and the embedding is a triangulation. This result was generalized by Ringel and Youngs [4] for K_n with $n \equiv 9 \pmod{24}$.

It seems that the coloring problem on the modified surface $S_{0,1}$ is not solved. The *conjecture* reads: If G is a graph embeddable into $S_{0,1}$, then the chromatic number is $\chi(G) \leqslant 5$. Certainly a slight modification of the proof of the well known 5 color-theorem should settle the matter.

The reader should know that in such a short report one cannot mention all details and difficulties encountered in the determination of the genus of graphs.

References

1. Ringel, G., Über drei kombinatorische Probleme am n-dimensionalen Würfel and Würfelgitter, *Abh. Math. Sem. Univ. Ham.*, 20 (1955), 10-19.

2. Ringel, G., Das Geschlecht des vollständigen paaren Graphen, *Abh. Math. Sem. Univ. Ham.* 28 (1965), 139-150.

3. Ringel, G., and Youngs, J.W.T., Solution of the Heawood Map-Coloring Problem, *Nat. Acad. Sci.* 60 (1968), 438-445.

4. Ringel, G., and Youngs, J.W.T., Die Lösung des Problems der Nachbargebiete, *Archiv der Math.* (in press)

5. Ringel,G., and Youngs, J.W.T., Das Geschlecht des symmetrischen vollständigen drei-färbbaren Graphen, *Commentarii Helvetici Mathematici* (to appear).

6. White, A.T., The Genus of Cartesian Products of Graphs, Ph.D. Thesis to be submitted to Michigan State Univ. (1969).

A CHARACTERIZATION OF CLIQUE GRAPHS*

Fred S. Roberts and Joel Spencer

The RAND Corporation, Santa Monica, CA, U.S.A.

If G is a graph, $V(G)$ will denote the set of vertices of G and $E(G)$ the set of edges. We denote the adjacency relation by I, i.e., if $x, y \in V(G)$, then xIy iff $(x,y) \in E(G)$. A *clique* of G is a maximal complete subgraph. The *clique graph* of G is the graph $H = K(G)$ whose vertices are the cliques of G and so that two cliques are adjacent if and only if they intersect. The main problem we are concerned with is this: given a graph H, is it the clique graph of some G?

Let K be a collection of complete subgraphs of a graph H. We shall say K has property I (for intersection) if whenever $L_1, L_2, \ldots, L_p$ are in K and $L_i \cap L_j \neq \emptyset$ for all i, j then the total intersection $\bigcap_{i=1}^{p} L_i \neq \emptyset$. Hamelink [1] showed that if the collection of all cliques of the graph H satisfies property I, then H is a clique graph. It is possible to show that the converse of this Theorem is false. Our basic result is the following.

Theorem. *A graph H is a clique graph iff there is a collection K of complete subgraphs of H which satisfies the following two properties:*

(1) *K covers all the edges of H, i.e., if $x, y \in H$ and xIy, then $\{x,y\}$ is contained in some element of K.*

(2) *K satisfies property I.*

---

* A more detailed version of this paper will appear in the Journal of Combinatorial Theory.

Proof. The proof of sufficiency is essentially the same as the proof of Hamelink's result. Let $K = \{L_1, L_2, \ldots, L_p\}$. Define the graph G as follows.

$$V(G) = V(H) \cup K.$$
If $h \in V(H)$, then $h I L_i$ iff $h \in L_i$.
$L_i I L_j$ iff $i \neq j$ and $L_i \cap L_j \neq \emptyset$
If $h, h' \in V(H)$, then not $h I h'$.

The claim is that $H = K(G)$. To prove this, let $C(h) = \{h\} \cup \{L_i : h \in L_i\}$. It is easy to see that each $C(h)$ is a clique of G. Moreover, these are only cliques of G. For, let C be a complete subgraph of G. Then if C contains an element h of $V(H)$, we have $C \subseteq C(h)$. And otherwise, C is contained in some $C(h)$ by property I.

To prove the necessity of the conditions, suppose $H = K(G)$. Let $V(G) = \{g_1, g_2, \ldots, g_n\}$, let $V(H) = \{h_1, h_2, \ldots, h_m\}$, and let $K_1, K_2, \ldots,$ denote the cliques of G, labelled in such a way that $h_i I h_j$ iff $K_i \cap K_j \neq 0$. For $i = 1,2,\ldots, n$, define $L_i = \{h_j : g_i \in K_j\}$. Each L_i is complete, becaus if h_j and h_k are in L_i, then $g_i \in K_j \cap K_k$ and so $h_j I h_k$. The claim is that $K = \{L_1, L_2, \ldots, L_n\}$ satisfies properties (1) and (2). Property (1) is satisfied because if $h_j I H_k$ then $K_j \cap K_k \neq \emptyset$. Finally, K satisfies propert I. For, suppose $L_{i_1}, L_{i_2}, \ldots, L_{i_r}$ pairwise intersect. Then for all j, k, there is a point h_{jk} in $L_{i_j} \cap L_{i_k}$. Thus g_{i_j} and g_{i_k} are both in K_{jk} and therefore we have $g_{i_j} I g_{i_k}$. It follows that $\{g_{i_1}, g_{i_2}, \ldots, g_{i_r}\}$ is contained in some clique K_s of G and thus $h_s \in \bigcap_{t=1}^{r} L_{i_t}$. Q.E.D.

References

1. Hamelink, Ronald C., A Partial Characterization of Clique Graphs, *J. of Combinatorial Theory* 5 (1968), 192-197.

ON THE CHROMATIC NUMBER OF STEINER TRIPLE SYSTEMS

Alexander Rosa*

McMaster University, Hamilton, Ont., Canada

It is still an open question, how to construct effectively two or more Steiner triple systems (STS) of a given order so that we are assured about their non-isomorphism. The investigation of the chromatic number of STS constitutes one of the possible approaches to this problem.

Let $S(n)$ be an STS of order n. The chromatic number of $S(n)$, $chr\ S(n)$, is the least integer k such that the elements of S can be colored by k colors in such a way that there is no triple in S with all three elements having the same color.

We have trivially $chr\ S(3) = 2$, $chr\ S(n) \geqslant 3$ whenever $n \geqslant 7$.

Theorem 1. *Let $n \equiv 1$ or $3 \pmod 6$, $n \geqslant 7$. Then there exists an STS of order n having chromatic number 3.*

When $n \equiv 3 \pmod 6$, the well-known construction (see, e.g., [1], Theorem 15.3.2) provides a required STS. When $n \equiv 1 \pmod 6$, $n = 6k+1$, the construction of an STS with chromatic number 3 may be described as follows:

Denote the elements of the STS by $a_1, a_2, \ldots, a_{2k}, b_1, b_2, \ldots, b_{2k}, c_1, c_2, \ldots, c_{2k}, d$. Take a decomposition $L = \{L_i,\ i=1,2,\ldots 2k-1\}$ of the complete graph $\langle 2k \rangle$ with the vertices $1,2,\ldots, 2k$ into $2k-1$ linear factors such that

$$L_i = \{[k+i-1, 2k],\ [i+j,\ i-j-1] \pmod{2k-1},\ j = 0,1,\ldots k-2\}$$

*This research was supported by the National Research Council of Canada.

and define a system $L' = \{L'_i,\ i = 1,2,\dots,\ 2k\}$ of $2k$ subgraphs of $\langle 2k \rangle$ by

$$L'_i = \begin{cases} L_i & \text{if } i \text{ is odd} \\ L_i - [i-1,i] & \text{if } i \text{ is even}, i < 2k \\ [j-1,j],\ j = 2,4,\dots,\ 2k-2 & \text{if } i = 2k. \end{cases}$$

For all $\alpha,\beta \in \{1,2,\dots,2k\}, \alpha \neq \beta$ define $\phi(\alpha,\beta)$:

$$\phi(\alpha,\beta) = i \text{ if and only if } [\alpha,\beta] \in L'_i\ .$$

Now include into the STS the following triples:

$$\left.\begin{array}{ll} (1) & (a_\alpha, a_\beta,\ b_{\phi(\alpha,\beta)}) \\ & (b_\alpha, b_\beta,\ c_{\phi(\alpha,\beta)}) \\ & (c_\alpha, c_\beta,\ a_{\phi(\alpha,\beta)}) \end{array}\right\} \qquad \text{for all } \alpha,\beta \in \{1,2,\dots,2k\},\ \alpha < \beta$$

(2) $(a_{2i}, b_{2i}\ c_{2i})$ for $i = 1,2,\dots,\ k$

(3) (a_i, b_{i+1}, d), (b_i, c_{i+1}, d), (c_i, a_{i+1}, d), $i = 1,3,5,\dots,2k-1$.

It is easy to verify that the system of triples (1), (2), (3) is an STS of order $n = 6k+1$, and evidently, its chromatic number is 3, since we can color all the a's (b's and c's) blue (red and green, respectively) and the element d by either color and there will be no triple of the system with all three elements having the same color.

Denote by $S^*(2^n-1)$ the well-known "totally associative" STS of order 2^n-1 (see e.g., [2]), and denote further by $S_1(49)$ the STS of order 49 obtained by the well-known "product rule" (see e.g., [1]) as a product of two STS of order 7.

Lemma. (a) $chr\ S^*(31) = 4$, (b) $chr\ S_1(49) = 4$.

Immediately from this lemma follows:

Theorem 2. *If S_1, S_2 are STS of order p, q, respectively, both containing a subsystem of order 7, then there exists a Steiner triple system S of order pq such that $chr\ S \geqslant 4$.*

It seems that $n = 31$ is the smallest order for which an STS with a chromatic number greater than 3 exists.

Let us remark, that Lemma, Theorem 2, the product rule and the well-known theorem about constructing from systems of order n_1, n_2, n_3

a new STS of order $n = n_3 + n_1(n_2 - n_3)$ (see e.g., [1],Theorem 15,4.2) are alone sufficient to prove by a recursive method that there exists a constant n_0 such that for all $n > n_0$, $n \equiv 1$ or 3 (mod 6), there is an STS of order n having chromatic number $\geqslant 4$, and consequently, at least two non-isomorphic STS of order n.

References

1. Hall, Jr., Marshall, Combinatorial Theory, Ginn-Blaisdell, Toronto 1967.

2. di Paola, Jane W., When is a Totally Symmetric Loop a Group?, *Amer. Math. Monthly* 76 (1969), 249-252.

ON TAIT COLORING OF CUBIC GRAPHS

M. Rosenfeld

University of Washington, Seattle, WA, U.S.A.

The purpose of this note is to show that every cubic graph with n vertices is homomorphic to a cubic graph with $\frac{6n+5}{5}$ vertices at most, that admits a Tait coloring. M. Watkins [2], has given an example of a cubic graph with 20 vertices that is 3-edge colorable (Tait coloring) and contains a homomorphic image of the Peterson graph, thus indicating that a converse of a conjecture of Tutte is not true. In view of our result this is not a surprise, as a matter of fact, it can be shown that the Peterson graph is homomorphic to a cubic graph on 12 vertices that admits a Tait coloring.

Let $E = (a,b)$ and $E' = (a',b')$ be two disjoint edges in a graph G. We denote by $G(E',E)$ the graph obtained from G by subdividing the edges E and E' by a pair of new vertices s and s' and joining them by an edge. Let such an operation be called an "H" operation. The following facts can easily be established: if G is a cubic graph, any graph G', obtained from G by a sequence of "H" operations will be homomorphic to G, cubic and with connectivity not smaller than the connectivity of G. We will show that by few "H" operations one can obtain from a given cubic graph G, a graph G' that admits a Tait coloring. Our result follows from the following simple lemma:

<u>Lemma</u>. If $G(n,k)$ is a graph on n vertices, regular of degree $k \geq 3$, then either G is k-edge colorable or in any coloring of G by $k+1$ colors every color class contains at least two edges.

<u>Proof</u>. Observe first that by Vizing's theorem [1] G is always k+1-edge colorable. Since every color class can contain $[\frac{n}{2}]$ edges at most, the lemma follows immediately if n is odd. If n is even, let a coloring of

G by $k + 1$ colors be given and assume that one of the colors is used only once. Since G contains $\frac{n \cdot k}{2}$ edges it follows that $k - 1$ color classes contain $\frac{n}{2}$ edges. The edges belonging to these classes determine a $(k-1)$-factor of G and therefore the remaining edges determine a 1-factor of G that can be colored by one additional color. Hence G is k-edge colorable and the lemma is proved.

Remark: The lemma is best possible in the sence that if for a given k a graph $G(n, k)$ is not k-edge colorable, then a graph $G(2n, k)$ exists that admits a coloring of the edges by $k + 1$ colors with one of the classes containing exactly two edges. This is obtained by taking two disjoint copies of $G(n, k)$, coloring similarly the edges of both graphs by $k + 1$ colors in such a way that one of the classes contains the least possible number of edges, m, in any $(k+1)$-coloring of G. The graph obtained by deleting $m - 1$ of this edges in both graphs and joining by an edge the corresponding vertices in both copies, will have the desired property.

Theorem: *Every cubic graph G, with n vertices, is homomorphic to a cubic graph with $\frac{6n+5}{5}$ vertices at most that admits a Tait coloring.*

Proof: If G is 3-edge colorable there is nothing to prove. Assume therefore that G is not 3-edge colorable and let a coloring of the edges of G by 4 colors be given. We denote the four color classes by A_1, A_2, A_3, A_4. Without loss of generality, we may assume that A_4 contains the least possible number of edges, m, in any 4-coloring of the edges of G. The edge $E = (a, b)$ is in A_4 iff the two other edges incident with a are in A_α and A_β, the edges incident with b are in A_α and A_γ and $\{\alpha, \beta, \gamma\} = \{1, 2, 3\}$. Such an edge will be called an α-edge. By our lemma, $m \geqslant 2$. Let $E_1, E_2 \in A_4$, and assume that both are α-edges. In the graph $G(E_1, E_2)$ we denote by S_1 and S_2 the two additional vertices. In $G(E_1, E_2)$ we color all the edges except the edges E_1 and E_2 by the same colors used in the coloring of G, color the edges: (a_1, s_1)-β, (b_1, s_1)-γ, (a_2, s_2)-β, (b_2, s_2)-γ and (s_1, s_2)-α. (Fig. 1).

$G(E_1,E_2)$ is obtained from G by a single "H" operation, and in this graph, A_4 contains $m - 2$ edges. If m is even, since A_4 contains 3 distinct types of edges, by a sequence of $\frac{m-2}{2}$ "H" operations we obtain a graph G' in which A_4 contains exactly two edges. If the two edges are of the same type an additional "H" operation will yield a graph F that is obtained from G by $\frac{m}{2}$ "H" operations and admits a Tait coloring. If the two edges are of distinct type an additional "H" operation between them (Figure 2) will yield a graph F with A_4 containing exactly one edge and therefore by the lemma, admits a Tait coloring.

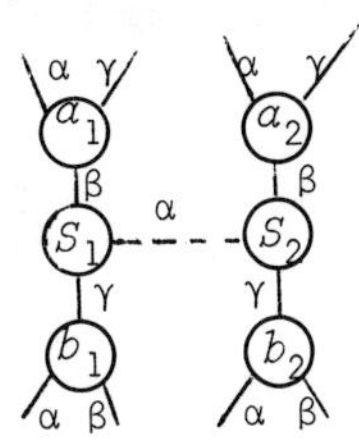

Figure 1.

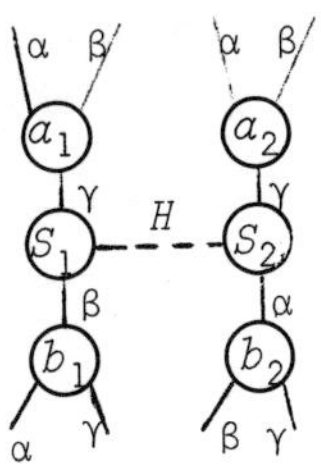

Figure 2.

Hence, if m is even, by $\frac{m}{2}$ "H" operations on G we obtain a graph that admits a Tait coloring. If m is odd, after $\frac{m-3}{2}$ "H" operations we obtain a graph G' in which A_4 contains 3 edges. If two of these edges are of the same type an additional "H" will yield a graph with only one edge in A_4 and therefore this graph will be 3-edge colorable. If all three edges are of distinct types, an "H" operation between an α-edge and β-edge yields a single γ-edge (Figure 2) and an "H" operation between this edge and the third edge will yield a 3-edge colorable graph. Therefore in this case $\frac{m+1}{2}$ "H" operations yield the desired graph. To complete the proof we have to estimate m. Simple considerations show that $m \leqslant \frac{n}{5}$. For example, if G is at least two connected and does not contain a triangle it contains a 1-factor. We color the edges of the 1-factor by one color. The remaining edges constitute a family of disjoint circuits of length $\geqslant 4$. We color the edges of the even circles by 2 colors, use the same colors to color all the edges of the odd circuits except one, and color the exceptional edges by the fourth color. The number of these edges is at most $\frac{n}{5}$. If G contains a triangle, we can shrink it to a single point, use induction on n, and rebuild the triangle, color its edges without using the fourth color. Similar considerations in case G is 1-connected yield the same result.

Since by $[\frac{m+1}{2}]$ "H" operations we obtained a cubic graph homomorphic to G that admits a Tait coloring, the number of vertices of this graph will be $n + m + 1$ at most, and since $m \leqslant \frac{n}{5}$ the theorem is proved.

Remark. It can be easily seen that the Peterson graph can be colored by 4 colors with any preassigned pair of disjoint edges, as the only edges in A_4. Hence a single arbitrary "H" operation on the Peterson graph will yield a cubic graph with 12 vertices that admits a Tait coloring.

References

1. Vizing, V. G., On an estimate of the chromatic class of a p-graph, *Diskret. Analiz.* 3 (1964) 25-30.

2. Watkins, M. E., A theorem on Tait Colorings with an Application to the Generalized Peterson Graphs, *J. of Combinatorical Theory*, 6 (1969).

ON THE BERGE CONJECTURE CONCERNING PERFECT GRAPHS*

Horst Sachs

Technische Hochschule, Ilmenau, DDR

All graphs considered are finite indirected graphs without loops and multiple edges.

Obviously, the chromatic number χ is never smaller than the density ω (i.e., the number of vertices of a maximum clique contained in the graph G):

$$\chi(G) \geq \omega(G) . \tag{1}$$

The question may be raised whether or not there is any other relation between χ and ω besides (1). The answer is: No, for, as is well known, Tutte, Kelly-Kelly, and Zykov were the first to prove that there exist graphs of any given chromatic number without containing a triangle.

In view of this result another question arises, namely: Which are those graphs for which equality holds in (1)? Or, more precisely, in order to make it a reasonable mathematical problem:

Problem. Characterize all those graphs G for which $\chi = \omega$ holds not only for G itself but also for each of its subgraphs G':

$$\chi(G') = \omega(G') \quad \text{for all} \quad G' \subset G. \tag{2}$$

(G' is a subgraph of G if every vertex of G' is a vertex of G and if two vertices of G' are connected by an edge if and only if these vertices are also connected in G.)

Graphs with property (2) have been called χ-perfect graphs. There exist such graphs, for instance every bipartite graph is χ-perfect with

---

* Results partly due to E. Olaru from Iaşi (now at Ilmenau).

$\chi = \omega \leqq 2$; other non-trivial classes of χ-perfect graphs have been given by Gallai, Berge, and other authors.

If G is replaced by its complement $\overline{G}$ then, obviously,

$$\chi(G) = \theta(\overline{G}), \qquad \chi(\overline{G}) = \theta(G) \tag{3}$$

where $\theta(G)$ is the partition (or covering) number of G, i.e., the minimum number of cliques covering all vertices of G, and

$$\omega(G) = \alpha(\overline{G}), \qquad \omega(\overline{G}) = \alpha(G) \tag{4}$$

where $\alpha(G)$ is the stability number of G, i.e., the maximum number of vertices of G no two of which being connected by an edge.

Obviously,

$$\theta(G) \geqq \alpha(G) \tag{$\overline{1}$}$$

and if $\overline{G}$ is χ-perfect then G has the property

$$\theta(G') = \alpha(G') \quad \text{for all} \quad G' \subset G. \tag{$\overline{2}$}$$

A graph with property $(\overline{2})$ is called α-perfect.

Shannon and Gilmore noticed that α-perfect graphs correspond to perfect communication channels considered in information theory.

According to a famous theorem of König all vertices of a bipartite graph G without isolated vertices can be covered by $\alpha(G)$ pairwise disjoint edges which implies that every bipartite graph is also α-perfect.

In accordance with this observation Berge conjectured that a graph is χ-perfect if and only if it is α-perfect. In fact, Berge conjectured a bit more, namely:

BC: A graph G is α-perfect if and only if neither G nor $\overline{G}$ contains an odd hole.

(A hole is a circuit with more than 3 edges having no chord.) Since the condition occurring in the *BC* is symmetric in G and $\overline{G}$ the *BC* is equivalent with the complementary conjecture.

$\overline{BC}$: A graph G is χ-perfect if and only if neither G nor $\overline{G}$ contains an odd hole,

which, if true, would give a perfectly satisfactory answer to our problem formulated above and would indeed be an extremely interesting result.

It is easy to see that the condition mentioned is necessary since for a circuit C of odd length > 3 or its complement $\overline{C}$,

$$\theta = \alpha + 1,$$

consequently neither C nor $\overline{C}$ can be a subgraph of an α-perfect graph; all we have to show is

BC^1: If neither G nor $\overline{G}$ contains an odd hole then G is α-perfect.

A very useful concept is the concept of criticity: Let $G = (X,U)$ where X and U denote the sets of vertices and edges of G, respectively.

Definition 1. Let β stand for χ or α. G is called β-*critical* if G is β-imperfect but all of its proper subgraphs are β-perfect;

G is called vertex-β-critical if G is β-imperfect but $G - x$ is β-perfect for every $x \in X$.

Obviously, G is β-critical if and only if G is vertex-β-critical.

Definition 2. The graph G without isolated vertices is called *edge-α-critical* if G is α-imperfect but $G - u$ is α-perfect for every $u \in U$.

Obviously,

(A) Every edge-α-critical graph is α-critical.

The converse is *not* true as the graph $\overline{C}_7$ (the complement of a 7-gon) shows which is α-critical but not edge-α-critical.

(B) If G is α-critical then

$$\theta(G) = \alpha(G) + 1,$$
$$\left.\begin{aligned}\theta(G-x) &= \theta(G) - 1\\ \alpha(G-x) &= \alpha(G)\end{aligned}\right\} \quad x \in X;$$

If G is edge-α-critical then

$$\theta(G) = \alpha(G) + 1,$$
$$\left.\begin{aligned}\theta(G-u) &= \theta(G)\\ \alpha(G-u) &= \alpha(G) + 1\end{aligned}\right\} \quad u \in U.$$

(B) enables us to apply results of the theories of graphs which are vertex - critical with respect only to χ (Dirac, Gallai, ...) or which are edge-critical with respect only to α (Andrásfai, Wessel, Berge) to our problems, but unfortunately does not help us too much.

(C) Every α-imperfect graph G contains an α-critical graph as a subgraph.

For a subgraph $G' \subset G$ with $\theta(G') > \alpha(G')$ having minimum number of vertices is obviously α-critical.

If the BC is true then an α-critical graph G contains a circuit $C_{2\lambda+1}$ ($\lambda \geqq 2$) or its complement $\overline{C}_{2\lambda+1}$ as a subgraph and necessarily is identical with it. On the other hand, if we can show that every α-critical graph is a $C_{2\lambda+1}$ or a $\overline{C}_{2\lambda+1}$ then the BC is true: For if G does not contain a $C_{2\lambda+1}$ or a $\overline{C}_{2\lambda+1}$ then our assumption implies that G does not contain an α-critical graph and therefore (according to (C)) must be perfect.

We conclude that the BC is equivalent with

BC^2: Every α-critical graph is either a $C_{2\lambda+1}$ or a $\overline{C}_{2\lambda+1}$ ($\lambda \geqq 2$).

In the following, G is supposed to be an α-critical graph.

If $\alpha(G) = 2$ then $\alpha(\overline{G}) = 2$, i.e., $\overline{G}$ does not contain a triangle. If we assume that $\overline{G}$ does not contain an odd hole then $\overline{G}$ cannot contain any circuit of odd length at all, i.e., $\overline{G}$ is bipartite; consequently, $\overline{G}$ is χ-perfect and G is α-perfect with $\alpha = 2$ - contradicting the hypothesis that G be α-critical. Therefore, $\overline{G}$ necessarily contains an odd hole $C_{2\lambda+1}$ which implies $G = \overline{C}_{2\lambda+1}$ with some $\lambda \geqq 2$.

As $\alpha(\overline{C}_{2\lambda+1}) = 2$ for all $\lambda \geqq 2$, for $\alpha \geqq 3$ an α-critical graph cannot be a $\overline{C}_{2\lambda+1}$, and what remains to be proved is the following statement equivalent with the BC.

BC^3: For $\alpha \geqq 3$, every α-critical graph is a $C_{2\lambda+1}$.

The difficulties in proving or disproving this conjecture are due to the fact that essential use must be made of the hypothesis $\alpha \geqq 3$ for otherwis[e] all properties of α-critical graphs we may derive from our assumptions ar[e] common properties of the $C_{2\lambda+1}$ and $\overline{C}_{2\lambda+1}$ (if the BC is true).

E. Olaru succeeded in proving a result which is a bit weaker than BC^3, namely:

Every edge-α-critical graph is a $C_{2\lambda+1}$. (see Theorem 2)

In the following some properties of α-critical graphs G are listed: Some of these should be well known to everybody familiar with the proble[m] some of them are nearly trivial and can be proved in a few lines, others need a longer proof which in some cases is a bit more involved.

(D) G is connected; $\overline{G}$ is connected.

(E) For every clique $Q \subset G$:

$$\alpha(G-Q) = \alpha(G).$$

(F) For every clique $Q \subset G$: $G - Q$ is connected; for every clique $Q^* \subset \overline{G}$: $\overline{G} - Q^*$ is connected.

Because of $\alpha(G-x) = \alpha(G)$ for every $x \in X$, we find a system of $\alpha(G) = \alpha$ cliques $Q_i = (X_i, U_i)$ $(i = 1,2,\ldots,\alpha)$ covering $G - x$ with

$$X_i \cap X_j = \emptyset \ (i \neq j), \qquad \bigcup_{i=1}^{\alpha} X_i = X - \{x\}.$$

<u>Conclusions</u> (using (E)):

(G) To every $x \in X$ there is a maximum stable set of G not containing x, and there are at least $\alpha(G)$ different maximum stable sets each of which containing x.

(A vertex-set is called *stable* in G if no two of its vertices are F adjacent in G.)

Let X_i^1 be the subset of all vertices of X_i not connected with x, $X^1 = \bigcup_{i=1}^{\alpha} X_i^1$ (= set of all "non-neighbours" of x), $X^2 = X - \{x\} - X^1$ (= set of all neighbours of x). For $Y \subset X$, let $[Y]_G$ denote the subgraph of G having vertex-set Y. Then

(H) $X_i^1 \neq \emptyset$ $(i = 1,2,\ldots,\alpha)$.

(I) For every $T \subsetneq \{1,2,\ldots,\alpha(G)\}$,

$$\alpha([\bigcup_{i \in T} X_i^1]_G) = |T|.$$

Let $G^1(x) = [X^1]_G$, $G^2(x) = [X^2]_G$.

(J) $\theta(G^1(x)) = \alpha(G^1(x)) = \alpha(G) - 1$ for all $x \in X$.

(K) If $G^1(x)$ is covered with $\alpha(G) - 1$ cliques then each of these contains at least two vertices.

Consequently, if $v(x) = |X^2|$ is the valence of x and $|X| = n(G)$:

(L) $v(x) \leqslant n(G) - 2\alpha(G) + 1$ for all $x \in X$.

(M) $G^1(x)$ is connected; $\overline{G^2(x)}$ is connected.

From the second statement of (M) a famous theorem of Dilworth can easily by derived:

<u>Theorem</u>. (Dilworth). *If the graph H is transitively orientable then H is α-perfect.*

In a very similar way, from the first statement of (M) follows

Theorem. (Berge) *If the graph H is transitively orientable then H is χ-perfect.*

Using the preceding statements, E. Olaru was able to prove the following theorems.

Theorem 1. (main result) *The α-critical graph G is an odd circuit of length ≥ 5 if and only if G has the following property: There exist three vertices x, x^1, x^2 in G such that*

$$(*)\quad \begin{cases} (x,x^1),\ (x,x^2) \in U \\ (x^1,x^2) \notin U \\ \alpha([X^1 \cup \{x^j\}]_G) = \alpha(G) \quad \text{for} \quad j = 1,2. \end{cases}$$

As a consequence, the *BC* is equivalent with

BC^4: If G is α-critical with $\alpha \geq 3$, then G contains three vertices x, x^1, x^2 satisfying (*).

If G is edge-α-critical then $\alpha([X^1 \cup \{y\}]_G) = \alpha(G)$ for *every* y adjacent to x (for the proof use (B)), and as a consequence of Theorem 1 we have

Theorem 2. *G is edge-α-critical if and only if G is an odd circuit $C_{2\alpha+1}$ (True for $\alpha \geq 2$).*

Therefore we have another equivalent formulation of the *BC*:

BC^5: If G is α-critical with $\alpha \geq 3$ then G is edge-α-critical.

The fact that this statement is not true if $\alpha = 2$ is admitted (cf. the remark following (A)) clearly shows us the origin of all trouble.

Dropping the hypothesis (*) in Theorem 1, Olaru could only prove the following:

Theorem 3. *Every vertex x of an α-critical graph G belongs to some odd circuit of length ≥ 5 all of whose chords in G are issuing from x.*

As a consequence we have

Theorem 4. *If the α-critical graph G is not an odd circuit $C_{2\lambda+1}$ ($\lambda \geq 2$), then every vertex $x \in G$ belongs to some triangle.*

(In fact we should show in this case that G is a $\overline{C}_{2\lambda+1}$ with $\lambda \geq 3$.)

Theorem 4. yields another equivalent formulation of the *BC*:

BC^6: If G is α-critical with $\alpha \geq 3$ then G has at least one vertex which is contained in no triangle.

nother consequence of Theorem 3 is

heorem 5. *If every odd circuit of length* $\geqslant 5$ *in a graph H has two rossing chords then H is* α*-perfect.*

This theorem is particularly interesting when compared with a well-nown result of Gallai.

heorem. (Gallai) *If every odd circuit of length* $\geqslant 5$ *in a graph H is riangulated by some of its chords then H is* α*-perfect.*

References

1. Andrásfái, B., On critical graphs, Theory of Graphs, Intern. Symp. Rome, 1966, 9-20.
2. Berge, C., Färbung von Graphen, deren sämtliche bzw. deren ungerade Kreise starr sind (Zusammenfassung) *Wiss. Z. Martin-Luther-Univ. Halle-Wittenberg,* (1961), 114-115.
3. Berge, C., Some Classes of Perfect Graphs. Graph Theory and Theoretical Physics. (Academic Press, London and New York, 1967), Chapter 5, 155-165.
4. Descartes, Blanche, A three colour problem. *Eureka*, (April 1947 and March 1948).
5. Descartes, Blanche, Solution to Advance Problem No. 4526. *Amer. Math. Monthly,* 61 (1954), 352.
6. Dilworth, R. P., A decomposition theorem for partially ordered sets. *Ann. Math.* 51 (1950), 161.
7. Gallai, T., Graphen mit triangulierbaren ungeraden Vielecken. *Magyar Tud. Akad. Mat. Kutató. Int. Közl.* 7 (1962), 3-36.
8. Gilmore, Cf., *Referativnyĭ Žurnal Matematiki,* 1967/3/B 250.
9. Hajnal, A. and Surányi, J., Über die Auflösung von Graphen in vollständige Teilgraphen. *Annales Univ. Sci. Budapestinensis,* 1 (1958), 113-121.
10. Kelly, J. N. and Kelly, L. M., Path and circuits in critical graphs. *Am. J. Math.* 76 (1954), 791.
11. Shannon, C. E., The Zero-error Capacity of a Noisy Channel, *IRE Trans. Inf. Theory,* IT-3 (1956), 3.

12. Surányi, L., The covering of Graphs by cliques. *Studia Sciantiarum Mathematicarum Hungaria.* 3 (1968), 345-349.

13. Zykov, A. A., On some properties of linear complexes. (russ.) *Math. Sbornik,* N. S. 24 (1949), 163-188.

A NEW PERFECT SINGLE ERROR-CORRECTING GROUP CODE. MIXED CODE

J. Schönheim

The University of Calgary, Alta., Canada
and
Tel Aviv University, Israel

Consider the five component vectors having as first four components elements from $GF(2)$, while the fifth component is from $GF(4)$. Define accordingly vector addition as in $GF(2)$ for the first four components and as in $GF(4)$ for the fifth component. This set of vectors is under the above defined addition operation a group G. If we denote by α and β the elements of $GF(4)$ different from 0 and 1, then the following subgroup of G is a single error correcting perfect code:

$$(00000),\quad (11001),\quad (1010\alpha),\quad (1001\beta),$$

$$(0110\beta),\quad (0101\alpha),\quad (00111),\quad (11110).$$

ON THE SKELETONS OF A GRAPH OR DIGRAPH

J. Sedláček

The University of Calgary, Alta., Canada

0. Ore in [8] has used the notation *skeleton* and *scaffolding* for certain subgraphs of a given graph. Some authors are calling them *frame*. Most frequently these notations stand for the maximal tree which is contained in a given graph.

Let G be a finite undirected and connected graph without loops and multiple edges. The maximal tree subgraph of G is usually called the *spanning* tree of G. The number of all trees spanning a given graph G will be denoted by $k(G)$. For $k(G)$ J.W. Moon uses the name of *complexity* of the graph G. It is well known that the number $k(G)$ equals the value of a determimant, the elements of which depend on G.

In a former paper [11] we defined a set A_n as follows: A_n is the set of all such positive integers q that there exists a connected graph G of n vertices and q spanning trees. In this contribution we shall first add to this definition the condition that G be a regular graph of degree t. Thus we obtain the set $B_n^{(t)}$. The cases for $t = 0,1,2$ are trivial and the investigation of the set $B_n^{(t)}$ begins to be interesting for $t = 3$. We see that $B_n^{(t)} = \phi$ for $n \leqslant t$. By $|M|$ let us denote the number of elements of the (finite) set M. Evidently, $|B_{t+1}^{(t)}| = 1$ for any $t \geqslant 3$. Further, if t is odd, then $B_n^{(t)} = \phi$ for every odd n. For even t evidently $|B_{t+2}^{(t)}| = 1$.

We may deduce two theorems on the number of elements of the set $B_n^{(t)}$ for large values of n. Both theorems affirm what might be anticipated by intuition.

<u>Theorem 1</u>. *For a given odd number* $t \geq 3$

$$\lim_{a\to\infty} |B_{2a}^{(t)}| = +\infty$$

<u>Theorem 2</u>. *For a given even number* $t \geq 4$

$$\lim_{b\to\infty} |B_{b}^{(t)}| = +\infty.$$

We shall not give the proofs of Theorems 1 and 2 - the reader may find them in the paper [12], which is to be published. We shall limit ourselves to the statement that both proofs are constructive. For instance, in the proof of Theorem 1, for every sufficiently large number $2a$ we construct regular graphs $G_1, G_2, \ldots, G_r$, each graph is of degree t, has $2a$ vertices and the relation

$$k(G_1) < k(G_2) < \ldots < k(G_r)$$

holds. And the number r tends to infinity, if the number a does the same.

At the Conference in Manebach in 1967 we showed that there is a close connection between spanning trees and Lucas numbers, well known in the theory of numbers - see [9]. First, let us recall this concept. Let be given an equation $z^2 - Lz + M = 0$, $L > 0$ and M being integers and α, β, $\alpha \neq \beta$ the roots of this equation. Then the *Lucas number* is defined by

$$L_m = \frac{\alpha^m - \beta^m}{\alpha - \beta} \qquad (m = 1,2,3,\ldots). \tag{1}$$

We shall certainly recollect that the well known Fibonacci numbers are a special case of Lucas numbers.

We should like to describe here further graphs the number of spanning trees of which may be expressed in the form of (1). From the literature on graphs the so called wheel graphs are well known; they may be defined as follows:

A wheel W_n denotes a graph which consists of a circuit C_n of length $n \geq 3$ together with a vertex v not belonging to the circuit and joined to each vertex u_i $(i = 1,2,\ldots,n)$ of C_n by one edge. The W_n graphs are treated, for instance, in [1] and [2]. To determine the number $k(W_n)$ it is advantageous to introduce an auxiliary graph H_n, which is obtained from W_n by removing of the edge $u_1 u_n$. We may also extend the definition of the graph H_n to the case $n = 1$ (the graph H_1 has the vertices u_1, v and only one edge u_1v) and $n = 2$ (three vertices u_1u_2, v and three edges u_1u_2, u_1v, u_2v). Now we may prove

Theorem 3.

$$k(H_n) = \frac{(3 + \sqrt{5})^n - (3 - \sqrt{5})^n}{2^n \sqrt{5}} \tag{2}$$

for every positive integer n.

Proof. Let us put $a_n = k(H_n)$ and let b_n denote the number of subgraphs S of the graph H_n where S satisfies the following conditions: S contains all vertices of H_n and consists of two components which are trees, one of them containing v and the other u_n. We see that $a_1 = 1$, $b_1 = 1$ and the following equations hold:

$$a_{n+1} = 2a_n + b_n, \quad b_{n+1} = a_n + b_n.$$

By solving this system we obtain the formula (2) for $k(H_n)$.

Theorem 4. *If n is a given positive integer* $\geqq 3$, *then*

$$k(W_n) = \left(\frac{3 + \sqrt{5}}{2}\right) + \left(\frac{3 - \sqrt{5}}{2}\right)^n - 2.$$

Proof. The result follows from the preceding theorem and the formula

$$k(W_n) = k(H_n) + 2 \sum_{j=2}^{n} k(H_{j-1})$$

which is easy to be verified.

R.K. Guy and F. Harary in [6] investigated the so called *Möbius ladder* M_{2r} which may be defined as follows: The graph M_{2r} consists of a polygon of length $2r \geqq 4$ and the r chords joining opposite pairs of vertices. It can also be drawn as in Figure 1, which is reminiscent of the Möbius strip. The proof of the next theorem shall be left to the reader.

Theorem 5.

$$k(M_{2r}) = \frac{r}{2} ((2 + \sqrt{3})^r + (2 - \sqrt{3})^r + 2)$$

for every integer $r \geqq 2$.

To conclude I should like to mention the finite directed graphs G without loops and multiple edges and one theorem which we found with M. Fiedler in 1958 and published in Czech in the paper [4]. Let G have vertices $1, 2, 3, \ldots, n$. Label each edge ik by a number x_{ik} and put

(i) $a_{ik} = -x_{ik}$ for $i \neq k$ and $ik \varepsilon G$, (i) $a_{ik} = 0$ for $i \neq k$ and $ik \notin G$,

(iii) $a_{ii} = \sum_{k, ik \varepsilon G} x_{ik}$.

For $\phi \neq M \subset \{1,2,\ldots,n\}$ we denote by $S(M)$ the set of all rooted subforests of G whose set of roots is exactly M. Then, the principal minor of the matrix $A = (a_{ik})$ obtained by omitting the rows and columns corresponding to M is

$$\det A(M) = \sum_{S \varepsilon S(M)} \pi(S)$$

where $\pi(S)$ denotes $\prod_{ik \varepsilon S} x_{ik}$.

As an application of this theorem we gave in 1958 the results $a^{b-1} b^{a-1}$ for the number of trees spanning the complete bipartite graph $\langle a,b \rangle$.

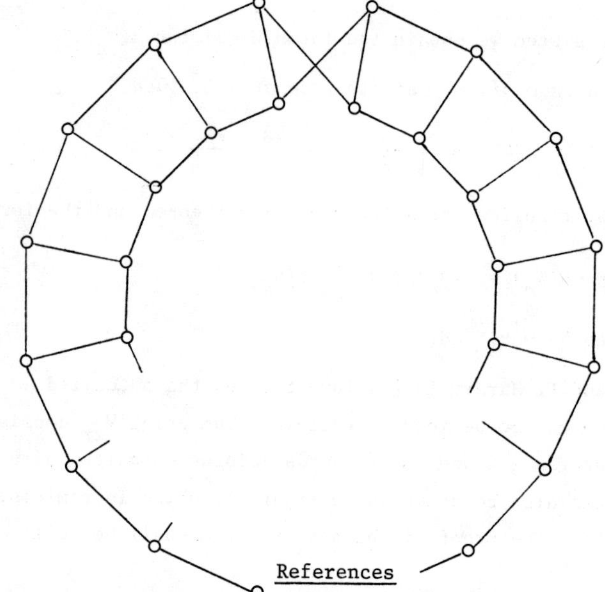

References

1. Dirac, G.A., Some results concerning the structure of graphs, *Canad. Math. Bull.*, 6(1963), 183-210.

2. Erdös, P., Harary, F., Tutte, W.T., On the dimension of a graph, *Mathematatika*, 12 (1965), 118-122.

3. Fiedler, M., Graphs and linear algebra, Theory of graphs (Internat. Sympos; Rome 1966), 131-134, Paris 1967.

4. Fiedler, M., Sedláček, J., O W-basích orientovaných grafů, *Casopis Pest. Mat.,* 83 (1958), 214-225.

5. Glicksman, S., On the representation and enumeration of trees, *Proc. Camb. Phil. Soc.,* 59 (1963), 509-517.

6. Guy, R.K., Harary, F., On the Möbius ladders, The University of Calgary, Research Paper No. 2, November 1966.

7. Moon, J.W., Enumerating labelled trees, Graph Theory and Theoretical Physics (edited by F. Harary), 261-272, London - New York, 1967.

8. Ore, O., Theory of graphs, Amer. Math. Soc. Colloq. Publ. 38, Providence 1962.

9. Rotkiewicz, A., On Lucas numbers with two intrinsic prime divisors, *Bulletin de L'Acadêmie polonaise des sciences,* 10 (1962), 229-232.

10. Scoins, H.I., The number of trees with nodes of alternate parity, *Proc. Camb. Phil. Soc.,* 58 (1962), 12-16.

11. Sedláček, J., On the spanning trees of finite graphs, *Casopis Pest. Mat.,* 91 (1966), 221-227.

12. Sedláček, J., Regular graphs and their spanning trees, *Casopis Pest. Mat.* (to appear).

ON AN ALGORITHM FOR ORDERING OF GRAPHS

Milan Sekanina

University of Manitoba, Winnipeg, Man., Canada

Let (G,ρ) be a finite connected (undirected) graph without loops and multiple edges. So x, y being two elements of G (vertices of the graph (G,ρ)), $\langle x,y\rangle \in \rho$ means that x and y are connected by an edge. Two vertices $x, y \in G$ have the distance $\mu(x,y)$ equal to n, if n is the smallest number with the following property: there exists a sequence $x_1, x_2, \ldots, x_{n-1}$ of vertices such that $\langle x,x_1\rangle \in \rho, \ldots, \langle x_{n-1}, y\rangle \in \rho$. If $x \in G$, we put $\mu(x,x) = 0$.

In [1] the following theorem was proved:

If a, b are vertices of (G,ρ), $a \neq b$, then it is possible to order G in a sequence $a = x_1, x_2, \ldots, x_g = b$ (where g = card G) such that

$$\mu(x_i, x_{i+1}) \leqslant 3.$$

Let us consider our graph as a collection of points with edges considered as ways connecting the points. We shall give an algorithm according to which one can successively proceed from one point to another using each way in one direction at most once and labelling some of the passed points successively with members $1, 2, \ldots, g$ in such a manner that between two labellings at most two points can be passed without labelling. The last label will be given to some neighboring point to the starting point and after that every point has some label.

First, let us suppose (G,ρ) is a tree.

Algorithm I.

1. Choose some point a and label it with 1.

2. Let us be in a point x which has just been labelled with n.

2.1. If there is a way $\langle x,y_1\rangle$ not passed in any direction as yet, we proceed to y_1.

2.2. If there exists no such y_1 described in 2.1., choose some $\langle x,y_1\rangle$ not passed in the direction from x to y_1 and we proceed to y_1.

3.1. If 2.1 occurred then

3.1.1. in the case there exists $\langle y_1,y_2\rangle$ not passed in any direction we proceed to y_2 and label it with $n+1$.

3.1.2. If y_2 from 3.1.1 does not exist, we label y_1 with $n+1$.

3.2. If 2.2 occurred then

3.2.1. if 3.1.1 is valid we pass to y_2.

3.2.2. If 3.1.1 does not hold then:

3.2.2.1. if y_1 is not labelled we label it with $n+1$.

3.2.2.2. If y_1 is labelled we proceed to some y_2 where $\langle y_1,y_2\rangle$ has not been passed in the direction from y_1 to y_2.

4. If 3.2.1 occurred then if there exists $\langle y_2,y_3\rangle$ not passed in any direction we proceed to y_3 and label it with $n+1$. In other case y_2 is labelled with $n+1$.

If 3.2.2.2 occurred and if there exists $\langle y_2,y_3\rangle$ not passed in any direction, we pass to y_3 and label it with $n+1$. Otherwise y_2 is labelled with $n+1$.

Let (G,ρ) now be quite arbitrary finite connected (undirected) graph without loops and multiple edges with card $G > 1$.

Let Algorithm II differ from Algorithm I by adding the following rule:

Let us be in the point x and let $\langle x,y\rangle$ be such a way which was not passed before in any direction but in some previous step we were already in y (y can be still without label). Then $\langle x,y\rangle$ cannot be used.

So by this rule the cases 3.1.1, 3.2.1, 4. (first part) of Algorithm I are modified.

Algorithm II has the same effect as Algorithm I for trees with one exception: some ways may not be passed.

References

1. Sekanina, Milan, On an ordering of the set of vertices of a connected graph, Publ. Fac. Sci. Univ. Brno., No. 412, 137 - 142 (1960).

FIXING SUBGRAPHS

J. Sheehan

University of Waterloo, Ont., Canada

Let G be a finite graph with no loops or multiple edges. Let $A(G)$ denote the automorphism group of G. Let $\pi(G)$ be the set of all labelled graphs obtainable from G by labelling the vertices of G with a fixed set L of labels ($|L| = |V(G)|$). Let $\pi(G) = \{G_1, G_2, \ldots, G_p\}$. Let $r_i(\alpha,\beta)$ be the number of edges in G_i which have end-vertices labelled α and β, $\alpha,\beta \in L$. Let $r_{ij}(\alpha,\beta) \equiv r_i(\alpha,\beta) + r_j(\alpha,\beta)$. Then $G_i \cap G_j$ is the graph defined by: (i) $V(G_i \cap G_j) \equiv L$, (ii) α and β are adjacent if and only if $r_{ij}(\alpha,\beta) = 2$, $\alpha,\beta \in L$. Let $\text{Int}(G) \equiv \{G_i \cap G_j : i,j \in \{1,2,\ldots,p\}\}$. Let U be a spanning subgraph of G.

Definition. U is a *fixing subgraph* of G if G is the only element of $\text{Int}(G)$ containing a subgraph isomosphic to U. Let $F(G)$ denote the set of fixing subgraphs of G.

Theorem 1. *$U \in F(G)$ if and only if $A(U) \subseteq A(G)$ and G contains exactly $|A(G)|/|A(U)|$ subgraphs isomosphic to U.*

Remark. The fixing subgraphs of G possess many interesting properties. In this talk however we shall restrict attention to the following question. Can we characterize those graphs G such that $|F(G)| = 1$ (obviously $G \in F(G)$)?

Examples ($|F(G)| = 1$)

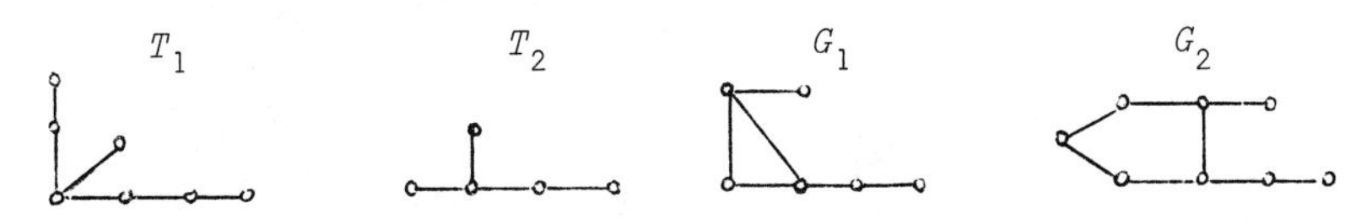

Definition. G is *pseudo asymmetric* if $|A(G)| = 1$ and $\forall\, \lambda \in E(G)$ $|A(G-\lambda)| \neq 1$.

Theorem 2. *Let G be a finite non-trivial tree or asymmetric monocyclic graph. Then* $|F(G)| = 1$ *if and only if* $G \varepsilon \{T_1, T_2, G_1, G_2\}$.

A very obvious result is contained in Theorem 3.

Theorem 3. *Let G be a finite non-trivial regular graph then* $|F(G)| > 1$.

Conclusion. In order to prove Theorem 2 we proved that the only pseudo-asymmetric tree was T_2. I am pleased to acknowledge that J.A. Zimmer Jr (see these proceedings) obtained a more general result than this a littl earlier than myself. Possibly the number of pseudo-asymmetric graphs is very small.

EXTREMAL GRAPH PROBLEMS

M. Simonovits

Budapest, Hungary

Notations. $v(G)$, $e(G)$, $\chi(G)$ denote the number of vertices, edges and the chromatic number of the graph G. Here the graphs have no directed, multiple or loop edges. $\overset{d}{\underset{i=1}{\times}} G_i$ denotes the product of graphs G_i, i.e. the graph, obtained by joining vertices of G_i to the vertices of the other G_i-s.

Generalizing a well-known theorem of Turán [1] Erdös and I have proved independently [3], [4] that for any given graph $M_1,\dots,M_k$ and fixed n if K^n has maximum number of edges among graphs of n vertices, not containing any M_i as a subgraph, then

Theorem A. *There exist graphs $N_1,\dots,N_d$, $(d+1 = \min \chi(M_i))$ such that K^n can be obtained from $\overset{d}{\underset{i=1}{\times}} N_i$ omitting $0(n^{2-\frac{1}{r}})$ edges from it. Here is an integer depending only on $M_1,\dots,M_\mu$ and*

(1) $v(N_i) = \frac{n}{d} + 0(n^{1-\frac{1}{r}})$, $e(N_i) = 0(n^{2-\frac{1}{r}})$

(2) *any vertex of N_i has valence $\geq \frac{n}{d}(d-1) + 0(n^{1-\frac{1}{r}})$*

(3) *the number of vertices of N_i joined to at least εn vertices of the same N_i is $0_\varepsilon(1)$.*

The graph K^n is called the *extremal graph* for $M_1,\dots,M_\mu$. Theorem A shows that the extremal graphs for $M_1,\dots,M_\mu$ are fairly well determined by $\min \chi(M_i)$, they depend loosely on the structure of M_i-s.

How the structure of M_i-s influence the structure of the extremal graphs? Erdös and I have proved [5] that the extremal graphs for $K(3,r_1,\dots,r_d)$ are products: $K^a = \overset{d}{\underset{i=1}{\times}} N_i$ where $3 \leq r_1 \leq r_d$ and

(1) $v(N_i) = \frac{n}{d} + O(n^{2/3})$

(2) N_i is an extremal graph for $K(3,r_1)$.

(3) $N_2,\dots,N_d$ are extremal graphs for $(1,r_2)$.

Here 3 can be replaced by 2 or 1 as well.

I have found the following generalization of this latest theorem:

Notation.

(1) $f(n,M_1,\dots,M_k)$ denotes the number of edges of the extremal graphs for $M_1,\dots,M_\mu$.

(2) Let $\chi(M) = 2$ and colour both M and $K(n, n)$ by two colours: red and blue. We consider subgraphs G^{2n} of $K(n,n)$ such that if M is the subgraph of G^{2n}, then the class of blue vertices of M is not contained by the class of blue vertices of $K(n,n)$. The maximum of $e(G^{2n})$ will be denoted by $h(n,G^{2n})$.

Definition. $x \in M_1$ is a *weak point* for $M_1,\dots,M_\mu$ if $\chi(M_1) = 2$ and $h(n;M_1 - x) = o(f(n;M_1,\dots,M_\mu))$.

Remark. If there exists an automorphism of $M_1 - x$ changing the colours, then our condition with $f(n;M_1 - x) = o(f(n;M_1,\dots,M_\mu))$.

Examples.

(1) $K(r_0,\dots,r_d)$ has weak points if either $r_0 \leq 3$, or if $r_0^2 - 3r_0 + 3 > r_1$. [5] Probably it always has.

(2) If M is not a tree, but $M - x$ is, $\chi(M) = 2$ then $x \in M$ is a weak point of it.

(3) Let $C(2l)$ be a circuit of $2l$ vertices, $x \notin C(2l)$ and let x be joined to 5 or more vertices of $C(2l)$ so that the obtained graph M be two-chromatic. Then $x \in M$ is a weak point of it.

(4) Let M be a graph, obtained from two $C(2l)$ or from two $K(r,r)$ by joining them by a path of length 2. Then M has *no* weak point.

Theorem 1. *Let M be a d+1 chromatic graph and let us colour it by* 1,2,...,d+1. *L_{ij} denotes the subgraph of M spanned by the vertices of the ith and jth colours. If $x \in L_{ij}$ is a weak point of $\{L_{ij}\}$ and K^n is an extremal graph for M, then K^n can be obtained from a suitable product $N^n = \times_{i=1}^{d} N_i$ omitting $o(n)$ edges from it. Here*

(1) $v(N_i) = \frac{n}{d} + o(n)$

(2) N_i *is almost an extremal graph for* $\{L_{ij}\}$ *it has* $f(n;\ldots,L_{ij},\ldots) + O(n)$ *edges, but it does not contain any* L_{ij}.

(3) *The vertices of* N_i $(i=2,\ldots,d)$ *are joined to less than* s *other vertices of* N_i, *if* x *is joined to* s *vertices of the 3rd colour.*

Theorem 2. *If in Theorem* 1. $r \leq 3$, *then* $o(n)$ *can be replaced by* $o(1)$. *If* $r \leq 2$, *then there exists an extremal graph* K^n *such that*

$$K^n = \mathop{\times}_{i=1}^{d} N_i$$ *whenever* n *is large enough.*

Remarks.

(1) Similar theorems hold if M is replaced by $M_1,\ldots,M_\mu$. The only change is that L_{ij}-s must be replaced by those subgraphs of $M_1,\ldots,M_\mu$, for which $\chi(M_j - L_t) = \min \chi(M_j) - 2$ if $L_t \subseteq M_j$.

(2) Theorem 1 has "assymptotic" character, but it has many corollaries of "exact" character. One of them is the theorem of Erdös and mine about the extremal graphs for $K(3,r_1,\ldots,r_d)$. Another one is

Theorem 3. *Let* $\Gamma(3k)$ *be the graph, having the vertices* $x_1,\ldots,x_k$; $y_1,\ldots,y_k$; $z_1,\ldots,z_k$ *and defined by*

(i) $x_i \to y_i \to z_i \to x_i$ *is an automorphism of* $\Gamma(3k)$.

(ii) $x_1,\ldots,x_k$, $y_1,\ldots,y_k$ *determine a* $C(2l)$.

Then for $n > n_0$ *any extremal graph* K^n *for* $\Gamma(3k)$ *is a product:* $K^n = k_1 \times k_2$ *where* $v(K_i) = \frac{n}{2}$, $e(K_2) = 0$ *and* K_1 *is an extremal graph for* $\{\ldots,C(2l),\ldots\}$ $\frac{k}{2} \leq l \leq k$.

References

2. Turán, P., *Matematikai Lapik*, 48 (1941), 436-452. (in Hungarian).

3. Erdös, P., On some new inequalities concerning extremal properties of graphs. Theory of Graphs, Proc. Coll. held at Tihany, Hungary, 1966.

4. Simonovits, M., A method for solving extremal problems. Stability problems. Theory of Graphs, Proc. Coll. held at Tihany, Hungary, 1966.

5. Erdös, P. and Simonovits, M., An extremal graph problem. *Acta Math. Acad., Sci. Hungar.* (forthcoming).

SOME PROPERTIES OF THE SPECTRUM OF A GRAPH

John H. Smith

Boston College, Chestnut Hill, MA, U.S.A.

All graphs will be undirected, without loops or multiple edges. By subgraph we mean a graph obtained by deleting a set of vertices and all incident edges. We consider real valued functions on the vertices and the adjacency operator, A, which replaces the value at a vertex by the sum of the values and adjacent vertices. (It can be represented by the adjacency matrix). By eigenvalue of a graph we mean eigenvalue of A.

The author thanks A.J. Hoffman for several suggestions concerning Theorem 5.

Theorem 1. *A graph has exactly one positive eigenvalue if and only if the non-isolated points form a complete multipartite graph.*

Proof. We may ignore isolated points. If the graph is not complete multipartite it contains one of the following:

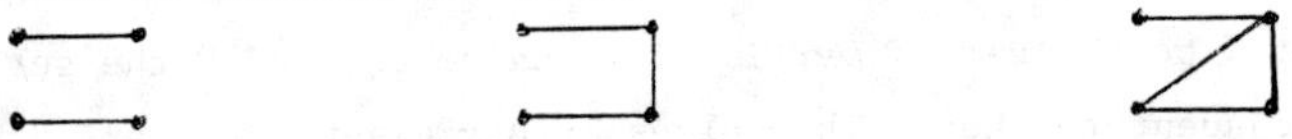

All of these have two positive eigenvalues, hence the original graph has at least two.

Conversely if G is complete multipartite and λ a positive eigenvalue let the function x be a corresponding eigenvector. It is constant on the parts of G. Suppose $(x)_a > 0$ $(x)_b < 0$. Then $(\lambda x)_a < \sum_{\text{all } c} (x)_c < (\lambda x)_b$, a contradiction. Hence the sign of x is constant and λ is the largest eigenvalue by [2].

The next result uses repeatedly Lemma 2, p. 57 of [2].

Theorem 2. *The largest eigenvalue of G is ≤ 2 (<2) is and only if each connected component of G is a subgraph (proper subgraph) of one of the* *following:*

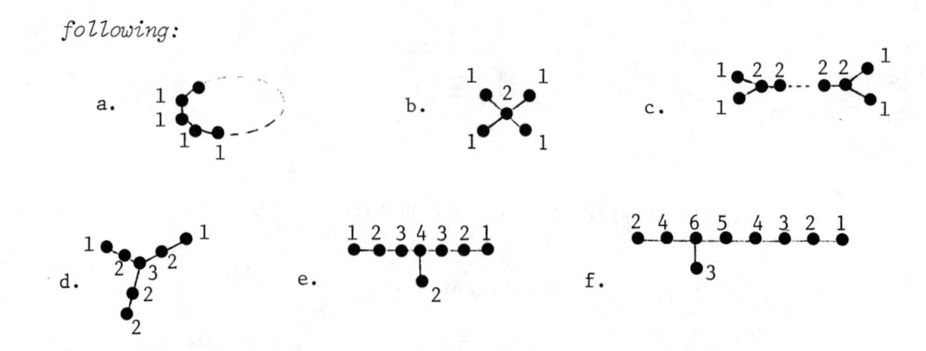

Proof. We may assume G is connected. The numbers associated with the vertices give eigenvectors with eigenvalue 2. Since all entries are positive, 2 is the largest eigenvalue.

Conversely let G have largest eigenvalue less than 2. Then G is a tree by (a) with no vertices of degree > 3 by (b) and at most one of degree 3 by (c). If there are none of degree 3,G is a proper subgraph of (a); if there is one, at least one branch from it has length one by (d). If two have length one G is a proper subgraph of (c); if not at least one other branch has length two by (e) and the remaining branch has length less than 5 by (f).

If the largest eigenvalue is 2, the proof is similar.

Theorem 3. *Let G be a tree with s vertices of degree one and t vertices adjacent to vertices of degree one. Then the multiplicity of any eigenvalue (non-zero eigenvalue) is less than s (t, for $t \neq 1$).*

Proof. Let S be the set of vertices of degree one and T the set of vertices adjacent to them. The values of any eigenvector are determined by its values on S (by values on T for a non-zero eigenvalue). Hence the restriction {eigenvectors for λ} $\rightarrow$ {functions on S (T)} is one to one. It is not onto (for $t \neq 1$) for no eigenvector can be 1 at one point of S (T) and 0 elsewhere.

Theorem 4. *If Γ, the automorphism group of G (connected), is transitive on edges then the only non-zero simple eigenvalue is the largest ($\pm$ the largest if G is bipartite).*

Proof. Let $\lambda \neq 0$ be a simple eigenvalue with eigenvector x. If x takes on the same non-zero value at some two adjacent vertices then it is constant and we are done. If not let $|(x)_a|$ be maximal. Then x

is constant and non-zero on the vertices adjacent to a since if $x_b \neq x_{b'}$, $a \sim b$, $a \sim b'$ then $\sigma(x)$ is independent of x, $\sigma \in \Gamma$, $\sigma(a,b) = (a,b')$. If $a \sim b$ then x is constant on vertices adjacent to b; otherwise x and $\tau(x)$ are independent where $\tau(a,b) = (a',b)$ $a' \sim b$. Continuing, we get G is bipartite and x is constant on the parts. Q.E.D.

If G is connected, for any two vertices a and b let $N_{rs}(a,b)$ be the number of vertices at distance r from a and s from b. If $N_{rs}(a,b)$ depends only on r,s and $d(a,b)$ we get an association scheme in the sense of [1]. We consider graphs satisfying a weaker hypothesis.

Theorem 5. *If G is connected and $N_{r1}(a,b)$ depends only on r,a, and $d(a,b)$ then*

(i) *the number of distinct eigenvalues is $D+1$ ($D = diam(G)$)*

(ii) *for all a, $\max_b d(a,b)$ is D or $D-1$*

(iii) *either $\max_b d(a,b)$ is D for all a or it gives a bicoloring of G*

(iv) *the only simple eigenvalue is the largest (± the largest if G is bipartite).*

Remark. If G is regular (i) is contained in [3].

Proof. For a vertex a, let S_a be the operator which replaces the value of a function at a point by the average of the values at all points having the same distance from a. Then the condition that $N_{r1}(a,b)$ depend only on r and $d(a,b)$ is equivalent to $S_aA = AS_a$. Hence if λ is an eigenvalue with eigenvector x, $AS_ax = S_aAx = \lambda S_ax$. Further if $x_a \neq 0$, $(S_ax)_a = X_a \neq 0$. If in addition $\lambda \neq 0$ and b is adjacent to a then $(S_a\boldsymbol{x})_a \neq 0$ implies $(S_ax)_b \neq 0$ which simplies S_ax is an eigenvector of AS_b. Hence every eigenvalue of A is an eigenvalue of some AS_a and every non-zero eigenvalue an eigenvalue of all of them.

Since A^D is not a linear combination of lower powers of A, the degree of the minimal polynomial of A (i.e. the number of distinct eigenvalues) is $\geq D+1$. Hence $1 + \max_b d(a,b) = \text{rank}(S_a) \geq \text{rank}(AS_a)$ $\geq$ number of distinct non-zero eigenvalues of $A \geq D$. This gives (i) and (ii).

For (iii) note that if $\max_b d(a,b) = D$ and $\max_b d(a',b) = D-1$, then $d(a,a') = 2$ for all x, $Ax = 0$ implies $(x)_{a'} = 0$ but for some y, $Ay = 0$, $(y)_a \neq 0$. Then $(S_ay)_{a'} = 0$ which is impossible.

Finally let $\lambda \neq 0$ be a simple eigenvalue with eigenvector x. If $(x)_a \neq 0$ all vertices adjacent to a have the same (non zero) value; otherwise $S_a x$ would be independent of x. Hence either $(x)_a$ is constant or G is bipartite and $(x)_a$ depends only on the part. In either case (iv) follows.

References

1. Bose, R.C. and Mesner, Dale M., On linear associative algebras corresponding to association schemes of partially balanced designs, *Ann. Math. Stat.*, 30 (1959), 21-38.

2. Gantmacher, F.R., Matrix Theory, Vol. II, Chelsea, 1959, New York.

3. Hofmann, A.J., On the polynomial of a graph, *Amer. Math. Monthly*, January, 1963, 30-36.

ON EXTREMAL PROBLEMS IN GRAPH THEORY

Vera T. Sós

Eotvös L. University, Budapest, Hungary

In this talk I am going to speak on results and problems in connection with the investigations concerning distance distribution in a point set A, discussed by P. Turán in his talk.

I shall use the following notations:

$G(P;E)$ is the graph with the set of vertices P and set of edges E

$C_k = C_k(P;E_c)$ is the complete graph, where $|P| = k$.

$G^r(P;G_1,G_2,\ldots,G_r)$ is a so-called r-class graph.

$G_i = G_i(P,E_i)$, where $E_i \cap E_j = \phi$ for $i \neq j$ and $\bigcup_{i=1}^{r} E_i = E_c$.

$H(k_1,\ldots,k_r)$ is the class of G^r r-class graphs, with the property, that $C_{k_\nu} \not\subseteq G_\nu$ for every $1 \leq \nu \leq r$.

$H_n(k_1,\ldots,k_r)$ is the subclass of it containing G^r with $|P| = n$.

Let n have the form $n = (k-1)m + \ell$ (n,k,m,ℓ non negative integers) and

$$t(n,k) \stackrel{\text{def}}{=} \frac{k-2}{2(k-1)}(n^2-\ell^2) + \binom{\ell}{2}$$

the Turán-numbers.

Let $R(k_1,\ldots,k_r)$ the Ramsey-number, defined as the smallest number having the property, that if

$$G^r(P;G_1,\ldots,G_r) \subset H(k_1,\ldots,k_r)$$

then $|P| \leq R(k_1,\ldots,k_r)$.

In order to obtain *lower* bound for the number of large distances in a point set, we recall the definition of the k'th covering-constant of a point-set A on the plane.

Let $K(r,P)$ be the disk with radius r and center P. We call $K_1(r,P_1),\ldots,K_k(r,P_k)$ a *covering-system* of A, if $P_i \in A$ for $1 \leq i \leq k$ and $A \subset \bigcup_{i=1}^{k} K_i(r,P_i)$. The *covering-constant* c_k of a set A is defined usually as the infimum of the r's with the above property.

From our point of view it's more appropriate to use the equivalent definition:

$$c_k \overset{\text{def}}{=} \inf_{(P_1,\ldots,P_k)} \sup_{Q} \min_{1 \leq i \leq k} \overline{QP_i}$$

where $Q \in A$, $P_i \in A$.

<u>Theorem</u>. *If $A = \{Q_1,\ldots,Q_n\}$ having k^{th} covering constant c_k, then at least*

$$e(n,k) \overset{\text{def}}{=} (k-1)(n-1) + \left[\frac{n-k+2}{2}\right]$$

among the distances $\overline{Q_iQ_j}$ $(1 \leq i < j \leq n)$ are not less, than c_k.

The result is sharp, as it is shown by the following example:

Let $n - k + 1$ even, $Q_1,\ldots,Q_{n-k+1}$ the vertices of a regular n-k+1-gon inscribed into $K(r,0)$, and let $Q_{n-k+2},\ldots,Q_n$ additional points with the property

$$\overline{Q_\nu 0} > 2r \qquad \text{for } n\text{-}k+2 \leq \mu,\ \nu \leq n.$$
$$\overline{Q_\nu Q_\mu} > r$$

For this point set $c_k = r$ and the number of distances, which are $\geq r$ is exactly $e(n,k)$.

The proof follows easily using the following theorem of P. Erdös - L. Moser:

<u>Theorem</u>. (P. Erdös-L. Moser): *If the graph $G(P,E)$ has the property, that for every $\{P_1,\ldots,P_k\} \in P$ there exists a $P^* \in P$ such that the edges $P_iP^* \in E$ for $1 \leq i \leq k$, then*

$$|E| \geq (k-1)(n-1) + \left[\frac{n-k+2}{2}\right].$$

The unique extreme graph has the property, that it has k-1 points with degree n-1.

From the point of view of application the following question arises: how fast the minimal possible value of $|E|$ increases, if we restrict the maximal degree. More exactly, if the graph $G(P,E)$ has the above property and the maximal degree in G is δ, what is the minimal possible value for $|E|$ depending on δ (and on k,n).

E.g. in the simplest case, when $\delta = n-2$, the following holds:

Theorem. *If the graph $G(P,E)$ has the property as in theorem of P. Erdős - L. Moser, and every vertex in P has degree $\leq n-2$, then*

$$|E| \geq k(n-k) + \binom{k}{2}.$$

The above graph problems are strongly connected with the following one for (0,1) matrices:

Let $A = \{a_{ij}\}$ an $n \times n$ symmetric (0,1) matrix with $a_{ii} = 0$ and with the property, that for $A^2 = \{b_{ij}\}$ we have $b_{ij} \geq k$. What is the minimal possible value of $\sum_{i=1}^{n} \sum_{i=1}^{n} a_{ij}$? Further, the same question under the additional condition $\sum_{i=1}^{n} a_{ij} \leq \delta$?

In order to get some results for the distribution of distances of a point set A using the packing-constants, refining the ideas of P. Turán, we proceed as follows:

For the case of simplicity we consider a point set A with diameter 1.

We divide the interval $[0,1]$ by the packing-constants*
$1 = d_2 \geq d_3 \geq \ldots > d_{r+1} > 0$. We define the r-class graph $G^r = G^r(P,G_1,\ldots,G_r)$ with $P = \{1,2,\ldots,n\}$ belonging to a point-set $\{Q_1,\ldots,Q_n\} \subset A$ by:

$$(i,j) \in E_\nu \Longleftrightarrow \overline{Q_iQ_j} \in [d_{\nu+1}, d_{\nu+2}) \quad \text{for } \nu \leq r-1$$
$$\in [d_{r+1}, 0] \quad \text{for } \nu = r.$$

Now from the definition of the numbers d_ν we get simultaneous structural conditions for the graph $G_1,\ldots,G_r$. The general question is, what are the possible values of $|E_1|,\ldots,|E_r|$ (under certain condition). This leads to the following simplest special but important graph problem:

Let $k_1,\ldots,k_r$ given positive integers. If G^r is an r-class graph having the property, that

---

*For the definition of the packing constants $d_2,d_3,\ldots$ see P. Turán: Applications of graph th. etc. in this volume.

$$C_{k_\nu} \not\subseteq G_\nu \qquad \text{for } 1 \leq \nu \leq r$$

then what is max $(E_1 + \ldots + E_{r-1})$?

In the case, when $k_r = n+1$ (i.e. there is no condition for G_r) it is easy to see, that

$$E_1 + \ldots + E_{r-1} \leq t(n, R(k_1,\ldots,k_{r-1}))$$

and the upper bound is sharp.

E.g. for $r = 3$, $k_1 = k_2 = 3$ the 3-class graph in $H_n(3,3, n+1)$ having the minimal number of edges in G_3 is the following: (for the sake of symplicity let 5 decide n).

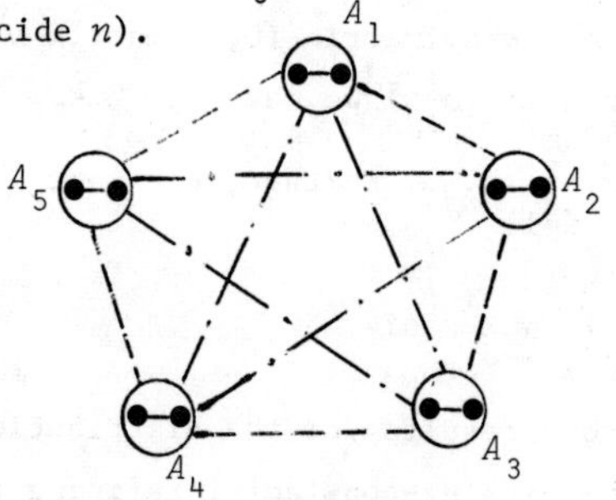

Let A_i $(1 \leq i \leq 5)$ be disjoint sets of the n vertices each having $\frac{n}{5}$ points. E_3 consists of all the edges with both vertices in the same set A_ν $(1 \leq \nu \leq 5)$. E_2 consists of all the edges (i,j) for which $i \in A_\nu$ and $j \in A_{\nu+1}$ $(A_6 \equiv A_1)$ and E_1 consists of all the remaining edges.

Already in the case $k = 2$ it seems to be very difficult to determine

$$f_n(k_1,k_2) \overset{\text{def}}{=} \max_{G^2 \subset H_n(k_1,k_2)} E_1 .$$

The remark, that (i) $f_n(k_1,n+1) = t(n,k_1)$ and (ii) $f_n(k_1,k_2) = o(n^2)$ for fixed k_1,k_2 shows the connection between Turán's and Ramsey's theorem. Namely the statement (i) is just Turán's theorem. While statement (ii) implies that for n large the class $H_n(k_1,k_2)$ must be empty, which is just Ramsey's theorem. So having some information on $f_n(k_1,k_2)$ Turán's and Ramsey's theorems are the consequences of the two extreme cases.

For the case, when k_1 is fixed and $k_2 = cn$ or $k_2 = o(n)$ for large enough, we have some results with P. Erdös which will appear later.

A FAMILY OF BIBDS

R.G. Stanton and D.G. Gryte

York University, Toronto, Ont., Canada

1. Introduction. While the results hold for a prime power, we shall for convenience restrict discussion to the case of a prime p. We represent the Galois Field on p^2 symbols, $GF(p^2)$, by adjoining a mark w to $GF(p)$ such that $w^2 = c$ (a non-square) for $p > 2$ (for $p = 2$, take $w^2 = w + 1$). Then the plane projective geometry $PG(2,p^2)$ consists of all triples with co-ordinates from $GF(p^2)$, proportional triples being identified; thus the first non-zero element in a triple may be taken as 1. We shall study the set of points on the curve C with equation $\Sigma\, x_i^{p+1} = 0$.

Since $(d+ew)^{p+1} = (d+ew^p)(d+ew) = (d-ew)(d+ew)$ for $d, e, \varepsilon\, GF(p)$, $p > 2$, with a suitable modification for the case $p = 2$, it follows that if $a \in GF(p^2)$, then $a^{p+1} \in GF(p)$. Also, if α is a primitive element of $GF(p^2)$, it follows that, for g fixed, $\alpha^{g+h(p-1)}$ produces the same $(p+1)$-th power for $0 \leqslant h \leqslant p$. We thus obtain

Lemma 1. Every non-zero element of $GF(p^2)$ is the $(p+1)$-th power of $p+1$ elements of $GF(p^2)$.

Now the curve C consists of points $(1,0,x_3)$, $(0,1,x_3)$, $(1,x_2,0)$, where $x_3^{p+1} = x_2^{p+1} = -1$, as well as points $(1,x_2,x_3)$, where $x_2^{p+1} = b \in GF(p)$, with $b \neq 0$, $b \neq -1$ (in this case, $x_3^{p+1} = -b-1$). By Lemma 1, there are $3(p+1)$ points of the first kind and $(p-2)(p+1)^2$ points of the second kind. Adding, we obtain

Lemma 2. There are p^3+1 points on the curve C.

2. Secants to the Curve C. If $A_1(x_{11},x_{12},x_{13})$ and $A_2(x_{21},x_{22},x_{23})$ are points of C, then $A_1 + \delta A_2$, $\delta \neq 0$, will lie on C if and only if $S + \delta^{p-1} S^p = 0$, where $S = \Sigma\, x_{2j}x_{1j}^{\ p}$. If $S = 0$, then every point $A_1 + \delta A_2$ lies on C, that is, the line A_1A_2 lies in C. Otherwise, there may be no

solution for δ. However, if there is a solution δ^* for δ, then there are exactly p-1 solutions, that is, $\delta^{*1+i(p+1)}$. We thus have

Lemma 3. If a line L through A_1 on C meets C again at A_2, then it contains exactly p-1 other points of C or lies entirely in C.

The second possibility in Lemma 3 disappears by a purely combinatorial argument. Suppose that all p^2+1 points of L lie on C; take D on C but not on L, and join D to all points A_i on L; the lines DA_i contain all points in the geometry. Suppose that r of these lines contain 2 points of C, t of them contain p+1 points of C, z of them contain p^2+1 points of C. Then we have:

$$r + t + z = p^2+1,\ 1 + r + pt + p^2z = p^3+1.$$

By subtraction, $z(p^2-1) + t(p-1) = p^2(p-1)-1$. Thus p-1 divides 1, and thi is not possible for $p > 2$. The case $p = 2$ is easily checked separately (there are just 9 points forming an $EG(2,3)$), and so we may state

Lemma 4. A line L through A_1 on C may meet C in 0,1, or p other points.

Now let r and t retain the same meaning as above, but suppose that th are u lines through A_1 which do not meet C again (these may be called the *tangents* at A_1). It is easy to check that $\Sigma\, x_{1j}^{\ p} x_j = 0$ is a tangent at A_1; hence $u \geq 1$. Also, we have:

$$r + t + u = p^2+1,\ 1 + r + tp = p^3+1.$$

It follows that $r+tp = p^3$, $r+t \leq p^2$. Hence, we find that $t(p-1) \geq p^2(p-1)$, and thus $t \geq p^2$. As a result, we have $t = p^2$, $r = 0$, $u = 1$. This gives

Lemma 5. Each of the p^2 secants through A_1 on C meets C in exactly p othe points.

3. Design Properties of the Curve C. We now form a balanced incomplete block design by selecting the points of C as varieties and the secants to C as the blocks. The results already established show that we have

Lemma 6. If points are selected as varieties, secants as blocks, then C generates a BIBD with parameters $(p^3+1,\ p^2(p^2-p+1),\ p^2,\ p+1,\ 1)$.

The above methods, with variations, produce interesting configuratio by the use of other fields such as $GF(p^3)$. J.J. Seidel has drawn our attention to a paper by R.C. Bose, *On the Application of Finite Projective Geometry for Deriving a Certain Series of Balanced Kirkman Arrangements*, Golden Jubilee Commem. Vol., Calcutta Math. Soc. (1958-59), 341-354, in wh by somewhat different arguments, the results given for $GF(p^2)$ are also established.

SOME COMBINATORIAL PROBLEMS IN GENERAL TOPOLOGY

A.H. Stone

University of Rochester, Rochester, N.Y., U.S.A.

Many of the combinatorial problems in point-set topology are connected with the notion (due to Alexandroff) of the "nerve" of a cover. For simplicity I shall take "cover" to mean "finite open cover" throughout, though generalizations to infinite open covers or closed covers usually present no difficulty. Sometimes the 1-dimensional part of the nerve (consisting of the 0- and 1-cells) suffices, and the problem is one in graph theory. This is often the case, for example, in the theory of unicoherent and multicoherent spaces (see for instance [4, Th. 8], [5, Ths. 2 and 3]). Another case is conjectured to arise in the theory of "density", as follows.

A topological space X is said to have "density $\leqslant d$" if it has arbitrarily fine covers U such that each set in U meets at most d others. There is an obvious resemblance here to Lebesgue's definition of dimension, and it is an old conjecture [3] that an n-dimensional normal space has density at least $2^{n+1} - 2$. This conjecture has been proved (by Boltyanskii [1]; see also [6] for further results) only for $n \leqslant 2$. But it would follow from the following graph-theoretic conjecture.

(*) For each $n = 1,2,\ldots$, there exists $k(n)$ with the following property: if G is a (proper, finite) linear graph in which each vertex has degree $\leqslant 2^{n+1} - 3$, then G can be vertex-colored so that (a) whenever $n + 1$ vertices form a complete subgraph of G, at least two of them have the same color, (b) no color is used more than $k(n)$ times. (The number of colors is unrestricted.)

For suppose (*) holds, for a particular n, and let X be a normal space of dimension n but with density $d < 2^{n+1} - 2$. There is a cover V

of X such that for every refinement of V there is some $x \in X$ belonging to $n+1$ sets of the refinement. Take a $(k(n)+\ )$-fold star-refinement U of V in which each set meets at most d others, and let G be the 1-dimensional part of the nerve of U. Thus each vertex of G has degree $\leqslant d$. According to (*) we obtain a vertex coloring of G, using (say) N colors 1,2,... N. We may suppose that, for each $i = 1,2,\ldots, N$, the vertices colored i generate a connected subgraph of G; for if not, we replace this subgraph by its components and color each component with a new color; this does not affect (a) or (b). Let W_i ($i = 1,2,\ldots, N$) be the union of those members of U which correspond to vertices in the ith color class. Then $\{W_1,\ldots, W_N\}$ is a cover of X which (because of (b) and the connectedness of the color classes) refines V. But (because of (a)) no point of X can belong to $n+1$ of the sets W_i, contradicting the choice of V.

The conjecture (*) is trivial for $n = 1$, but unproved for $n = 2$. It is at any rate "best possible"; there exist graphs G in which the vertices have degrees $\leqslant 2^{n+1} - 2$ which cannot be colored as in (*). We may obtain such G's from the nerves of "efficient" covers (roughly speaking) of the unit cube in Euclidean n-space.

A rather different, and very well known, application of combinatorial methods to topology is Sperner's Lemma. Several generalizations of it have been given (for instance in [2]). The following result is a partial generalization in a different direction.

Theorem. *Given positive integers n, p, k, let an $(n-1)$-simplex $\sigma = (v_1, v_2,\ldots, v_n)$ be subdivided simplicially; and suppose that to each vertex w of the subdivision we assign a set of p or fewer of the numbers $1,2,\ldots, n$ in such a way that if w lies in a face $(v_{i_1},\ldots, v_{i_r})$ of σ then its assigned numbers are among $i_1,\ldots, i_r$. Then there is a $(k-1)$-simplex of the subdivision such that no two of its k vertices have a number assigned in common, provided $n \geqslant 4p^3(k-1)^2$.*

The final proviso could be somewhat improved, but the method does not seem to give the "best" result, which I conjecture should be $n \geqslant kp$. The case $p = 1$ would then give the original Sperner lemma. (The conjecture is also true when $p = 2 = k$, $n \geqslant 4$.)

Proof. For each vertex w of the subdivision K of σ, let $f(w)$ denote the assigned subset of $\{1,2,\ldots, n\}$, with $|f(w)| \leqslant p$. Consider the barycentr subdivision K^* of K. Each vertex c of K^* is the centroid of a cell (say

$(w_1, w_2, \ldots, w_r)$ of K. Consider the corresponding subsets $f(w_i)$ $(1 \leq i \leq r)$ of $\{1,2,\ldots, n\}$. For each $j = 1,2,\ldots, n$, let λ_j = number of i's $\leq r$ for which $j \in f(w_i)$; pick j so that λ_j is as large as possible, and define $g(c) = j$. Then g satisfies the conditions of Sperner's Lemma (applied to K^*), so there exists an $(n-1)$-cell $(c_1, c_2, \ldots, c_n)$ such that the numbers $g(c_i)$, $i = 1,\ldots\ldots, n$, are all different.

From the nature of the barycentric subdivision, there is an $(n-1)$-cell $(w_1, w_2, \ldots, w_k)$ of K such that (if the c_i's are suitably renumbered) $c_1 = w_1$, c_2 is the centroid of (w_1, w_2), and so on; finally, c_n is the centroid of $(w_1, w_2, \ldots, w_n)$.

We may assume $n = 4p^3(k-1)^2$, for otherwise we merely replace σ by a lower-dimensional face. We write $s = 4p^2(k-1) - 2$, and consider two cases.

(i) If none of $1,2,\ldots, n$ belongs to $f(w_i)$ for more than s values of i, we observe that each $f(w_i)$ can meet at most $p(s-1)$ others. But, writing $q = p(s-1) + 1$, we see that $n > q(k - 1)$. Thus we can select successively k pairwise disjoint sets $f(w_i)$, and obtain the desired $(k-1)$-cell of K.

(ii) In the remaining case, we first simplify the notation (by permuting $v_1, v_2, \ldots, v_n$) so as to obtain $g(c_i) = i$ $(i = 1,2,\ldots, n)$. In particular, $g(c_n) = n$; by definition this means that, if j belongs to $f(w_i)$ for λ_j i's then $\lambda_n \geqslant \lambda_j$ $(j = 1,2,\ldots, n)$. Because (i) is false, $\lambda_n \geqslant s + 1 = 4p^2(k-1) - 1$. Then n belongs to at least s of the sets $f(w_1), \ldots, f(w_{n-1})$; hence, by definition of $g(c_{n-1})$, so does $n-1$. Similarly we see that, for $r = 1,2,\ldots, n-1$, each of the $r+1$ numbers $n, n-1, \ldots, n-r$ belongs to at least $s - r + 1$ of the sets $f(w_1), f(w_2), \ldots, f(w_{n-r})$. Since each of these $n-r$ sets has at most p elements, we have $(r + 1)(s - r + 1) \leqslant p(n - r) < pn$. Apply this with $r = s/2 = 2p^2(k - 1) - 1$; we obtain $4p^4(k - 1)^2 < pn$, contradicting our assumption $n = 4p^3(k-1)^2$. Thus case (ii) cannot arise, and the proof is complete.

References

1. V. Boltyanskii, On a property of 2-dimensional compacta, *Doklady Akad. Nauk SSSR* 75 (1950), 605-608.

2. Ky Fan, Extensions of two fixed point theorems of F.E. Browder, a talk given at the Conference on Set-Valued Mappings, Selections, and Topological Properties of 2^X, S.U.N.Y. at Buffalo, May 1969.

3. L. Lichtenbaum, Sur un invariant topologique, *C.R. Acad. Sci. Paris* 193 (1931),1307-1308.

4. A.H. Stone, Incidence relations in unicoherent spaces, *Trans. Amer. Math. Soc.* 65 (1949), 427-447.

5. A.H. Stone, Incidence relations in multicoherent spaces II, *Canadian J. Math.* 2 (1950), 461-480.

6. A.H. Stone, On coverings of 2-dimensional spaces, *Proc. London Math. Soc.* (3) 3 (1953), 338-349.

COMBINATORIAL STRUCTURES IN GALOIS DOMAINS

Thomas Storer

University of Michigan, Ann Arbor, MI, U.S.A.

Consider the primes 5 and 13, respectively, so that g.c.d. (5-1, 13-1) = 4. Noting that 2 is a common primitive root of both 5 and 13, we order the multiplicative cosets of the fourth-power in $\mathbb{Z}_5$ and $\mathbb{Z}_{13}$ on 2 as follows:

$$C_0^{(5)} = \{1\} \qquad\qquad C_0^{(13)} = \{1, 3, 9\}$$

$$C_1^{(5)} = 2 \cdot C_0^{(5)} = \{2\} \qquad\qquad C_1^{(13)} = 2 \cdot C_0^{(13)} = \{2, 5, 6\}$$

$$C_2^{(5)} = 2^2 \cdot C_0^{(5)} = \{4\} \qquad\qquad C_2^{(13)} = 2^2 \cdot C_0^{(13)} = \{4, 10, 12\}$$

$$C_3^{(5)} = 2^3 \cdot C_0^{(5)} = \{3\} \qquad\qquad C_3^{(13)} = 2^3 \cdot C_0^{(13)} = \{7, 8, 11\}.$$

Now, if $C_4^{(5)}$ is the (4 by 4)-matrix whose (i,j) entry is the number of times an element of $C_i^{(5)}$ is followed by an element of $C_j^{(5)}$, we directly verify that

$$C_4^{(5)} = \begin{bmatrix} 0&1&0&0\\ 0&0&0&1\\ 0&0&0&0\\ 0&0&1&0 \end{bmatrix}, \qquad C_4^{(13)} = \begin{bmatrix} 0&1&2&0\\ 1&1&0&1\\ 0&1&0&1\\ 1&0&1&1 \end{bmatrix}.$$

We now define the *star-product* $C_4^{(5)} * C_4^{(13)}$ of these two matrices; leaving $C_4^{(13)}$ alone, we operate on $C_4^{(5)}$ by sending its initial column "right" to become its last column (in which case the new matrix is placed to the right of the old) or by sending its initial row "below" to become its last row (in which case the new matrix is placed below the old). By combinations of these two operations, we generate the following *scheme*:

$$C_4^{(5)} = \begin{array}{ccccccc}
\begin{bmatrix}0100\\0001\\0000\\0010\end{bmatrix} & \rightarrow & \begin{bmatrix}1000\\0010\\0000\\0100\end{bmatrix} & \rightarrow & \begin{bmatrix}0001\\0100\\0000\\1000\end{bmatrix} & \rightarrow & \begin{bmatrix}0010\\1000\\0000\\0001\end{bmatrix} \\
\downarrow & & \downarrow & & \downarrow & & \downarrow \\
\begin{bmatrix}0001\\0000\\0010\\0100\end{bmatrix} & \rightarrow & \begin{bmatrix}0010\\0000\\0100\\1000\end{bmatrix} & \rightarrow & \begin{bmatrix}0100\\0000\\1000\\0001\end{bmatrix} & \rightarrow & \begin{bmatrix}1000\\0000\\0001\\0010\end{bmatrix} \\
\downarrow & & \downarrow & & \downarrow & & \downarrow \\
\begin{bmatrix}0000\\0010\\0100\\0001\end{bmatrix} & \rightarrow & \begin{bmatrix}0000\\0100\\1000\\0010\end{bmatrix} & \rightarrow & \begin{bmatrix}0000\\1000\\0001\\0100\end{bmatrix} & \rightarrow & \begin{bmatrix}0000\\0001\\0010\\1000\end{bmatrix} \\
\downarrow & & \downarrow & & \downarrow & & \downarrow \\
\begin{bmatrix}0010\\0100\\0001\\0000\end{bmatrix} & \rightarrow & \begin{bmatrix}0100\\1000\\0010\\0000\end{bmatrix} & \rightarrow & \begin{bmatrix}1000\\0001\\0100\\0000\end{bmatrix} & \rightarrow & \begin{bmatrix}0001\\0010\\1000\\0000\end{bmatrix}
\end{array}$$

We now take the *inner-product* of each of these (4 by 4)-matrices with $C_4^{(13)}$, to obtain

$$\begin{array}{ccccccc}
3 & \rightarrow & 0 & \rightarrow & 2 & \rightarrow & 4 \\
\downarrow & & \downarrow & & \downarrow & & \downarrow \\
0 & \rightarrow & 4 & \rightarrow & 2 & \rightarrow & 2 \\
\downarrow & & \downarrow & & \downarrow & & \downarrow \\
2 & \rightarrow & 2 & \rightarrow & 2 & \rightarrow & 2 \\
\downarrow & & \downarrow & & \downarrow & & \downarrow \\
4 & \rightarrow & 2 & \rightarrow & 2 & \rightarrow & 0
\end{array}$$

and this ordered (4 by 4)-array we write as

$$C_4^{(5)} * C_4^{(13)} = \begin{bmatrix} 3 & 0 & 2 & 4 \\ 0 & 4 & 2 & 2 \\ 2 & 2 & 2 & 2 \\ 4 & 2 & 2 & 0 \end{bmatrix}$$

To see that this bizarre matrix product has some relation to reality, we look at the domain $\mathbb{Z}_{5\cdot 13} = \mathbb{Z}_{65}$. Here we note that

$$27 \equiv 2 \pmod 5), \qquad 27 \equiv 1 \pmod{13}$$

and we order the multiplicative cosets of the powers of 2 in $\mathbb{Z}_{65}$ as follows:

$$C_0^{(65)} = \{1, 2, 4, 8, 16, 32, 33, 49, 57, 61, 63, 64\}$$
$$C_1^{(65)} = 27 \cdot C_0^{(65)} = \{11, 19, 21, 22, 23, 27, 38, 42, 43, 44, 46, 54\}$$
$$C_2^{(65)} = 27^2 \cdot C_0^{(65)} = \{7, 9, 14, 18, 28, 29, 36, 37, 47, 51, 56, 58\}$$
$$C_3^{(65)} = 27^3 \cdot C_0^{(65)} = \{3, 6, 12, 17, 24, 31, 34, 41, 48, 53, 59, 62\}.$$

Now, if $\mathrm{C}_4^{(65)}$ is the (4 by 4)-matrix whose (i,j)-entry is the number of times an element of $C_i^{(65)}$ is followed by an element of $C_j^{(65)}$ we directly verify that

$$\mathrm{C}_4^{(65)} = \begin{bmatrix} 3 & 0 & 2 & 4 \\ 0 & 4 & 2 & 2 \\ 2 & 2 & 2 & 2 \\ 4 & 2 & 2 & 0 \end{bmatrix},$$

and hence that $\mathrm{C}_4^{(65)} = \mathrm{C}_4^{(5)} * \mathrm{C}_4^{(13)}$.

The example above is no fluke, in that it is contained in a general theorem relating the class structure of $\mathbb{Z}_{pq}$ to the structure of $\mathbb{Z}_p$ and $\mathbb{Z}_q$:

<u>Theorem 1</u>. *If g is a fixed common primitive root of the odd primes p and q for which* g.c.d. $(p-1,q-1) = e$, *then the matrices* $\mathrm{C}_e^{(p)}$, $\mathrm{C}_e^{(q)}$ *and* $\mathrm{C}_e^{(pq)}$ *associated with g are related by*

$$\mathrm{C}_e^{(pq)} = \mathrm{C}_e^{(p)} * \mathrm{C}_e^{(q)}.$$

Further, the entries of $\mathrm{C}_e^{(p)}$ are known for

$$e = 2, 3, 4, 5, 6, 8, 10, 12, 14, 16, 18, 20 \quad \text{for all } p,$$

and hence, by Theorem 1, the entries of $\mathrm{C}_e^{(pq)}$ are also known for these (even) e and all corresponding p, q.

Now, writing

$$Q = \{0, q, 2q, 3q, \ldots, (p-1)\, q\}$$

for the multiples of q in $\mathbb{Z}_{pq}$, a theorem of Whiteman [1] states:

<u>Theorem 2</u>. *The set $D_e^{(pq)} = C_0^{(pq)} + Q$ is a difference set in $\mathbb{Z}_{pq}$ if and only if $q = (e-1)p + 2$ and the entries of the initial column of $\mathrm{C}_e^{(pq)}$ are constant.*

Thus the problem of the existence of difference sets $D_e^{(pq)}$ (e even) in the domains $\mathbb{Z}_{pq}$ is solved for $e = 2, 4, 6, 8, 10, 12, 14, 16, 18, 20$ by Theorem 1, modulo an horrendous computation.

Reference

1. Whiteman, A. L., A family of difference sets. *Illinois J. Math.*, 6 (1962), 107-121.

GENERALIZED CHROMATIC NUMBERS

Walter Taylor

University of Colorado, Boulder, CO, U.S.A.

Let R be an n-ary relation on the (finite or infinite) set A, and let ρ be an equivalence relation on $\{1,\ldots,n\}$. An equivalence relation $\equiv$ on A will be called *permissible* iff

$$(\forall x_1)\ldots(\forall x_n)\bigwedge_{i\rho j}(x_i \equiv x_j) \Rightarrow \sim R(x_1,\ldots,x_n).$$

The ρ-*chromatic number* $\chi_\rho(R)$ is the least cardinal M such that there is a permissible equivalence relation partitioning A into M blocks. If no permissible equivalence relation exists, then we say $\chi_\rho(R) = \infty$. A *coloring* of A in M colors is a map $C:A \to M$ such that the induced equivalence relation is permissible.

This notion of chromatic number was introduced in [3] to solve a problem in model theory. The notion also seems to have applications in combinatorics. For example, suppose that one is to seat a set of k people at various tables for a banquet, and that certain sets of 3 persons are inimical. The problem of finding the minimum number of tables required reduces to finding the chromatic number of the relation "inimical", where $1 \rho 2 \rho 3$. For further specific instantiation of this notion, consult the works of Hedetniemi, Kronk and Rosa [*].

We propose that in the finite case, at least, the problem of chromatic numbers can be replaced by a simpler problem in the study of equivalence relations. If a set E of equivalence relations on A contains with each θ, every refinement of θ, we say that E is *downward-closed*.

Proposition 1. E is downward-closed iff for some n, R, and ρ, E is the set of permissible equivalence relations on A.

Fixing A, n, R and ρ, we define $p(\lambda)$ to be the number of distinct ways of coloring A in λ colors. We let b_j ($1 \leq j \leq k = \overline{\overline{A}}$) denote the number of permissible equivalence relations partitioning A into j blocks. We let $x^{(m)}$ denote the polynomial $x(x-1)\ldots(x-m+1)$.

Proposition 2. $p(\lambda) = b_k\lambda^{(k)} + b_{k-1}\lambda^{(k-1)} + \ldots + b_1\lambda^{(1)}$.

Thus the number of colorings is a polynomial, the *chromatic polynomial* of A. Much of the theory of chromatic polynomials of graphs does *not* carry over to these chromatic polynomials. In particular, the chromatic polynomial in this case is not necessarily the characteristic polynomial of a geometric lattice [2]. For example, the polynomial $p(\lambda) = \lambda^3 - \lambda$ is the chromatic polynomial of a certain 3-element structure. One can, however, calculate these chromatic polynomials by coalescing vertices and adding "edges," as outlined in [1].

We do not know a good criterion for deciding when a polynomial is a chromatic polynomial. If θ is a partition into blocks B_i having b_i elements, we let $\lambda^\theta = \Pi\lambda^{b_i}$, where the product is formed by treating the polynomials $\lambda^{(j)}$ as powers. And we let $\lambda^{\theta_1} \cdot \lambda^{\theta_2}$ denote the polynomial $\lambda^{\theta_1 \wedge \theta_2}$.

Proposition 3. A polynomial is chromatic iff it is of the form $1 - \Pi(1 - \lambda^\theta)$, where the product is over any set of equivalence relations on A.

References

1. Read, R. C., An Introduction to Chromatic Polynomials, *J. Comb. Theory* 4 (1968), 52-71.

2. Rota, G. C., On the Foundations of Combinatorial Theory I. Theory of Möbius Functions, *Z. Wahrscheinlichkeitstheorie und Verwandte Gebiete* 2 (1963), 340-368.

3. Taylor, W., Compactness and Chromatic Number, *Fund. Math.* (to appear)

*. This volume.

APPLICATIONS OF GRAPH THEORY TO GEOMETRY AND POTENTIAL THEORY

Paul Turan

Eotvös L. University, Budapest, Hungary

Professor G. Sabidussi wrote in his review of the colloquium lectures held in Snolenice in 1963, printed in Canad. Math. Bull. 1966 the following remarkable lines.

"... Those who proclaim the usefulness of graphtheory for the rest of mathematics would perform a genuine service if they took the existing body of graphtheoretical theorems and made a list of instances of applicability or actual applications of such theorems in other parts of mathematics. I suspect that this list would be short ...".

The aim of this talk is not to furnish a required list but - in complete agreement of the intentions of Professor Sabidussi - to extend a bit the fields of applications even beyond those indicated in the title. As to the *existing* applications in the literature I confine myself to one remark only. These applications use mostly Ramsey's theorem; typical examples are the "geometrical Waring problem" solved by Erdös and Szekeres and another one in approximation theory found by Vera T. Sós. As the former one shows particularly clearly the application of Ramsey's theorem furnishes rather rough upper bounds on the quantities in question, very little beyond mere existence. In what follows we intend also to show that graphtheory can produce results in other subject which are *best possible* showing that their role in the subject is *genuine* (which is no wonder since the part of the graphtheory mainly relevant for applications belongs essentially to logic).

The first geometrical problem we have in mind first is the following. Let H be a geometrical condition for finite pointsets in E_k (k-dimensional euclidean space) satisfying the following postulates only*

a) boundedness (in the sense that the pointsets after suitable translation can be shifted into a universal bounded domain).

b) closedness (in the sense that if the pointsets $(P_{1\nu}, P_{2\nu}, \ldots, P_{l\nu})$ $\nu = 1,2,\ldots$ satisfy the condition H and $P_{j\nu} \to P_j$ $(j = 1,\ldots,l)$ then the set $(P_1,\ldots,P_l)$ satisfies it too).

c) continuity (in the sense that if the set $P_1,\ldots,P_l$ of different points satisfies the condition H and the integers $k_1,k_2,\ldots,k_l$ are given then to arbitrarily small neighbourhoods of $P_1,\ldots,P_l$ one can place $k_1,k_2,\ldots,k_l$ different points so that the new pointset satisfies the condition H too.**

The problem now is what can be said on the distribution of the mutual distances $\overline{P_\mu P_\nu}$ $(\mu \neq \nu)$ of our pointset $(P_1,P_2,\ldots,P_\mu)$?

In such generality the problem seems to be at first quite hopeless; one cannot even see through what quantities belonging to condition H this distribution can be hoped to be governed. Now we have found that such constants are the so called packing constants d_l belonging to the condition H, which are defined for $l \geq 2$ by

$$d_l = d_l(H) = \sup \min_{1 \leq \mu \leq \nu \leq l} \overline{P_\mu P_\nu} \tag{0}$$

where *sup* is taken to all $(P_1,\ldots,P_l)$ sets $(P_\mu \neq P_\nu)$ satisfying condition H. The name "packing constant" can be justified most easily in the case when the condition H means that our points are on the surface of the unitsphere. In this case namely when d_l is realized by the system $(P_1^*,\ldots,P_l^*)$, the spherical caps with centers at $P_\nu^*(\nu = 1,\ldots,l)$ and radii $\frac{1}{2} d_l$ realize obviously the densest packing of the surface of a unit sphere by l congruent spherical caps. The packing constants were investi gated since Newton and Gregory; their dispute boils down to the question whether $d_{13}^* < 1$ (Newton) or $d_{13}^* \geq 1$ (Gregory) where d_l^* mean the packing constants belonging to the surface of the unit sphere. The intensive

---

* I laid as usual no stress to state the postulates in their most general form

** Such conditions H are e.g. restriction to a bounded domain, or to an l-dimensional surface $(l < k)$, or max distance = 1, etc.

investigation of packing constants started with the work of Minkowski and Thu; their close influence to the distance distribution however seems to be unnoticed so far. We have clearly

$$d_2 \geqq d_3 \geqq \dots \geqq d_l \geqq \dots \to 0; \tag{1}$$

important role is played also by the indices $i_1 < i_2 < \dots$ defined by $i_1 = 1$ and

$$d_2 = d_3 = \dots = d_{i_2} > d_{i_2+1} = \dots = d_{i_3} > d_{i_3+1} = \dots \tag{2}$$

Then we can state the following

<u>Theorem 1</u>. *If $l \geqq 2$ and $n > i_l$ then, if the points $P_1, P_2, \dots, P_n$ satisfy the condition H, the number of distances $\overline{P_\mu P_\nu}$ $(\mu \neq \nu)$ satisfying the inequality*

$$\overline{P_\mu P_\nu} \leqq d_{i_{l+1}} \quad (= d_{i_l+1}) \tag{3}$$

is at least

$$\frac{n^2}{2i_l} - \frac{n}{2} \tag{4}$$

Increasing the quantity in (4) *by* 1 *the theorem is no more true.*

In order to give immediately an interesting application of this theorem let the condition H be that

$$\max_{1 \leqq \mu, \nu \leqq \mu} \overline{P_\mu P_\nu} = 1 \quad \text{in} \quad E_k. \tag{5}$$

Then we have obviously

$$d_2^{(k)} = d_3^{(k)} = \dots = d_{k+1}^{(k)} = 1$$

denoting the packing constants of this particular H-condition by $d_l^{(k)}$. As was shown recently by Schoenberg* and Seidel*

$$d_{k+2}^{(k)} = \begin{cases} \sqrt{\dfrac{k}{k+2}} & \text{if } k \text{ is even} \\[2ex] \sqrt{\dfrac{k^2+2k-1}{k^2+4k+3}} & \text{if } k \text{ is odd} \end{cases} \tag{6}$$

Hence in our case

$$i_1 = 1, \qquad i_2 = k+1$$

and applying Theorem 1 with $l = 2$ we get that for $n > k+1$ at least

---

* J. F. Schoenberg, Linkages and distance geometry I, II,
J. J. Seidel, Quasi-regular two distance sets, Proc. of the *Koningl. Nederlandse Akad. van Wetensch.*, Ser. A, Vol. LXXII, p. 43-63 resp. p. 64-70.

$$\frac{n^2}{2k+2} - \frac{n}{2}$$

distances are $\leqq d_{k+2}^{(k)}$ and this is best possible.

A perhaps more striking formulation of this fact is that having a system of n different points with maximal distance 1 so that more than

$$\frac{kn^2}{2k+2} \tag{7}$$

distances are $> d_{k+2}^{(k)}$ then the system cannot be isometrically embedded into E_k. Having just $\frac{kn^2}{2k+2}$ points the conclusion is no more true (as can be shown by putting m different points "near" to each vertex of a k-dimensional simplex with edge length 1 inside the simplex then we have

$$\binom{k+1}{2} m^2 = \binom{k+1}{2}\left(\frac{n}{k+1}\right)^2 = \frac{kn^2}{2k+2}$$

distances which are even greater than $1-\varepsilon$).

The theorem in (7) can perhaps be extended to more general classes of metric spaces.

For further aims we reformulate a bit theorem 1 as

Theorem 1'. *For integer $l \geqq 2$, arbitrarily small $\varepsilon > 0$ and $n > n_0(\varepsilon,H)$ at least $(\frac{1}{i_l} - \varepsilon)^{th}$ part of the mutual distances $\overline{P_\mu P_\nu}$ $(\mu \neq \nu)$ is $\leqq d_{i_{l+1}}$.*

Theorem 1' is again best possible in the sense that replacing $(\frac{1}{i_l} - \varepsilon)$ by $(\frac{1}{i_l} + \eta)$ with a *fixed* $\eta > 0$ the theorem is no more true.

One can give a still stronger form of the theorem 1'. For this sake let us define the "lower distance distribution function $f_H(\theta)$ belonging to condition H" defined for $0 < \theta \leqq d_2$ by

$$f_H(\theta) = \lim_{n\to\infty} \min \frac{1}{\binom{n}{2}} \sum_{\overline{P_\mu P_\nu} \leqq \theta} 1 \tag{8}$$

where the minimum refers to all systems $(P_1,\ldots,P_n)$ satisfying condition H with fixed n. Theorem 1' states that

$$f_H(d_{i_{l+1}}) = \frac{1}{i_l} \qquad l = 1,2,\ldots \tag{9}$$

We can state the

Theorem 1". *The function $f_H(\theta)$ is always a stepfunction, continuous from the right whose only jumps are at $\theta = d_{i_{l+1}}$ $(l = 1,2,\ldots)$ and given by (9*

In order to show that theorem 1 gives quite concrete geometrical theorems I shall enumerate some examples. I shall make use of theorem 1'; I shall use however the expression "$\frac{1}{i_l}$ th part" for abbreviation.

a) Condition H: let $k = 2$ and the points on the periphery of the unitcircle. Then we have evidently

$$d_l = 2 \sin \frac{\pi}{l} \qquad l = 2,3,\ldots$$

i.e., $i_l = l$. Hence we obtained that having n points on the periphery of the unitcircle, at least $\frac{1}{l}$ th part of the distances is $\leqq 2 \sin \frac{\pi}{l+1}$ and this is best possible especially for $l = 5$ at least $\frac{1}{5}$ th part of the distances is $\leqq 1$.

b) Condition H: let $k = 2$ and maximal distance is 1. Then Erdös and Bateman determined the first seven packing constants for completely different aims* in 1951. We mention among them only

$$d_2 = d_3 = 1,\ d_4 = \frac{\sqrt{2}}{2},\ d_7 = \frac{1}{2},$$

and also

$$i_v = v + 1 \qquad v = 2,3,4,5,6.$$

Hence in our case

$$\text{at least } \frac{1}{i_2} = \frac{1}{3} \text{ rd part of the distances is } \leqq d_{i_3} = d_4 = \frac{\sqrt{2}}{2}, \tag{10}$$

$$\text{at least } \frac{1}{i_5} = \frac{1}{6} \text{ th part of the distances is } \leqq d_{i_6} = d_7 = \frac{1}{2}. \tag{11}$$

and all best possible.(10) constitutes the only known result in this direction, it is due to P. Erdös, who published it as a problem in Elemente der Mathematik in 1955.

c) Perhaps the most classical case is when the condition H is that the points are located on the surface of the unitsphere in E_3. Beyond the classically known values for $l \leqq 6$ (which show that

$$d_2^* > d_3^* > d_4^* > d_5^* = d_6^*) \tag{12}$$

van der Waerden and Schutte found the exact values up to 9 (which show incidentally

$$d_6^* > d_7^* > d_8^* > d_9^*). \tag{13}$$

It is known from estimations that

$$d_9^* > d_{10}^* > d_{11}^*, \text{ further } d_{12}^* > d_{13}^*.$$

---

* Bateman, P. T. and Erdös, P., Geometrical extrema suggested by a lemma of Besicovitch, *Amer. Math. Monthly*, Vol. 58, (1951), 306-314.

It is an upward conjecture that

$$d^*_{11} = d^*_{12}. \tag{14}$$

From all these one has in the case

$$i_1 = 1,\ i_2 = 2,\ \ldots,\ i_4 = 4$$

$$i_5 = 6,\ i_6 = 7,\ \ldots,\ i_{10} = 11$$

and

$$i_{11} = \begin{cases} 13 \text{ if the conjecture is true} \\ 12 \text{ if not.} \end{cases}$$

Hence theorem 1 gives for $l = 10$ that for $n \geqslant 12$ at least

$$\frac{n^2}{2i_{10}} - \frac{n}{2} = \frac{n^2}{22} - \frac{n}{2} \tag{15}$$

distances are

$$\leqslant d^*_{i_{11}} = \begin{cases} d^*_{13} \text{ if the conjecture (14) is true} \\ d^*_{12} \text{ if not.} \end{cases} \tag{16}$$

A perhaps not uninteresting consequence of (15) - (16) is that if for an $n \geqslant 12$ suitably chosen $(P^*_1,\ldots,P^*_n)$ system more than $\frac{10}{22}n^2$ distances are $> d^*_{13}$ then the conjecture (14) is false*. It is perhaps not without interest to realise that graphtheory has in such a problem to say something at all.

d) Let the condition H be restriction to the unitsquare resp. unitcube. For these cases as well as to the cases when the points are located on the periphery of the unitsquare resp. unitcube Professor A. Meir communicated to me the exact values of several packing constants belonging to these H-conditions; with their aid one can prove among others, e.g., that in the case of unitsquare

at least $\frac{1}{3}$ rd of the distances are $\leqslant 1$

at least $\frac{1}{8}$ th of the distances are $\leqslant \frac{1}{2}$

and in the case of unitcube

at least $\frac{1}{7}$ th of all distances are $\leqslant 1$,

all these being best possible.

---

* Since d^*_{13} is not exactly known, only upper and lower bounds for it, it is better to replace d^*_{13} by $\tilde{d}_{13}$ in the assertion where $\tilde{d}_{13}$ means any known upper bound for d^*_{13}.

All these referred to *lower* bounds for the number of distances $\lessgtr v$. In order to obtain possibly *upper* bounds it was plausible to think that covering-constants will play an essential role instead of packing constants. I - and some geometers too - did not succeed in finding the *proper* definition of these general covering constants; it was Vera P. Sós who succeeded in it. Her results in this direction are contained in her talk.

Incidentally the above mentioned theorems suggest many purely geometrical problems concerning packing constants. All these amount to the problem how far packing constants can be prescribed beyond (1) (problem of existence); if there is a condition H with prescribed packing constants at all, what is the minimal dimension in which it can be realised; problems of uniqueness in the obvious sense, various problems with equality sign, e.g., does there exist in prescribed positive integer $l_1 + 2 \lessgtr l_2$ a condition H such that $d_{l_1}(H) < d_{l_1+1}(H) = \ldots = d_{l_2}(H) < d_{l_2+1}(H)$, etc.

Leaving aside temporarily the geometrical applications we turn to potentials. Putting in the points $P_1, \ldots, P_n$ in E_2 unitmasses the potential energy of the system under Newton law of force is given by

$$\sum_{\mu \neq \nu} \frac{1}{r_{\mu\nu}} \tag{17}$$

$r_{\mu\nu}$ denoting the distance of P_μ and P_ν. In E_2 the role of $\frac{1}{x}$ is taken by $\log \frac{1}{x}$, in E_k for $k \gtrless 3$ by x^{2-k}. Replacing these special functions by a monotonically decreasing $g(x)$ and having uniformly distributed mass with density 1 over a k-dimensional Jordan measurable domain V the generalized potential energy of the system is given by

$$I(g,V) = \int_V \int_V g(\overline{PQ}) d_{v_P} d_{v_Q} \tag{18}$$

denoting by d_{v_P} the k-dimensional volume element at P. It is intuitively clear from (17) that splitting the sum into partial sums of type

$$\sum_{d_{l-1} < r_{\mu\nu} \lessgtr d_l} \frac{1}{r_{\mu\nu}}$$

and applying theorem 1' we can obtain a lower estimation of the potential energy. Passage to limit gives the

Theorem 2*. *If the domain V is bounded and Jordan measurable with k-dimensional volume $|V|$ then the inequality*

$$\frac{1}{|V|^2} \int_V \int_V g(\overline{PQ}) d_{v_P} d_{v_Q} \gtrless \sum_{l=2}^{\infty} \frac{g(d_l)}{(l-1)l}$$

---

* For $g(x) \equiv 1$ we have obviously equality.

holds. Here $g(x)$ is monotonically decreasing and d_l stands for the packing constants of V.

Some interest is given to this inequality by the fact that the lower bound depends only on values of g taken on places independent of g reminding one to formulae of mechanical quadrature.

Let us consider as condition H the points being restricted to a bounded closed set S in E_2. As well known the transfinite diameter $\Delta(S)$ is defined by

$$\lim_{n\to\infty} \max \left(\prod_{1\leq\mu<\nu\leq n} \overline{P_\mu P_\nu}\right)^{\frac{1}{\binom{n}{2}}}$$

where the maximum is taken over all $(P_1,\ldots,P_n)$ systems belonging to S. Then theorem 1 leads to the result that the divergence of the series

$$\sum_l \frac{1}{l^2} \log \frac{1}{d_l} \tag{19}$$

implies $\Delta(S) = 0$. As Erdös showed by an ingenious example the result is best possible in the sense that the divergence of the series

$$\sum_l \frac{1}{l^{2-\varepsilon}} \log \frac{1}{d_l} \qquad \varepsilon > 0 \tag{20}$$

does *not* imply $\Delta(S) = 0$. Further if S is a simply connected closed domain then the so called conformal radius R being equal to $\Delta(S)$ satisfies the inequality

$$R \leq \prod_{l=2}^{\infty} d_l^{\frac{1}{(l-1)l}} \tag{21}$$

and dropping the d_l's with $d_l < 1$ we have *upper* bounds for the conformal radius by the first few packing constants.

We may ask now what all these have to do with graphtheory? The essential point in the proof of theorem 1 is the following graphtheorem* I found in 1940.

Let $3 \leq k \leq n$ and the integer t and h are uniquely determined by

$$n = (k-1)\, t + h, \qquad 0 \leq h \leq k-2. \tag{22}$$

Then the maximal number of edges in a graph G_n with n vertices not containing a complete subgraph of order k is

---

* Egy gráfelméleti szélsöértékfeladat, *Matematikai és Fizikai Lapok*, vol. 49, (1941) 436-452 (in Hungarian with German abstract). An English translation of my original proof is given in the Appendix of my paper "On the theory of graphs", Colloq. Math. (1954).

$$\frac{k-2}{2(k-1)} (n^2-h^2) + \binom{h}{2}; \tag{23}$$

equality is only for the graph G_n^* where the vertices are distributed into $(k-1)$ disjoint classes possibly equally and two vertices are connected by an edge if and only if they belong to different classes.

We shall not give detailed proofs of the results in this paper; these - together with proofs of other results in this connection - will be given in a later paper with P. Erdös, A. Meir and Vera P. Sós. But we may shortly indicate why the theorem in (22) - (23) is so successful in applications. The box principle of Dirichlet can be formulated as the assertion of a complete subgraph in a special graph; hence all theorems concluding from some data of a graph to the existence of a complete subgraph can be considered as more elastic than the box principle. So theorem (22) - (23) too.

Returning to the applications I mention two quite recent applications of the graph-theorem in (22) - (23), found by G. Katona. First he asked that given the d-dimensional vectors $w_1,\dots,w_n$ with $|w_j| \geqq 1$ for at least how many pairs (w_μ, w_ν) $\mu \neq \nu$ we may assert that

$$|w_\mu + w_\nu| \geqq 1, \tag{24}$$

or equivalently what is the probability of (24). He proved that denoting the minimal number of pairs by $E(n)$ one has

$$E(n) = \begin{cases} \frac{n}{2}\left(\frac{n}{2} - 1\right) & \text{if } n \text{ is even} \\ \left(\frac{n-1}{2}\right)^2 & \text{if } n \text{ is odd} \end{cases} \tag{25}$$

independently of d. That (25) cannot be improved can be shown easily choosing "half" of the vectors +1 and "half" as -1*. Suitable passage to limit led lim to the following two theorems.

If $f_\nu(x)$ $(\nu = 1,\dots,n)$ belong to $L_2(a,b)$ and

$$\int_a^b |f_\nu(x)|^2 \, dx \geqq 1 \qquad (\nu = 1,\dots,n) \tag{26}$$

---

* One might remark that - with above convention - for arbitrary positive integer $k \geqq 3$ at least $\left(\frac{1}{k-1}\right)^{\text{th}}$ part of pairs of plane vectors w_j with $|w_j| \geqq 1$ satisfies the inequality $|w_\mu + w_\nu| \geqq 2\cos\frac{\pi}{k}$. This is probably *not* best possible for $k \geqq 4$.

then for at least $E(n)$ pairs $(\mu,\nu,\mu\neq\nu)$ the inequality

$$\int_a^b |f_\mu(x) + f_\nu(x)|^2 \, dx \geqq 1 \tag{27}$$

holds.

If ξ and η are vector-valued independent random variables with the same distribution then

$$P(|\xi+\eta| \geqq x) \geqq \tfrac{1}{2} P(|\xi| \geqq x)^2. \tag{28}$$

Dr. Katona will publish his results in separate papers.

The last general geometrical problem we are going to discuss is the following. Let in E_k be given n different points $P_1,\ldots,P_n$ and a $\phi(P_\mu P_\nu P_\lambda)$ "triangle functional" for $1 \leqq \mu < \nu < \lambda \leqq n$, such as the area, the periphery, etc. Now if the points satisfy a geometrical condition H what can be said on the value-distribution of the functional ϕ? In this case it is plausible to define the lth "packing-constant" δ_l for $l \geqq 3$ by

$$\delta_l = \sup_{P_1\ldots P_l} \min_{1\leqq\mu<\nu<\lambda\leqq l} \phi(P_\mu P_\nu P_\lambda) \tag{29}$$

where the sup refers to all $(P_1,\ldots,P_l)$ systems satisfying the condition H. Then for continuous ϕ say

$$\delta_3 \geqq \delta_4 \geqq \ldots \to 0.$$

We assert then the

Theorem 3. *If the function ϕ is continuous with respect to all vertices and the condition H is as before then for all $l \geqq 3$ at least $\frac{1}{\binom{l}{3}}$ th part of the triangles Δ satisfies the inequality*

$$\phi(\Delta) \leqq \delta_l. \tag{30}$$

Here the $\frac{1}{\binom{l}{3}}$ is not best possible. In order to improve this percentage one has to extend the theorem in (22) - (23) to hypergraphs. This problem - which I raised already in my above quoted paper from 1941 - runs as follows. Let $m \leqq k \leqq n$; call a collection of combinations $(1 \leqq)\ i_1 < i_2 < \ldots < i_m\ (\leqq n)$ an m-graph, n as the order of our m-graph, each of the above combinations as an m-edge, a subcollection of the above m-edges an m-subgraph, further if all m-edges of k vertices occur in our m-graph we call it a complete m-subgraph of order k. Then the problem asks what is the maximal number of m-edges in an m-graph of order n if it does not contain any complete m-subgraph of order k. The problem is not solved until none for $m \geqq 3$; for $m = 2$ this is given by the theorem in (22) - (23). I conjecture that at least for

$$(m-1)/(k-1) \tag{31}$$

one "extremal m-graph" is given by the following construction. We divide the vertices into $\frac{k-1}{m-1}$ disjoint classes "possibly equally" and take all m-edges with *not all* vertices in the same class. In the proof of theorem 3 we used instead of theorem (22) - (23) the following weaker theorem of Katona - Nemetz - Simonovits*.

For $2 \leqq m \leqq k \leqq n$ all m-graphs of order n having at least

$$\left\{1 - \frac{1}{\binom{k}{m}}\right\} \binom{n}{m} \tag{32}$$

m-edges contains certainly a complete m-subgraph of order K.

In order to give some concrete geometrical content let us consider first the case when the condition H is that the

$$\sup_{1 \leqq \mu < \nu < \lambda \leqq n} \text{area of } P_\mu P_\nu P_\lambda = 1. \tag{33}$$

In this case evidently $\delta_3 = \delta_4 = 1$ whereas one can prove that

$$\delta_5 = \frac{\sqrt{5}-1}{2}. \tag{34}$$

Hence theorem 3 gives with $m = 3$, $l = 5$ that at least $(\frac{1}{10})^{th}$ of the triangles have an area $\leqq \frac{\sqrt{5}-1}{2}$ if only (33) is satisfied. If the conjecture (31) is true then $(\frac{1}{10})^{th}$ could be replaced by $(\frac{1}{4})^{th}$ but even this would not be best possible.

In the case when the condition H is that

$$\sup_{1 \leqq \mu < \nu < \lambda \leqq n} \text{periphery of } P_\mu P_\nu P_\lambda = 1 \tag{35}$$

the situation is a bit different as remarked by Professor Meir. Here one has again $\delta_3 = \delta_4 = 1$ and one can easily show, $\delta_5 < 1$. Theorem 3 gives again that at least $(\frac{1}{10})^{th}$ of the triangles have a periphery $\leqq \delta_5$. If conjecture (31) is true then again $\frac{1}{10}$ can be replaced by $(\frac{1}{4})$th and this would be best possible. Namely let $\varepsilon > 0$ be arbitrarily small and $\overline{P_1P_2} = \frac{1}{2} - 2\varepsilon$. Then we can put m points "near" to P_1 and m points "near"

---

* G. Katona, T. Nemetz and M. Simonovits, Ujabb Bizonyitás a Turán Féle Gráftételre és Megjegyzések Bizonyos Általánositásaira, (in Hungarian with English abstract) *Matematikai Lapok* (15), p., 228-238.

to P_2 (m large) so that (35) is satisfied and we have exactly (with $n = 2m$)

$$\binom{n}{3} - 2\binom{m}{3} \sim \frac{3}{4}\binom{n}{3}$$

triangles even with a periphery $> 1-\varepsilon$. This shows that even for triangle functionals the graphtheoretical method can lead to *best possible* results.

Analogous problems and theorems could have been formulated and proved for higher dimensions too.

Returning finally to the relationship of the graphtheory and its applications in most cases it is graphtheory → applications in other subjects. This conference illustrated among others that also the inverse effect can occur; the problems concerning best *upper* bounds for the distance distribution and energy integrals let to numerous interesting *graph* problems some of which were dealt with in the talk of Vera P. Sós. Not wishing to dwell much upon the interaction of graphtheory and its applications I want to remind you of the situation of some rich families which degenerates quickly having only intermarriages in order to keep the family fortune together. If a theory, whatever beautiful, becomes too introverted, there is a danger it will loose soon its freshness, its problems become of secondary importance. New relationships bring new blood in, breed significant new problems. Another reason to propogate the tendency to increase the interaction between graphtheory and other parts of mathematics is that this seems to be the best way to eliminate the old stale aperçu, "Graphtheory is the slum of topology" from the mathematical public life. A success of this tendency will be to a large extent the result of the fertile criticism of Professor Sabidussi.

COMPLEX HADAMARD MATRICES

Richard J. Turyn

Raytheon, Sudbury, MA, U.S.A.

We are concerned with square matrices H satisfying $HH^* = nI$, whose entries are fourth roots of 1. A theorem of Williamson [5] is equivalent to the following pair of assertions:

1. If S is a matrix with 0 main diagonal, ±1 entries elsewhere, such that $S = S^1$, $S^2 = (n-1)I$ then $S + iI$ is Hadamard. (Such matrices arise from the quadratic character in $GF(n-1)$ when $n-1$ is a prime power $\equiv 1 \pmod 4$; others have been constructed by Goethals and Seidel [1].)

2. If H is a complex Hadamard matrix of order n and H_m a real Hadamard matrix of order m, then there exists a real Hadamard matrix of order mn. (If T is a real monomial matrix of order m such that $T^1 = -T$, $T^2 = -I$, and $H = X + iY$, we take $H_m \times X + TH_m \times Y$.)

The famous conjecture that there exists a real Hadamard matrix of order $4t$ for any integer t would follow from the existence of a complex Hadamard matrix of order $2t$.

(If A and B are real Hadamard matrices which arise from complex ones A_1 and B_1, then a real matrix arises from $A_1 \times B_1$ whose order is half that of $A \times B$.)

<u>Theorem</u>. *There exist (symmetric) Hadamard circulants of orders* 2^t, $t = 1, 2, 3, 4$. *There are no Hadamard circulants of order* 2^t, $t > 4$ *or* $2p^n$, *p an odd prime.*

The sequences: $1, i$; $-1,1,1,1$; $1,1,i,-1,1,-1,i,1$ of lengths 2,4,8 and the sequence $(i^{F(m)})$, $0 \leq m \leq 15$, $F(m) = \left[\frac{m+2}{4}\right]\left(m + 2 - \left[\frac{m+2}{4}\right]\right)$ (the matrix i^{jk} read from the middle of the first row) define the circulants.

If (x_n) defines a circulant of order $2p^n$, p a prime then $|\sum x_m \zeta^{km}|^2 = 2p^n$, ζ a primitive $2p$-th root of 1, any k. $k = 0$ shows $p^n \equiv 1 \pmod 4$. If $(g,p) = 1$ we have $\sum x_m \zeta_p^m = w \sum x_m \zeta_p^{gm}$, w a root of 1, ζ_p a primitive p-th root of 1, since $\sum x_m \zeta_p^m$ and $\sum x_m \zeta_p^{gm}$ have the same absolute value and prime ideal factorization. The factor of 2 is not affected since $\sum y_m \zeta_p^m = \sum (y_m - y_0) \zeta_p^m$, $y_m = x_m + x_{p+m}$, and $1+i \mid y_m - y_0$, $|1+i| = \sqrt{2}$. We can assume $w^4 = 1$ by replacing x_m by x_{m-a} for suitable a. Let g be a primitive root mod p. If $w = 1$, we have $y_m = y_1$, $0 < m < p$, so $\sum x_m \zeta_p^m = y_0 - y_1$, and it is easily seen that this is impossible. If $w \neq 1$, $\sum_1^{p-1} (y_m - y_0) = 0$, so $\sum y_n = py_0$ which is impossible. If $n > 1$, we can again assume that $\sum y_m \zeta^m = w \sum y_m \zeta^{gm}$, ζ a primitive p^n-th root of 1, $w^4 = 1$. It follows easily that $\sum y_m \zeta^m \in Q(i,\zeta_p)$. By reducing this equation using the equation for ζ over $Q(i,\zeta_p)$, we find an expression of absolute value $\sqrt{2p^n}$ expressed as a difference of two terms of absolute value $\leqslant 2$, so $2p^n \leqslant 16$, $p^n \leqslant 8$, which proves the theorem. (2^t, $t > 4$ is in [4].)

Other non-existence theorems for complex circulants can be proved, similarly to those for binary circulants ([3], [4]). It is likely none exist with the exception of those of order 2, 4, 8, 16.

A "good" form for matrices is

$$\begin{bmatrix} A & B \\ -B^* & A^* \end{bmatrix} \tag{1}$$

This includes constructions of the complex numbers and quaternions from the reals and complex numbers respectively, the extremal determinants found by Yang [6] of orders $\equiv 2 \pmod 4$ with A, B ± 1 circulants, and the low order Hadamard matrices. We consider Hadamard matrices of form (1), with A, B in the group algebra of a finite abelian group, e.g., complex circulants. We must have $AA^* + BB^* = 2nI$, i.e., $\sum_n (a_n \bar{a}_{n+j} + b_n \bar{b}_{n+j}) = 0$ for all $j \neq 0$. If the group is cyclic and A and B are symmetric, then expanding the above matrix to real form using the 2×2 real Hadamard matrix gives the Williamson quaternion type (see [2]). The two forms are equivalent: if X, Y are two of the real matrices in the Williamson type the corresponding complex matrix is $\frac{1}{2}(X + Y + i(X+Y))$. (1) allows us to remove some of the symmetry requirements: the real sequences of length 5 and 13 for which $\sum x_i x_{i+j} = 1$ for $j \neq 0$, taken with the sequence defined by the quadratic character and $x_0 = i$ give examples of matrices of form (1); the one of length 13 is not symmetric.

If A and B are circulants of order n such that (1) is Hadamard, C is any of the four circulants of the theorem above, then $A \times C$ and $B \times C$ are also. If A and B are symmetric, so are $A \times C$ and $B \times C$, and if n is odd, the group on which $A \times C$ is defined is cyclic. This generalizes the theorem of Williamson [5] for $t = 1$.

References

1. Goethals, J. M. and Seidel, J. J., Orthogonal Matrices with Zero Diagonal, *Can. J. Math.* 19 (1967), 1001-1010.

2. Hall, Marshall, Jr., Combinatorial Theory, Waltham, Mass. 1967.

3. Turyn, R., Character Sums and Difference Sets, *Pac. J. Math.*, 15 (1965), 319-346.

4. Turyn, R., Sequences with Small Correlation, in Error Correcting Codes, ed. by H. B. Mann, Wiley, 1968.

5. Williamson, J., Hadamard's Determinant Theorem and the Sum of Four Squares, *Duke Math. J.*, 11 (1944), 65-81.

6. Yang, C. H., On Designs of Maximal (+1,-1) Matrices of Order $n \equiv 2 \pmod 4$, II. *Math. Comp.*, 23 (1969), 201-205.

MORE ABOUT CHROMATIC POLYNOMIALS AND THE GOLDEN RATIO

W. T. Tutte

The University of Waterloo, Ont., Canada

Summary. Let M be a triangulation of the 2-sphere, with k vertices. Let $P(M,n)$ be its chromatic polynomial with respect to vertex-colourings. In this paper it is shown that

$$P(M,\tau+2) = (\tau+2)\ \tau^{3k-10}\ P^2(M,\tau+1),$$

where τ is the "golden ratio" $(1 + \sqrt{5})/2$.

The paper includes an introduction to the theory of chromatic polynomials.

1. Vertex-colourings.

We denote the numbers of vertices, edges and components of a graph G by $\alpha_0(G)$, $\alpha_1(G)$ and $p_0(G)$ respectively.

An n-colouring of G is a mapping of the vertex-set $V(G)$ of G into a set of n distinct elements called "colours" in such a way that each edge of G is incident with vertices of two different colours. We assume G to be finite. The number of n-colourings of G is a non-negative integer dependent on both G and n. We denote it by $P(G,n)$. It is defined for all non-negative integers n.

We admit the existence of a null graph Ω, having no edges or vertices. We consider that this graph has exactly one n-colouring for each non-negative integer n.

1.1. $$P(\Omega,n) = 1.$$

Evidently only the null graph can have a 0-colouring. It is clear from the definitions that a graph has a 1-colouring if and only if it is edgeless.

The two following results are obvious from the definitions.

1.2. *If G has a loop, then*

$$P(G,n) = 0.$$

1.3. *If G is edgeless, then*

$$P(G,n) = n^{\alpha_0(G)}.$$

If $\alpha_0(G) = n = 0$ in 1.3 the expression on the right is to be interpreted as unity, to achieve consistency with 1.1.

Consider the complete m-graph K_m. By definition it has exactly m vertices, it has no loops and each pair of distinct vertices are joined by exactly one edge. Thus $K_0 = \Omega$. If $m \geq 1$ and we are constructing an n-colouring for K, then we have n choices for the colour of the first vertex, $\max(0,n-1)$ choices for the colour of the second, $\max(0,n-2)$ for the colour of the third, and so on. We deduce that

1.4. $$P(K_m,n) = n(n-1)(n-2) \ldots (n-m+1).$$

If $m = 0$ in 1.4 we interpret the empty product on the right as unity, and the formula reduces to 1.1.

Let H be a subgraph of G. Then any n-colouring X of G induces an n-colouring Y of H in the sense that each vertex of H has the same colour in Y as in X. If C is a particular n-colouring of H we write $N(G,H,C)$ for the number of n-colourings of G inducing C.

1.5. *Let G have a subgraph H that is a complete m-graph. Let n be an integer not less than m, and let C be any n-colouring of H. Then*

$$P(G,n) = P(H,n)\, N(G,H,C).$$

Proof. If $m = 0$ the theorem is trivially true since every n-colouring of G induces the single n-colouring of the null graph K_m. If $P(G,n) = 0$ then $N(G,H,C) = 0$ and again the theorem is trivially true.

In the remaining case $m \geq 1$ and $P(G,n) \geq 0$. We can find an n-colouring F of G and an n-colouring C of H induced by F. Let C' be any other n-colouring of H. We can find a permutation θ of the n colours that transforms F into an n-colouring θF of G inducing C'. Evidently θ sets up a 1-1 correspondence of the set of n-colourings of G inducing C onto the set of n-colourings of G inducing C'. Thus

$$N(G,H,C') = N(G,H,C).$$

We deduce that

1.5.1. *$N(G,H,C)$ has the same value for all n-colourings C of the complete m-graph H.*

Since each n-colouring of G induces one of H Proposition 1.5 follows.

In the next theorems we relate values of $P(G,n)$ for different graphs G. If a graph G has an edge A we write G'_A for the graph obtained from G by deleting A. We also write G''_A for the graph obtained from G by first deleting A and then identifying the two ends of A, if these do not already coincide. Thus if A is a loop we have $G'_A = G''_A$.

1.6. *Let G be the union of two subgraphs H and K whose intersection L is a complete m-graph. Then*

$$P(G,n)\ P(L,n) = P(H,n)\ P(K,n).$$

Proof. It is clear that the n-colourings of G are in 1-1 correspondence with the ordered pairs $\{C_H, C_K\}$, where C_H is an n-colouring of H, and C_K is an n-colouring of K inducing the same n-colouring of L as does C_H. Hence, applying 1.5.1 to K, we have

$$P(G,n) = P(H,n)\ N(K,L,C),$$

for an arbitrary n-colouring C of L. We may assume that such a C exists since otherwise $P(G,n) = P(H,n) = P(L,n) = 0$ and the theorem is trivially true. This implies $n \geq m$, by 1.4. Hence

$$P(G,n)\ P(L,n) = P(H,n)\ P(K,n),$$

by 1.5, applied to K.

1.7. *Let A be any edge of G. Then*

$$P(G,n) = P(G'_A,n) - P(G''_A,n).$$

Proof. It is clear from the definition of an n-colouring that the n-colourings of G are those n-colourings of G'_A for which the ends x and y of A have different colours. It is also clear that the n-colourings of G''_A are in 1-1 correspondence with those n-colourings of G'_A for which x and y have the same colour. The theorem follows.

1.7.1. *If G has a second edge B with the same ends as A, then*

$$P(G'_A,n) = P(G,n).$$

Proof. In this case G''_A has a loop B. Hence 1.7.1 follows from 1.7 and 1.2.

2. Chromatic polynomials.

Consider a graph G with k vertices.

2.1. *$P(G,n)$ can be expressed as a polynomial in n with integral coefficients depending only on G. If G is loopless the polynomial is of degree k, and its leading term is n^k.*

Proof. If G has a loop the theorem is a trivial consequence of 1.2. Assume therefore that G is loopless.

If $\alpha_1(G) = 0$ the theorem is a consequence of 1.3. Assume as an inductive hypothesis that it is true whenever $\alpha_1(G)$ is less than some positive integer q and consider the case $\alpha_1(G) = q$.

G has an edge A, and

$$P(G,n) = P(G'_A,n) - P(G''_A,n)$$

by 1.7. But G'_A and G''_A have each fewer edges than G. They satisfy the theorem by the inductive hypothesis. But

$$\alpha_0(G''_A) < \alpha_0(G) = \alpha_0(G'_A) = k.$$

Moreover G'_A has no loop. We deduce that $P(G,n)$ is a polynomial in n with integral coefficients depending only on G. It has the same degree k as $P(G'_A,n)$ and its leading term is the leading term n^k of $P(G'_A,n)$.

The theorem follows in general by induction.

As a polynomial $P(G,n)$ has a value for each real or complex number n. From now on we regard it as a polynomial function of a complex variable n, though in this paper we shall be concerned only with certain values of n on the real axis. We call it the *chromatic polynomial* of G.

The theorems of Section 1, other than 1.5 and its corollary 1.5.1, can be regarded as polynomial identities valid for all non-negative integral values of n. It follows that they hold for all values of n; otherwise we could construct an algebraic equation, of positive degree, having infinitely many solutions.

Hassler Whitney has determined the coefficients in $P(G,n)$ in terms of the subgraphs of G. For example the coefficient of n^{k-1}, for a loopless G, is found to be $-\alpha_1(G)$. Another result of this kind runs as follows.

2.2. *If G is non-null the constant term of $P(G,n)$ is zero.*

Proof. The constant term is $P(G,0)$, the number of 0-colourings of G.

2.3. *$P(G,n)$ divides by $n^{p_0(G)}$.*

Proof. If G is null then $p_0(G) = 0$ and the theorem is trivial. In the remaining case let the components of G be enumerated as G_1, G_2, ..., G_p, where $p = p_0(G) \geq 1$. By repeated application of 1.6, with $m = 0$, we have

$$P(G,n) = \prod_{j=1}^{p} P(G_j,n).$$

But each $P(G_j,n)$ divides by n, by 2.2. The theorem follows.

3. An inequality

Let G be a loopless graph. It is well-known that the coefficients in $P(G,n)$ alternate in sign. This can be proved by induction over $\alpha_1(G)$, using 1.7. From this we can deduce that when n is negative $P(G,n)$ is non-zero and has positive or negative sign according as $\alpha_0(G)$ is even or odd. Here we establish a somewhat stronger result.

We observe first that there is a polynomial $Q(G,n)$ in n such that

$$P(G,n) = n^{p_0(G)} Q(G,n).$$

This is merely a restatement of 2.3.

3.1. *If $n < 1$, then for a loopless graph G*

$$(-1)^{\alpha_0(G)+p_0(G)} Q(G,n) > 0.$$

Proof. If possible let G be such that the theorem fails and $\alpha_1(G)$ has the least value consistent with this condition.

We have $\alpha_1(G) > 0$, by 1.3. Choose an edge A of G. There are two possibilities according as A is or is not an isthmus, that is an edge whose deletion increases $p_0(G)$. We have $p_0(G_A') = p_0(G)$ if A is not an isthmus, and $p_0(G_A') = p_0(G) + 1$ if A is an isthmus. However G_A'' always has the same number of components as G.

Write

$$R(G) = (-1)^{\alpha_0(G)+p_0(G)} Q(G,n),$$

and similarly for G_A' and G_A''. If G_A'' has a loop we write $Q(G_A'',n) = 0$.

Suppose first that A is not an isthmus. Then

$$P(G,n) = P(G_A',n) - P(G_A'',n),$$

by 1.7. Since $p_0(G) = p_0(G_A') = p_0(G_A'')$ we infer that

$$Q(G,n) = Q(G_A',n) - Q(G_A'',n).$$

Since $\alpha_0(G) = \alpha_0(G'_A) = \alpha_0(G''_A) + 1$ it follows that

$$R(G) = R(G'_A) + R(G''_A).$$

But the expression on the right of this equation is positive by the choice of G, since G'_A is loopless. Hence the theorem is true for G, which is a contradiction.

We may now assume that A is an isthmus. Then G can be represented as the union of two disjoint subgraphs H and K, together with a complete 2-graph L, with edge A, having one vertex x in H and one vertex y in K. Applying 1.6 we deduce that

$$n^2P(G,n) = P(H,n)\ P(K,n)\ P(L,n),$$

$$nP(G''_A,n) = P(H,n)\ P(K,n).$$

Since $P(L,n) = n(n-1)$, by 1.4, we deduce that

$$P(G,n) = (n-1)\ P(G''_A,n),$$

$$Q(G,n) = (n-1)\ Q(G''_A,n),$$

$$R(G) = (1-n)\ R(G''_A).$$

But G''_A is now loopless. Hence the expression on the right of the last equation is positive, by the choice of G. But $R(G)$ is not positive, again by the choice of G. This contradiction establishes the theorem.

3.1.1. *If G is non-null, loopless and connected, then the coefficient of n in $P(G,n)$ is non-zero.*

Proof. If this corollary were false we would have $Q(G,0) = 0$, contrary to the main theorem.

It is possible to make inferences from 3.1 to some real values of n exceeding 1. An example follows.

Let τ denote the "golden ratio" $(1 + \sqrt{5})/2$. It is the positive solution of the quadratic equation

$$\tau^2 = \tau + 1.$$

From this equation we deduce

3.2. $$\tau + 1 = \tau^2,\ \tau - 1 = \tau^{-1},\ \tau - 2 = -\tau^{-2}.$$

In later sections we shall be interested in the case $n = \tau + 1$, and we shall apply the following result.

3.3. *If G is loopless, then $P(G,\tau+1) \neq 0$.*

Proof. If the theorem fails we have

$$P(G(3 + \sqrt{5})/2) = 0.$$

But $P(G,n)$ is a polynomial with integers as coefficients. Hence we must have also

$$P(G(3 - \sqrt{5})/2) = 0.$$

But this is contrary to 3.1 since

$$0 < (3 - \sqrt{5})/2 < 1.$$

4. Maps on the sphere.

In this section we suppose G to be a planar graph, realized in the 2-sphere without crossings. If G is connected it defines a map M on the sphere. The vertices and edges of G are called vertices and edges of M respectively and the components of the remainder of the sphere are called the *faces* of M. If each face of M is bounded by a simple closed curve, necessarily made up of the edges and vertices of a circuit of G, we shall call M a *proper* map. A proper map in which the boundary of each face is made up of the edges and vertices of a triangle in G is a *triangulation* of the sphere.

We shall be interested chiefly in triangulations. However we shall find it necessary in some proofs to consider graphs in the sphere that may not define proper maps, and need not even be connected.

A *double triangle* in G is a subgraph H of G defined by four distinct vertices v_1, v_2, v_3, v_4 and five edges v_1v_2, v_2v_3, v_3v_4, v_4v_1, v_1v_3, the last of which we also denote by A. The first four edges define a quadrilateral Q of H. One further condition is imposed: that residual domain of Q in the sphere that contains the diagonal A must contain no other edge or vertex of G. We call this domain the *inside* of Q and the other one the *outside*.

Given such a double triangle H we define certain operations transforming G into other graphs on the sphere. We denote them by θ_A, ϕ_A and ψ_A.

θ_A replaces A by another diagonal B joining v_2 and v_4 across the inside of Q. It transforms H into a double triangle K of $\theta_A(G)$ with diagonal B.

ϕ_A contracts the inside and simple closed curve of Q into a single arc $v_2v_1v_4$ in such a way as to identify v_1 with v_3, v_2v_1 with v_2v_3 and v_1v_4 with v_3v_4. The operation ψ_A similarly contracts Q and its inside into a single arc $v_1v_2v_3$. Each of these operations leaves the surface topologically a sphere.

The effect of each of the three operations on the subgraph H is shown in Figure 1.

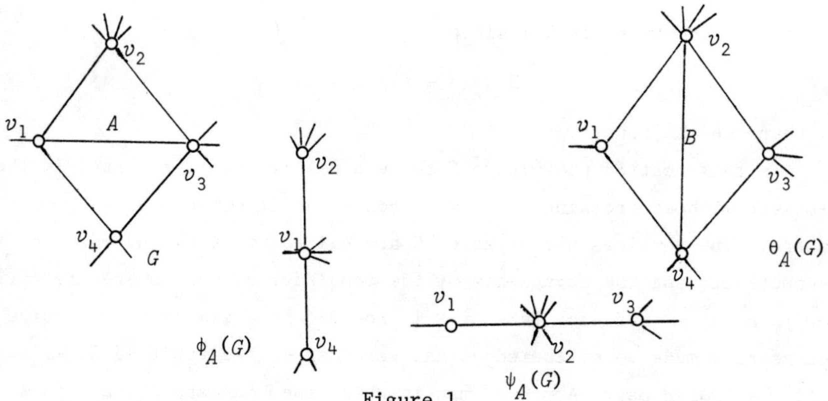

Figure 1.

By 1.7.1 we have

4.1. $$P(\phi_A G,n) = P(G''_A,n),\ P(\psi_A G,n) = P((\theta_A G)''_B,n).$$

We may operate similarly on $\theta_A G$ using the double triangle K and its diagonal B. We may suppose θ_B to replace B by the diagonal A. We then have

4.2. $$\theta_B(\theta_A G) = G,$$
$$\phi_B(\theta_A G) = \psi_A G,$$
$$\psi_B(\theta_A G) = \phi_A G.$$

4.3. *Let A be the diagonal of a double triangle H in G. Then, in the above notation,*

$$P(G,n) - P(\theta_A G,n) = P(\psi_A G,n) - P(\phi_A G,n).$$

Proof.

$$\begin{aligned} P(G,n) + P(\phi_A G,n) &= P(G'_A,n), && \text{by 1.7 and 4.1,}\\ &= P((\theta_A G)'_B,n) \\ &= P(\theta_A G,n) + P((\theta_A G)''_B,n), && \text{by 1.7,}\\ &= P(\theta_A G,n) + P(\psi_A G,n), && \text{by 4.1.} \end{aligned}$$

This formula can be used for the recursive calculation of the chromatic polynomials of triangulations. For if G is a triangulation then each of the graphs $\theta_A G$, $\phi_A G$ and $\psi_A G$ either has a loop or is a triangulation.

When $n = \tau + 1$ there is a second linear relation between the four polynomials.

4.4. *Let A and H be as in* 4.3. *Then*

$$P(G,\tau+1) + P(\theta_A G,\tau+1) = \tau^{-3}(P(\psi_A G,\tau+1) + P(\phi_A G,\tau+1)).$$

<u>Proof</u>. If possible let G be such that the theorem fails, $\alpha_0(G)$ has the least value consistent with this condition, and $\alpha_1(G)$ has the least value consistent with this.

Assume first that G has a vertex v not belonging to H.

The vertex v may be isolated, that is have no incident edge. In this case let J be the graph obtained from G by deleting v. Then J is realized on the sphere without crossings and it has H as a double triangle. Moreover the deletion of v from $\theta_A G$, $\phi_A G$ and $\psi_A G$ gives the graphs $\theta_A J$, $\phi_A J$ and $\psi_A J$ respectively. Hence

$$P(G,\tau+1) + P(\theta_A G,\tau+1) - \tau^{-3}(P(\psi_A G,\tau+1) + P(\phi_A G,\tau+1))$$

$$= (\tau+1) \{P(J,\tau+1) + P(\theta_A J,\tau+1) - \tau^{-3}(P(\psi_A J,\tau+1) + P(\phi_A J,\tau+1))\},$$

by 1.6,

$$= 0$$

since the theorem is true for J.

If v is not isolated let C be one of its incident edges. The contraction of C to a single point can be carried out by a deformation of G confined to the neighbourhood of C. It can be arranged that this operation, like the simple deletion of C, introduces no crossings and does not alter Q or its inside (except that one of the vertices v_i of Q may now have to be regarded as comprising all the identified points of C and its pair of ends). The graphs G'_C and G''_C each have H as a double triangle, and the theorem is true for them by the choice of G.

$$P(G,\tau+1) + P(\theta_A G,\tau+1) - \tau^{-3}(P(\psi_A G,\tau+1) + P(\phi_A G,\tau+1))$$

$$= \{P(G'_C,\tau+1) + P((\theta_A G)'_C,\tau+1) - \tau^{-3}(P((\psi_A G)'_C,\tau+1) + P((\phi_A G)'_C,\tau+1))$$

$$- \{P(G''_C,\tau+1) + P((\theta_A G)''_C,\tau+1) - \tau^{-3}(P((\psi_A G)''_C,\tau+1) + P((\phi_A G)''_C, +1))\},$$

by 1.7,

$$= \{P(G'_C,\tau+1) + P(\theta_A(G'_C),\tau+1) - \tau^{-3}(P(\psi_A(G'_C),\tau+1) + P(\phi_A(G'_C),\tau+1))\}$$

$$- \{P(G''_C,\tau+1) + P(\theta_A(G''_C),\tau+1) - \tau^{-3}(P(\psi_A(G''_C),\tau+1) + P(\phi_A(G''_C),\tau+1))\},$$

since the operations on A and C do not interfere with one another,

$$= 0,$$

since the theorem holds for G'_C and G''_C.

But we are assuming that the theorem fails for G. We therefore deduce from the preceding arguments that G has in fact no vertices other than those of H.

We next observe that G has no loop. For any such loop would be a loop also of $\theta_A G$, $\phi_A G$ and $\psi_A G$, and the theorem would be trivially true for G by 1.2.

Moreover G has no two edges X and Y, both distinct from A, with the same ends. For suppose two such edges exist. Then we may suppose X is not an edge of H. If $J = G'_X$ then the deletion of X from G, $\theta_A G$, $\phi_A G$ and $\psi_A G$ transforms these graphs into J, $\theta_A J$, $\phi_A J$ and $\psi_A J$ respectively without altering their chromatic polynomials (1.7.1). But the theorem is true for J, by the choice of G. Hence it must be true also for G, which is a contradiction.

There are now only three possibilities. Either $G = H$, or G is the graph H_1 obtained from H by adjoining an edge v_1v_3 outside Q, or G is the graph H_2 obtained by adjoining an edge v_2v_4 outside Q. (Figure 2). We cannot have two new joins v_1v_3 and v_2v_4 since these would cross outside Q. In view of 4.2 the third possibility can be reduced to the second by a change of notation in which A and B interchange symbols and $\theta_A G$ is renamed G. (If the theorem holds for one of G, $\theta_A G$, it holds for the other).

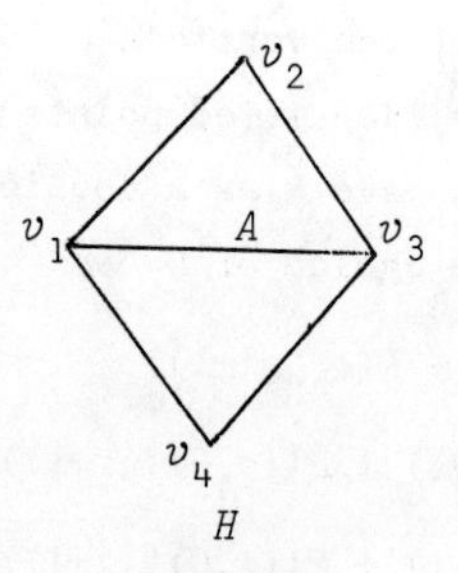

H_1

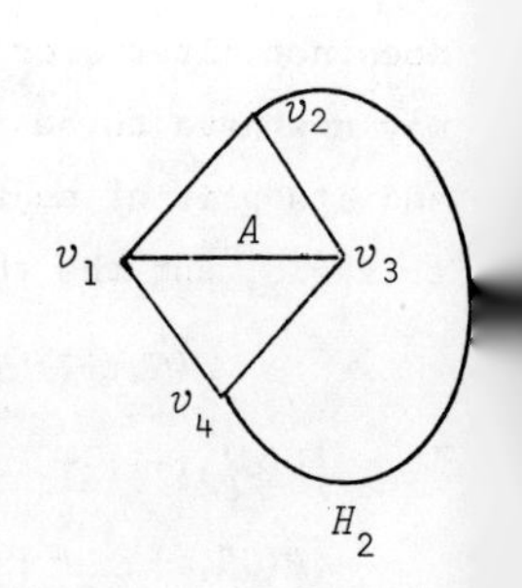

Figure 2.

We may now assume that G is H or H_1.

If $G = H$ we have

$$P(G,\tau+1) = P(\theta_A G,\tau+1)$$
$$= \{(\tau+1)\ \tau(\tau-1)\}^2/\{(\tau+1)\tau\}, \qquad \text{by 1.4 and 1.6,}$$
$$= \tau, \qquad \text{by 3.2.}$$

Each of the graphs $\phi_A G$ and $\psi_A G$ is an arc-graph of two edges. Hence

$$P(\phi_A G,\tau+1) = P(\psi_A G,\tau+1)$$
$$= \{(\tau+1)\tau\}^2/(\tau+1), \qquad \text{by 1.4 and 1.6,}$$
$$= \tau^4, \qquad \text{by 3.2.}$$

Accordingly G satisfies the theorem.

Suppose next that $G = H_1$.

$$P(G,\tau+1) = P(H,\tau+1) = \tau,$$

by 1.7.1 and the preceding argument. Since $\theta_A G$ is a complete 4-graph we have

$$P(\theta_A G,\tau+1) = (\tau+1)\ \tau(\tau-1)(\tau-2) = -1,$$

by 1.4 and 3.2. Since $\phi_A G$ has a loop and $\psi_A G$ is a complete 3-graph we have also

$$P(\phi_A G,\tau+1) = 0,$$
$$P(\psi_A G,\tau+1) = (\tau+1)\ \tau(\tau-1) = \tau^2,$$

by 1.2, 1.4 and 3.2. Hence

$$P(G,\tau+1) + P(\theta_A G,\tau+1) - \tau^{-3}(P(\psi_A G,\tau+1) + P(\phi_A G,\tau+1))$$
$$= \tau - 1 - \tau^{-1} = 0,$$

by 3.2.

We conclude that G, contrary to its definition, satisfies the theorem. This contradiction establishes the theorem.

Writing $n = \tau + 1$ in 4.3 and eliminating $P(\theta_A G,\tau+1)$ by 4.4 we obtain

4.5. $$P(G,\tau+1) = \tau^{-1}\ P(\psi_A G,\tau+1) + \tau^{-2}\ P(\phi_A G,\tau+1),$$

by 3.2. This formula would simplify recursive calculations if we were interested only in this particular value of n.

5. The main theorem.

The theorem stated in the Summary can be proved in the following slightly generalized form.

5.1. *Let M be a proper map of k vertices in which every face, with at most one exception, is triangular. Then*

$$P(M,\tau+2) = (\tau+2)\,\tau^{3k-10}\,P^2(M,\tau+1).$$

Here we write $P(M,n)$ for $P(G,n)$, where G is the defining graph of M.

Proof. If possible choose M so that the theorem fails and the number $\alpha_2(M)$ of faces has the least value consistent with this condition.

Choose a face F of M, the non-triangular one if such exists. We write Q for the boundary of F, referring to F as the *outside* of Q and to the complementary residual domain of Q as the *inside*.

Suppose first that M has no vertices other than those of F. Then the inside of Q is subdivided by (say) r diagonals into $r + 1$ triangular faces. The number of vertices is, by a simple induction, $r + 3$. It is possible for r to be zero. Repeated application of 1.4 and 1.6 yields

$$\begin{aligned} P(M,n) &= \{n(n-1)(n-2)\}^{r+1}/\{n(n-1)\}^r \\ &= n(n-1)(n-2)^{r+1} \\ &= n(n-1)(n-2)^{k-2}. \end{aligned}$$

Hence

$$P(M,\tau+1) = \tau^3\tau^{-k+2} = \tau^{5-k}, \qquad \text{by 3.2},$$

$$P(M,\tau+2) = (\tau+2)\,\tau^2\tau^{k-2} = (\tau+2)\,\tau^k,$$

by 3.2,

$$= (\tau+2)\,\tau^{3k-10}\,P^2(M,\tau+1).$$

We deduce that M has a vertex not incident with F. Let v be such a vertex of minimum valency.

Suppose M has two distinct edges A and B, not constituting together the boundary of F, which have the same two ends. (Clearly M can have no loop). The simple closed curve J that they define has two residual domains R_1 and R_2 in the sphere. We obtain a new map M_1 by deleting all the edges and vertices of M inside R_1 from the graph and then likewise deleting A. We define M_2 analogously, deleting within R_2. It is easily verified that M_1 and M_2 are proper maps, each with at most one non-triangular face. Each has fewer faces than M, and therefore each satisfie the theorem. Let M_i have k_i vertices ($i = 1, 2$). Then

$$k = k_1 + k_2 - 2,$$

$$P(M,n) = \{P(M_1,n)\ P(M_2,n)\}/\{n(n-1)\},$$

by 1.4 and 1.6. together with 1.7.1. Hence

$$P(M,\tau+1) = P(M_1,\tau+1)\ P(M_2,\tau+1)\ \tau^{-3}, \qquad \text{by 3.2,}$$

$$P(M,\tau+2) = P(M_1,\tau+2)\ P(M_2,\tau+2)(\tau+2)^{-1}\ \tau^{-2},$$

by 3.2,

$$= (\tau+2)^2\ \tau^{3k_2+3k_3-20}\ P^2(M_1,\tau+1)\ P^2(M_2,\tau+1)(\tau+2)^{-1}\ \tau^{-2}$$

$$= (\tau+2)\ \tau^{3k-16}\ P^2(M_1,\tau+1)\ P^2(M_2,\tau+1)$$

$$= (\tau+2)\ \tau^{3k-10}\ P^2(M,\tau+1).$$

We deduce that, except possibly for the bounding circuit of F, M has no digon, that is no circuit of two edges only.

We now further restrict the choice of M by postulating that v must have the least possible valency λ consistent with the conditions already imposed on M. Clearly $\lambda \geqslant 2$, since M is proper.

Suppose $\lambda = 2$. Then v is incident with two faces only, both triangular. The sides of the triangles opposite v must constitute a digon, since a third face F exists. This digon must bound F. The deletion of one of its edges transforms M into a triangulation M_1, of fewer faces, without altering its chromatic polynomial, by 1.7.1. The theorem being true for M_1 it must also be true for M. This contradiction shows that $\lambda \geqslant 3$.

Choose an edge A incident with v. Both ends of A have valency at least 3, even if one of them is incident with F. It follows that the two triangular faces incident with A have no other common incident edge. By the absence of digons their vertices opposite A must be distinct. Hence A is the diagonal of a double triangle H of the graph G of M. Using the notation of Section 4 we find

$$P(G,\tau+2) - P(\theta_A G,\tau+2)$$

$$= P(\psi_A G,\tau+2) - P(\phi_A G,\tau+2), \qquad \text{by 4.3,}$$

$$= (\tau+2)\ \tau^{3k-13}\ \{P^2(\psi_A G,\tau+1) - P^2(\phi_A G,\tau+1)\},$$

since $\psi_A G$ and $\phi_A(G)$ each define a map having at most one non-triangular face, having fewer faces than M, and having k-1 vertices,

$$= (\tau+2)\ \tau^{3k-10}\ \{\tau^{-3}(P(\psi_A G,\tau+1) + P(\phi_A G,\tau+1))\}$$
$$\times\ \{P(\psi_A G,\tau+1) - P(\phi_A G,\tau+1)$$
$$= (\tau+2)\ \tau^{3k-10}\ (P^2(G,\tau+1) - P^2(\theta_A G,\tau+1)),$$

by 4.3 and 4.4. Now $\theta_A G$ defines a map with at most one non-triangular face and with the same number of faces as G. In it the vertex v has valency $\lambda-1$ and is still not incident with the non-triangular face F. Hence $\theta_A G$ satisfies the theorem, that is

$$P(\theta_A G,\tau+2) = (\tau+2)\ \tau^{3k-10}\ P^2(\theta_A G,\tau+1),$$

since $\theta_A G$ has k vertices. Hence

$$P(G,\tau+2) = (\tau+2)\ \tau^{3k-10}\ P^2(G,\tau+1),$$

by the preceding result, contrary to the choice of M.

This contradiction establishes the theorem.

5.1.1. $$P(M,\tau+2) > 0,$$

(by 3.3 and 5.1).

Of course Theorem 5.1 remains valid when τ is replaced by the second root $(1 - \sqrt{5})/2 = 1 - \tau = -\tau^{-1}$ of the equation $x^2 = x + 1$.

6. Note on the discovery of Theorem 5.1.

D. W. Hall has kindly made available to the author his list of chromatic polynomials of some 900 triangulations (expressed in dual form) He and his colleagues calculated these in the course of their work on the chromatic polynomial of the truncated icosahedron. The polynomials are expressed in terms of $u = n - 3$ and are divided by the common factor

$$(u+3)(u+2)(u+1)u.$$

Inspection of this list revealed many cases of polynomials differing by

$$u^4 + 4u^3 + 3u^2 - 2u - 1,$$

or simple multiples thereof. It was thought that the zeros of this polynomial, two of which correspond to $n = \tau + 1$ and $n = \tau + 2$, might be specially significant values of u. Accordingly P. A. Kelly evaluated some of Hall's polynomials at these zeros on the Waterloo computer. The results suggested that $P(M,\tau+2)$ was proportional to the square of $P(M,\tau+1)$.

References

1. Berman, G. and Tutte, W. T., The golden root of a chromatic polynomial, *J. Combinatorial Theory,* 6 (1969), 301-302.

2. Birkhoff, G. D. and Lewis, D. C., Chromatic polynomials, *Trans. Amer. Math. Soc.,* 60 (1946), 355-451.

3. Hall, D. W., Siry, J. W. and Vanderslice, B. R., The chromatic polynomial of the truncated icosahedron, *Proc. Amer. Math. Soc.,* 16 (1965), 620-628.

4. Read, R. C., An introduction to chromatic polynomials, *J. Combinatorial Theory,* 4 (1968), 52-71.

5. Tutte, W. T., On chromatic polynomials and the golden ration, *J. Combinatorial Theory,* (to appear).

6. Whitney, H., A logical expansion in mathematics, *Bull. Amer. Math. Soc.,* 38 (1932), 572-579.

7. Whitney, H., The coloring of graphs, *Ann. of Math.,* 33, (1932), 688-718.

The generalized form of the main theorem stated in section 5 is not valid. The proof does not deal adequately with the case of the wheel, in which each double triangle has two edges in common with the non-triangular face. The proof is believed to be valid insofar as it applies to the ungeneralized theorem as stated in the Summary.

A PROBLEM IN COMBINATORIAL GEOMETRY ON THE LINE

H. Tverberg

Universitetet i Bergen, Bergen, Norway

Let A be a subset of $\{1, 2, \ldots, n\}$, and B its complement, with $A \neq \emptyset$, $B \neq \emptyset$.

In the set $\{\prod_{b \varepsilon B} |a-b|;\ a \varepsilon A\} \cup \{\prod_{a \varepsilon A} |a-b|; b \varepsilon B\}$ there is a greatest number, $f(A)$. Put $g(n) = \min\{f\,A;\ A \subset \{1, 2, \ldots, n\}\}$. The function $g(n)$ was discussed, together with its application to a problem in irreducibility of polynomials. Details will be pusblished elsewhere.

A CHARACTERIZATION OF MINIMAL GRAPHS WHICH ARE EMBEDDABLE (NON-EMBEDDABLE) IN A GIVEN 2-DIMENSIONAL MANIFOLD

Walter Vollmerhaus

The University of Calgary, Alta., Canada

A graph G has the Eulerian characteristic $\chi(G) = \nu$ iff ν is the largest integer with the property that G is embeddable in a 2-dimensional manifold of Eulerian characteristic ν. G is minimal of Eulerian characteristic ν iff $\chi(G) = \nu$ and for any edge e of G, $\chi(G - e) \geq \nu + 1$.

As a direct consequence of Kuratowski's theorem, it follows that there are two homeomorphism classes of minimal graphs of characteristic 1, which are represented by the two Kuratowski graphs $K_{3,3}$ and K_5. A generalization of this theorem is

Theorem 1. *For any integer $\nu \leq 1$ there are only finitely many homeomorphism classes of minimal graphs of characteristic ν.*

A better survey of these graphs gives

Theorem 2. *Any minimal graph of characteristic ν can be covered by $2 - \nu$ of the Kuratowski graphs $K_{3,3}$ and K_5.*

By restriction on embeddings only in orientable (non orientable) manifolds, the orientable (non orientable) genus $\gamma_0(G)$ $(\gamma_n(G))$ of a graph G is the smallest integer g with the property that G is embeddable in an orientable (non orientable) manifold of genus g. G is minimal of the orientable (non orientable) genus g iff $\gamma_0(G) = g(\gamma_n(G) = g)$ and for any edge e of G, $\gamma_0(G - e) < \gamma_0(G)$ $(\gamma_n(G - e) < \gamma_n(G))$.

Also for these restricted embeddings, a theorem analogous to Theorem 2 and therefore also a theorem analogous to Theorem 1, holds:

Theorem 3. *Any minimal graph of the orientable (non orientable) genus g can be covered by at most $2g - 1$ (g) of the Kuratowski graphs $K_{3,3}$ and K_5.*

A graph G is minimal non embeddable with characteristic ν, iff $\chi(G) < \nu$, and for any edge e of G $\chi(G - e) \geq \nu$. For these graphs again a theorem analogous to Theorem 2, holds:

Theorem 4. *Any minimal graph non embeddable with characteristic* ν *can be covered by* $3 - \nu$ *of the Kuratowski graphs* $K_{3,3}$ *and* K_5.

Again by restriction on embeddings only in orientable (non orientable) manifolds, a graph G is minimal non embeddable of the orientable (non orientable) genus g, iff $\gamma_0(G) > g$ ($\gamma_n(G) > g$) and for any edge e of G $\gamma_0(G - e) \leq g$ ($\gamma_n(G - e) \leq g$). For those graphs again a theorem like Theorem 2 holds:

Theorem 5. *Any minimal graph, which is non embeddable in the orientable (non orientable) manifold of genus* g *can be covered by at most* $2g + 1$ $(g+$ *of the Kuratowski graphs* $K_{3,3}$ *and* K_5.

A precise description of how these Kuratowski graphs in each of the theorems 2 - 5 have to be connected, will be given in another paper together with the proofs of these theorems.

THE LOWER BOUND CONJECTURE FOR 3- AND 4-MANIFOLDS

David W. Walkup

Boeing Scientific Research Laboratories, Seattle, WA, U.S.A.

For any closed connected d-manifold M, let $f(M)$ denote the set of vectors $f(K) = (f_0(K),\ldots,f_d(K))$, where K ranges over all triangulations of M and $f_k(K)$ denotes the number of k-simplices of K. The principal results announced here are Theorems 1 through 5 below, which, together with generalizations of the Dehn-Sommerville equations for manifolds, yield a characterization of $f(M)$ for some of the simpler 3- and 4-manifolds. The results for the 3- and 4-spheres, given in Theorems 1 and 5, have immediate and obvious implications for simplicial polytopes, i.e., closed bounded convex polyhedra all of whose proper faces are simplices. In particular, they provide a strong affirmative resolution in dimensions 4 and 5 of the so-called lower bound conjecture for simplicial polytopes. Theorem 3, which is concerned with triangulations of projective 3-space, also has an immediate implication for polytopes, specifically for a special subclass of centrally symmetric simplicial polytopes. This corollary of Theorem 3 is stated as Theorem 6.

The proofs of these results are available in [1] and will be published elsewhere.

Theorem 1. *There exists a triangulation K of the 3-sphere S^3 with f_0 vertices and f_1 edges if and only if*

$$4f_0 - 10 \leq f_1 \leq f_0(f_0-1)/2 \ .$$

Moreover, $f_1(K) = 4f_0(K) - 10$ for some triangulation K of S^3 if and only if $K \in H^3(0)$, i.e., if and only if K can be obtained from the boundary complex of a 4-simplex by a sequence of central retriangulations of 3-simplices.

Theorem 2. *Let M be either the orientable 3-handle $H^3_+ = S^2 \times S^1$ or the nonorientable 3-handle H^3_- obtained from $S^2 \times [0,1]$ by an antipodal identification of $S^2 \times 0$ and $S^2 \times 1$. There exists a triangulation K of M with f_0 vertices and f_1 edges if and only if*

$$4f_0 \leq f_1 \leq f_0(f_0-1)/2 \quad ,$$

except that $(f_0,f_1) = (9,36)$ is impossible if $M = H^3_+$. Moreover, $f_1(K) = 4f_0(K)$ for some triangulation of M if and only if $K \in \mathrm{H}^3(1)$, i.e., if and only if K can be obtained as a legitimate simplicial complex from a member of $\mathrm{H}^3(0)$ by removing two 3-simplices and identifying their boundaries only.

Theorem 3. *There exists a triangulation K of projective 3-space P^3 with f_0 vertices and f_1 edges if and only if*

$$4f_0 + 7 \leq f_1 \leq f_0(f_0-1)/2.$$

Moreover, $f_1(K) = 4f_0(K) + 7$ for some triangulation K of P^3 if and only if K can be obtained from K_0 by a sequence of central retriangulations of 3-simplices, where K_0 is the triangulation of P^3 with 11 vertices and 51 edges described following Theorem 6.

Theorem 4. *Suppose M is any closed connected 3-manifold distinct from S^3, H^3_+, H^3_-, and P^3. Then there exists an integer $\gamma(M) > 7$ such that*

$$f_1(K) \geq 4f_0(K) + \gamma(M)$$

for any triangulation K of M. Conversely, there exists $\gamma^*(M) \geq \gamma(M)$ such that for every (f_0,f_1) satisfying

$$4f_0 + \gamma^*(M) \leq f_1 \leq f_0(f_0-1)/2$$

there is a triangulation of M with f_0 vertices and f_1 edges.

Theorem 5. *If K is a triangulation of a closed connected topological 4-manifold, then*

$$f_1(K) \geq 5f_0(K) - \frac{15}{2}\chi(|K|),$$

where $\chi(|K|) = \chi(K)$ is the Euler characteristic of K. Moreover equality holds if and only if $K \in \mathrm{H}^4(1-\frac{1}{2}\chi(K))$, i.e., if and only if K can be obtained from the boundary complex of a 5-simplex by a sequence of centr retriangulations of 4-simplices, followed by removal of $2 - \chi(K)$ 4-simpl and identification of their boundaries in pairs.

Theorem 6. *Let P^d_p denote the class of centrally symmetric simplicial d-polytopes such that no centrally symmetric pair of vertices can be* *connected by a path consisting of fewer than three edges of the polytope. If* $P \in \mathcal{P}_p^4$ *then*

$$f_1(P) \geq 4f_0(P) + 14.$$

Moreover, $f_1(P) = 4f_0(P) + 14$ *if and only if* P *can be obtained by successively adding pyramids in centrally symmetric pairs to the faces of some member of* $\mathcal{P}_p^4$ *combinatorially equivalent to the polytope* P_0 *defined in the following paragraph.*

Consider the set V consisting of the following 22 points on the unit 3-sphere in E^4:

$$(0, 0, 0, \pm 1)$$
$$(\pm\beta, 0, 0, \pm\beta) \qquad \alpha = \sqrt{3}/3$$
$$(0, \pm\beta, 0, \pm\beta) \qquad \beta = \sqrt{2}/2$$
$$(0, 0, \pm\beta, \pm\beta)$$
$$(\pm\alpha, \pm\alpha, \pm\alpha, 0).$$

It is trivial that V is exactly the vertices of a centrally symmetric 4-polytope P_0. Let $H(\gamma)$ denote the hyperplane $\{x \mid x_4 = \gamma\}$. The five hyperplanes $H(-1)$, $H(-\beta)$, $H(0)$, $H(\beta)$, $H(1)$ contain all members of V, and as H ranges over these five in order $\operatorname{conv}(H \cap V)$ is successively a point, an octahedron, a cube, a second octahedron, and a second point. It can be verified that $\operatorname{conv}(H \cap V) = H \cap P$ in each of these five cases and moreover P_0 is simplicial with $f_0(P_0) = 22$, $f_1(P_0) = 102$, $f_2(P_0) = 160$, $f_3(P_0) = 80$. It can be further verified that no pair of centrally symmetric vertices of P_0 can be connected by a path consisting of fewer than three edges of P_0. It follows that identifying centrally symmetric faces in the boundary complex of P_0 yields a triangulation K_0 of projective 3-space.

The results for 4-manifolds in Theorem 5, for 3-manifolds in Theorems 1 through 4, and the corresponding facts concerning 2-manifolds obviously invite comparison and conjecture. It can be shown that there is a unique triangulation of the torus H_+^2 with 7 vertices, and this triangulation is a member of $\mathcal{H}^2(1)$. However, there is no triangulation of the Klein bottle H_-^2 having 7 vertices. These facts parallel the results for orientable and nonorientable 3-handles covered in Theorem 2 except that the roles of orientability and nonorientability are reversed.

The Dehn-Sommerville equations for triangulated 2-manifolds include

$$f_1(K) = 3f_0(K) - \frac{6}{2}\chi(|K|), \tag{1}$$

which is a natural analogue of the inequality in the first part of Theorem 5. However, no analogue of the second part of Theorem 5 holds for 2-manifolds. Thus, for example, if K is the boundary complex of an icosahedron, then K satisfies (1), but K cannot be obtained from the boundary complex of a 3-simplex or any other complex except itself by a sequence of central retriangulations of stars of faces of any dimension.

For any closed connected manifold M of dimension d which admits triangulations, let $\gamma(M)$ denote the supremum of all integers k such that every triangulation K of M satisfies $f_1(K) \geq (d+1)\, f_0(K) + k$. It is immediate from this definition that $\gamma(M) < +\infty$, and theorems 1 through 5 show $\gamma(M)$ is finite if $d \leq 4$, but it is not at all clear that $\gamma(M) > \infty$ if d is large also. Although Theorem 5 shows $\gamma(S^4) = -15$ and thus proves the lower bound conjecture for 5-polytopes, it leaves open the question whether $\gamma(M) \geq -15$ for all 4-manifolds.

The regular behavior of γ for 2-manifolds suggests a number of questions about its behavior for 3-manifolds: Can $\gamma(M)$ be related in some natural way to other algebraic invariants than $\chi(M)$? For manifolds of dimension d, let $\bar{\gamma}(M) = \gamma(M) - \gamma(S^d)$. Then $\bar{\gamma}$ is additive with respect to manifold addition for 2-manifolds. (The manifold sum of two d-manifolds is obtained by removing a d-cell from each and identifying boundaries.) Is the same true for manifolds of dimension 3 and higher? Projective 3-space P^3 is the simplest example of a lense space $L(p,q)$. What is the next larger value of $\gamma(L(p,q))$?

References

1. Walkup, D., The lower bound conjecture for 3- and 4-manifolds, document D1-82-0856, Boeing Scientific Research Laboratories, May, 1969.

A RÉSUMÉ OF SOME RECENT RESULTS ON HADAMARD MATRICES, (v,k,λ)-GRAPHS AND BLOCK DESIGNS

Jennifer Wallis

La Trobe University, Bundoora, Victoria, Australia

Hadamard Matrices: I have found many new classes of Hadamard matrices using an extension of a method of Williamson including 612 which is new. My methods depend heavily on the existence of skew-Hadamard matrices for which I have new constructions and on the existence of symmetric C-matrices. I have found skew-Hadamard matrices of orders 552 and 3304 which were previously not known. In addition I have found that if there is a skew-Hadamard matrix (symmetric C-matrix) of order n then there is a skew-Hadamard matrix (symmetric C-matrix) of order $(n-1)^r + 1$ where $r = 3$, 5 or 7. Most of my results may be found in [7], [8], [9] and [10].

By using subgroups of index four (4) in the group of powers of a primitive root of a prime $q \equiv 5 \pmod 8$ George Szekeres has constructed skew-Hadamard matrices of orders $2^t(q+1)$ with $t \geq 1$ an integer and t as above. This has obvious connections with tournaments and is important to a theory of twin difference sets. These and other similar results may be found in [5].

Goethals and Seidel in [3] have exhibited skew-Hadamard matrices of order 36 and 52, Szekeres has another of order 52 [2] and by a similar construction to that in [3] I have $4p$ where $p = 3, 5, 7, 9, 11, 13, 15$ and 19.

Theorem. (W. D. and J. Wallis [12]), *If there is an Hadamard matrix H of order $4p$ where p is odd and square free, then H is equivalent over the ring of integers to a diagonal matrix the modulus of whose entries are*

$$1, \underbrace{2, \ldots, 2}_{(2p-1)\text{ times}}, \underbrace{2p, \ldots, 2p}_{(2p-1)\text{ times}}, 4p.$$

(v,k,λ)-graphs. This concept originated with G. Szekeres and is defined as a graph on v points, each of valency k, and such that for any two points P and Q there are exactly λ points joined to both.

E. C. Johnsen's conditions give that $k - \lambda = m^2$ for some m/λ and Bose, in addition, has found that for a graph to exist $v - 1$ and $m + \frac{\lambda}{m}$ must have the same parity.

Theorem. (W. D. Wallis [13]): *The above conditions are not sufficient as there are infinitely many counter examples.*

Another necessary condition is that $m^4 \geq k + \lambda$ except for a complete graph, but it is not known if these conditions are now sufficient.

Theorem. (R. Ahrens): *If there is a* $(4k^2-1, 2k^2-k, 2k+1, k, 1)$*-configuration there is a* $(4k^2-1, 2k^2, k^2)$*-graph.*

W. D. Wallis has found there is always a $(2^{2k+2} - 1, 2^{2k+1}, 2k)^*$-graph for natural k and there is a $(4k^2-1, 2k^2, k^2)^{**}$-graph whenever $k = 2^a 3^b 5^c 7^d$, $a \geq b + c + d - 1$. The graphs found from * and ** for 63 and 255 are nonisomorphic.

Using a (36, 15, 6)-configuration of mine W. D. Wallis has found there are $(4k^2, 2k^2+k, k^2+k)$ and $(4k^2, 2k^2-k, k^2-k)$-graphs whenever $k = 2^a 3^b$, $a \geq b - 1$.

R. Ahrens [1] has used finite geometry to prove there is always a $(\lambda^2(\lambda+2), \lambda(\lambda+1), \lambda)$-graph for λ a prime power, giving a (45, 12, 3)-configuration.

Block Designs. In [6] Szekeres has shown a skew-design with parameters $(16t^2, 8t^2-2t, 4t^2-2t)$ exists whenever an H-design with parameters $(4t-1, 2t-1, t-1)$ (not necessarily skew) exists. Here skew design has the restricted definition that if S_i are the v subsets of a set $X = \{x_1, \ldots, x_v\}$ the design is skew if $x_i \in S_j \Rightarrow x_j \notin S_i$.

Using matrix properties I have shown in [11] that there exists a $(\lambda^2(\lambda+2), \lambda(\lambda+1), \lambda)$-configuration for λ prime or certain prime powers. This result was obtained independently but simultaneously with the resul of R. Ahrens and gives another, but non-isomorphic, (45, 12, 3)-configuration.

I have also found the following (b, v, r, k, λ)-configurations [11]:

(i) $(q(q+1), q^2, q+1, q, 1)$, q a prime power;

(ii) $(q(k^2+\lambda), qk, k^2+\lambda, k, \lambda)$, q a prime power and a (q, k, λ)-configuration must exist;

(iii) $(\frac{1}{2} q(3q-1), 3q, \frac{1}{2}(3q-1), 3, 1)$ q odd, $q \neq 1$;

(iv) $(2(n-1), n, n-1, n/2, (n-2)/2)$ n the order of a skew-Hadamard or a symmetric C-matrix.

(i) and (iv) give designs not listed in Hall [4].

References

1. Ahrens, R. and Szekeres, G., On a combinatorial generalization of 27 lines associated with a cubic surface, to appear in *J. Aust. Math. Soc.*

2. Blatt, D. and Szekeres, G., A skew-Hadamard matrix of order 52, to appear in *Can. J. Math.*

3. Goethals, J. M. and Seidel, J. J., A skew-Hadamard matrix of order 36, to appear in *J. Aust. Math. Soc.*

4. Hall, M., Jr., *Combinatorial Theory*, Blaisdell, Waltham, Mass, 1967.

5. Szekeres, G., Tournaments and Hadamard matrices, to appear in *L'Enseignment Math.*

6. Szekeres, G., A new class of Symmetric Block Designs, *J. Comb. Th.* 6 (1969) 219-221.

7. Wallis, Jennifer, A class of Hadamard matrices, *J. Comb. Th.* 6 (1969) 40-44.

8. Wallis, Jennifer, A note of a class of Hadamard matrices, *J. Comb. Th.* 6 (1969) 222-223.

9. Wallis, Jennifer, (v, k, λ)-configurations and Hadamard matrices, to appear in *J. Aust. Math. Soc.*

10. Wallis, Jennifer, Some (1,-1) matrices, to appear in *J. Comb. Th.*

11. Wallis, Jennifer, Some results on configurations, submitted to *J. Aust. Math. Soc.*

12. Wallis, W. D. and Wallis, Jennifer, Equivalence of Hadamard Matrices, to appear in *Israel Journal*.

13. Wallis, W. D., A non-existence theorem for (v,k,λ)-graphs, to appear in *J. Aust. Math. Soc.*

STILL ANOTHER PRODUCT FOR GRAPHS

Mark E. Watkins

Syracuse University, Syracuse, N.Y., U.S.A.

The automorphism group $\mathcal{A}(X)$ of a finite graph X is considered to be *regular* if its action on the set $V(X)$ of vertices of X is such that $u, v \in V(X)$ implies the existence of a unique graph-automorphism $\phi \in \mathcal{A}(X)$ with $\phi(u) = v$. A finite abstract group G shall be said to be *regular* if there exists a graph X such that $\mathcal{A}(X)$ is regular and isomorphic to G.

The regular abelian groups have been determined through the work of Chao [1] and McAndrew [2]. The author [5] has determined several classes of non-abelian groups and has further proved:

Theorem. *If G_i is regular and $G_i \neq Z_2$ $(i = 1,2)$, then the direct product $G_1 \times G_2$ is also a regular group.*

The proof of this theorem rests heavily upon a new kind of product for graphs. This note consists of several remarks concerning this new product.

Definition. *If X_1 and X_2 are graphs, the graph $X = X_1 \odot X_2$ is defined to have $V(X) = V(X_1) \times V(X_2)$ while $[(x_1,x_2), (y_1,y_2)] \in E(X)$ if and only if either*

$$x_1 = y_1 \text{ and } [x_2,y_2] \in E(X_2)$$

or

$$[x_1,y_1] \in E(X_1) \text{ and } x_2 \neq y_2.$$

It is not hard to see that the above theorem follows from the fact (also proved in [5]) that:

If $\mathcal{A}(X_i)$ is regular and $|V(X_i)| \geq 3$ $(i = 1,2)$, then

$$\mathcal{A}(X_1 \odot X_2) = \mathcal{A}(X_1) \times \mathcal{A}(X_2). \tag{1}$$

We remark that (1) does not necessarily hold if the operation $\odot$ is replaced by any of the other well-known graph products. For instance, the group of the lexicographic product of graphs is the wreath product of their groups (see [3]). If $\odot$ is replaced by the cartesian product, then (1) is not generally true unless X_1 and X_2 are "relatively prime" in the sense of Sabidussi [4].

On the other hand, $\odot$ fails to have various properties generally satisfied by other graph products. In particular, if $\overline{X}$ denotes the complement of X, then $\overline{X_1 \odot X_2}$ and $\overline{X}_1 \odot \overline{X}_2$ are generally distinct graphs. Also, like the lexicographic product, $\odot$ is not commutative. Since $\mathcal{G}(X) = \mathcal{G}(\overline{X})$ for any graph X and since groups do commute with respect to direct product, one infers that if G_1 and G_2 are regular and $\neq Z_2$, then there can be as many as eight distinct graphs whose automorphism group is $G_1 \times G_2$ and acts regularly. (There are fewer if $X_i = \overline{X}_i$, but the author knows of no graph having this property whose automorphism group is regular.)

Finally, the operation $\odot$ is not associative. In fact if $E(X_i) \neq \emptyset$ $(i = 1,2,3)$, then the graph $(X_1 \odot X_2) \odot X_3$ always has fewer edges than does $X_1 \odot (X_2 \odot X_3)$. It follows that any group which is the direct product of a large number of distinct regular groups $\neq Z_2$ is the regular automorphism group of a very substantial number of non-isomorphic graphs.

References

1. Chao, C.-Y., On a theorem of Sabidussi, *Proc. Amer. Math. Soc.*, 15 (1964), 291-292.

2. McAndrew, M.H., On graphs with transitive automorphism groups, *Notices Amer. Math. Soc.*, 12 (1965), 575.

3. Sabidussi, G., The composition of graphs, *Duke Math. J.*, 26 (1959), 693-696.

4. Sabidussi, G., Graph multiplication, *Math. Z.*, 72 (1959/60), 446-457.

5. Watkins, M.E., On the action of non-abelian groups on graphs, (to appear).

THE GENUS OF CARTESIAN PRODUCTS OF COMPLETE BIPARTITE GRAPHS

Arthur T. White

Michigan State University, East Lansing, MI, U.S.A.

In 1965 Ringel [2] showed that the genus of the complete bipartite graph $K_{p,q}$ is given by $\gamma(K_{p,q}) = \{\frac{(p-2)(q-2)}{4}\}$. We show that the genus of the cartesian product of $K_{2m,2m}$ with itself is given by $\gamma(K_{2m,2m} \times K_{2m,2m}) = 1 + 8m^2(m-1)$, with corresponding formulae for related graphs. Concepts to be used are defined in [1] and [3].

The following readily established lemmas are useful in computing the genus of $K_{2m,2m} \times K_{2m,2m}$ and related graphs.

Lemma 1. The cartesian product of two bipartite graphs is a bipartite graph.

Lemma 2. If the bipartite graph G with V vertices and E edges has a quadrilateral imbedding, then that imbedding is minimal, and $\gamma(G) = 1 + \frac{E}{4} - \frac{V}{2}$.

It therefore suffices to produce a quadrilateral imbedding for $K_{2m,2m} \times K_{2m,2m}$. We begin with the imbedding given by Ringel for $K_{2m,2m}$, which has $F = F_4 = 2m^2$ and $\gamma = (m-1)^2$. For this imbedding, we can prove the following lemma:

Lemma 3. The $2m^2$ quadrilateral facts can be partitioned into $2m$ sets of m faces each, with each set containing all $4m$ vertices of the graph $K_{2m,2m}$.

We are now able to prove the main result of this note.

Theorem 1. *The genus of the graph* $K_{2m,2m} \times K_{2m,2m}$ *is given by* $\gamma(K_{2m,2m} \times K_{2m,2m}) = 1 + 8m^2(m-1)$.

Proof. Imbed $2m$ copies of $K_{2m,2m}$ in $2m$ closed orientable 2-manifolds or like orientation, each of genus $(m-1)^2$, in the fashion described by Ringel. Now imbed $2m$ additional copies of $K_{2m,2m}$ similarly, but in closed orientable 2-manifolds of the opposite orientation. This partition of the $4m$ 2-manifolds corresponds to the vertex-set partition of $K_{2m,2m}$. Between each pair of oppositely oriented 2-manifolds, $4m$ edges must be added joining pairs of corresponding vertices, in order to imbed $K_{2m,2m}\times K_{2m,2m}$. Each set of $4m$ edges is carried over m tubes attached to corresponding faces in the 2-manifolds of opposite orientation, with each tube carrying four edges. Lemma 3 tells us that Ringel's imbedding of $K_{2m,2m}$ is ideally suited for this purpose. Each new face produced by this method is a quadrilateral, so that we have constructed a quadrilateral imbedding for $K_{2m,2m}\times K_{2m,2m}$. Using Lemma 2, with $V = 16m^2$ and $E = 32m^3$, we obtain $\gamma(K_{2m,2m}\times K_{2m,2m}) = 1 + 8m^2(m-1)$.

We now use the construction of Theorem 1 to obtain the following generalization:

Theorem 2. *The genus of the graph $K_{2m,2m}\times K_{r,s}$ is given by*

$\gamma(K_{2m,2m}\times K_{r,s}) = 1+m((m-2)(r+s) + rs)$, for $r \leq 2m$ and $s \leq 2m$.

Proof: Given $K_{2m,2m}\times K_{2m,2m}$ minimally imbedded as in Theorem 1, remove $(4m-(r+s))$ 2-manifolds containing copies of $K_{2m,2m}$ together with all attached tubes and the edges they carry, so that an imbedding of $K_{2m,2m}\times K_{r,s}$ remains. As each initial imbedding of $K_{2m,2m}$ was quadrilateral, each face re-introduced by this process is a quadrilateral, so that $K_{2m,2m}\times K_{r,s}$ is also quadrilaterally imbedded. A routine application of Lemma 2 now gives the result.

Using the technique of Theorem 1, we can also show that $\gamma(K_{3,3}\times K_{3,3}) = 10$. In this case the imbedding of $K_{3,3}$ with which we begin is not that given by Ringel, but has $F = F_6 = 3$. It is trivial that $\gamma(K_{1,1}\times K_{1,1}) = 0$. Combining these two observations with Theorem 1, we can state

Theorem 3. *The genus of the graph $K_{s,s}\times K_{s,s}$ is given by*

$\gamma(K_{s,s}\times K_{s,s}) = 1 + s^2(s-2)$, if s is even or if $s = 1$ or 3.

Acknowledgement. I wish to thank Professor E.A. Nordhaus for the assistance he has given me during the research on this problem.

References

1. Harary, F., Graph Theory, Addison Wesley, Reading, 1969.

2. Ringel, G., Das Geschlecht des Vollständigen Paaren Graphen, *Abh. Math. Sem. Univ. Hamburg,* 28 (1965), 139-150.

3. Youngs, J.W.T., Minimal Imbeddings and the Genus of a Graph, *J. Math. Mech.,* 12 (1963), 303-315.

LINEAR LISTS FOR SPIRO GRAPHS*

Lee J. White and J.E. Rush

The Ohio State University, Columbus, OH, U.S.A.

A *spiro graph* can be used as a model for spiro-fused ring systems of organic chemistry in order to both name the compound and store the structural information for subsequent retrieval. A weighted *spiro graph G* is defined as a graph with nonnegative integer node and edge weights such that

(1) G is finite and connected;

(2) each node has degree two or four; if of degree two, the node weight is positive, but if of degree four, the node weight must be zero;

(3) parallel edges may exist in G;

(4) two distinct circuits (simple closed paths) have no edge in common.

Given a spiro graph G with N nodes, the problem is to efficiently construct a unique linear list, from which G can be unambiguously reconstructed. This is accomplished by inducing the linear list from an *Euler tour* of G (a path containing all the edges of G exactly once).

Define a *terminal circuit* of G as a circuit containing only one node of degree four in G. G is *unbranched* if it contains a simple path (*spanning chain*) which includes all nodes of G; otherwise G is branched.

Theorem 1. *A spiro graph G is branched iff G contains either:*

*This research was supported by the National Science Foundation, under Grant GN-534.1.

(a) *a circuit including two nodes of degree four, v_1, v_2, such that each of the two paths between v_1, v_2 within the circuit contain at least one node distinct from v_1 and v_2, or*

(b) *a circuit which includes more than two nodes of degree four.*

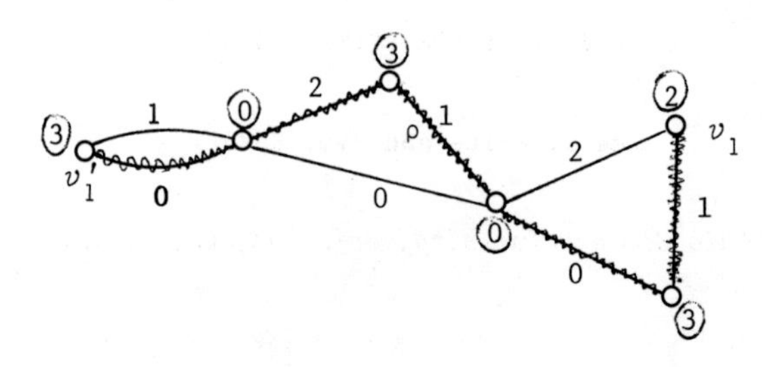

$$D = [\bar{2},1,\bar{3},0,1,\bar{3},2,0,\bar{3},1,0,2]$$

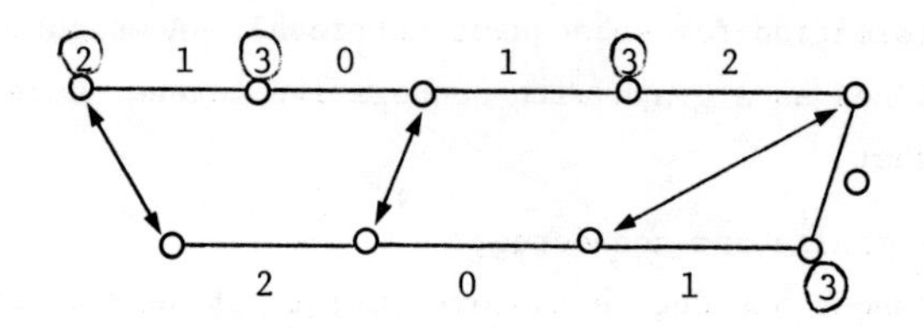

Figure 1. Unbranched Spiro Graph and Linear List D

The proof of Theorem 1 leads to the following algorithm which either denotes G as branched, or produces a spanning chain P. An upper bound on the computation grows linearly with N.

(1) Begin at any node v_1 of degree four in G, and search a circuit C containing v_1 to see if it is of type (a) or (b) of Theorem 1.

(2) If not, identify $v_2 \neq v_1$ as another node of degree four of C. If no v_2 exists, C is a terminal circuit, defining one end of the spanning chain P (see Fig. 1). Return to the initial node, and consider unexplored circuits in (1). Terminate when both ends of P are defined.

(3) Redefine v_2 as v_1, and repeat step (1).

Consider a list D with $2N$ entries of nonnegative integers, with bars over some entries, as in Fig. 1. Then list D_1 is *smaller* than

D_2 if the entries of D_1 are smaller lexicographically and where by convention, $\bar{k} > k$, for integer k.

Given a spanning chain P for an unbranched spiro graph G with N nodes, construct a list D as follows (see Fig. 1).

(1) For each terminal circuit C, C' of G, choose a starting node v_1, v_1' of P so that D will be a smallest list (this decision can be made by only considering C, C').

(2) Starting at v_1, construct list D_1 by entering positive node weights (with bars) and all edge weights as P is traversed.

(3) When v_1' is encountered, continue tracing the unique Euler tour and constructing D_1, terminating at v_1 when D_1 contains $2N$ entries.

(4) Repeat steps (2) and (3) with the roles of v_1, v_1' interchanged, constructing list D_2. Let D be the smaller of D_1, D_2. This algorithm grows as N^2 in general, but only grows linearly with N if G contains at least one node of degree four.

Theorem 2. *Two unbranched spiro graphs possess the same linear list D iff they are isomorphic.*

The method for reconstructing G given D is illustrated in Figure 1, and can be proven to yield the original spiro graph G (up to isomorphism). The growth of this reconstruction algorithm is linear with N.

These techniques are extended to branched spiro graphs, resulting in an algorithm for constructing linear lists which grows as N^2. Details of the results cited in this paper may be found in Technical Report 69-6, Computer and Information Science Research Center, The Ohio State University.

THE COLOURING OF VERY LARGE GRAPHS

M.R. Williams

The University of Calgary, Alta., Canada

The vertex colouring of very large graphs is a problem which often arises from attempting to solve scheduling and time-table construction problems. Several methods of solution have been suggested in the literature, some algorithmic (Berge, Williams) and some heuristic (Peck-Williams, Berge). The purpose of this note is to show an improvement to the Peck-Williams procedure which will result in potentially fewer colours being used.

The major assumption in the heuristic colouring procedures is: "Given a partially coloured graph, if an uncoloured vertex is adjacent to a large number of other vertices it will be harder to colour it than to colour a vertex which is adjacent to only a few other vertices". Assuming the first T-1 colours have been applied to the vertices of the graph, then to select the vertices which would have colour T the Peck-Williams heuristic procedure will procede as follows:

i. Find the uncoloured vertex with highest degree.

ii. Check to see if this vertex is adjacent to any vertex already given the colour T; if not then colour this vertex with colour T otherwise remove this vertex from further consideration in colour T and go to step 1.

By this heuristic the vertex coloured at any particular stage in the process will be the next colourable vertex of highest degree. However, complications arise when two or more vertices have the same degree. It can be shown (Williams) that the incorrect selection of a vertex can result in excess colours being used.

The adjacency matrix of a graph will be denoted by A, and a vector d will be defined as

$$d_i = \sum_j A_{ij}.$$

i.e. d_i is the degree of the ith vertex. It is this vector d which the Peck-Williams procedure uses to decide which vertex should be coloured next.

It is a well known result that the vector d is an approximation to the dominant eigenvector of the matrix A. In fact a closer approximation d^2 can be obtained by a matrix-vector multiplication, and in general the iterative process

$$d_i^{n+1} = \sum_j A_{ij} d_j^n$$

will converge to the dominant eigenvector of A as $n \to \infty$.

If we modify the Peck-Williams procedure to use d^n instead of d a significant improvement in the heuristic is noted. It has been found that the iteration need not be carried out a large number of times, in fact, if there are p vertices in the graph then

$$n = |\sqrt[3]{p}|$$

is generally sufficient. (See Williams).

This procedure has been used to colour graphs with over 700 vertices with very good results. One test case was done on a graph of unknown structure and resulted in the use of 28 colours, this graph was later found to contain a K_{26}, 27 vertices lacking only one edge to be a K_{27}, 28 vertices lacking only 3 edges to be a K_{28}, and several examples of 29 vertices lacking only 5 edges to be a K_{29}.

References

1. Berge, C., The Theory of Graphs and its Applications, John Wiley and Sons, New York 1962.
2. Peck, J.E.L. and Williams, M.R., Examination Scheduling, Algorithm 286 Communications of the Association for Computing Machinery, 9(6), 1966.
3. Williams, M.R., A Graph Theory Model for the Computer Solution of University Time-tables and Related Problems, PhD thesis, University of Glasgow, 1969.

REMARKS ON THE FOUR COLOR PROBLEM

J.W.T. Youngs

University of Santa Cruz, CA, U.S.A.

The recent confirmation of the Heawood conjecture by Ringel and Youngs prompts one to ask if the methods used by them can be employed to advantage in the most famous of all map coloring questions, namely the four color problem.

The methods they employed used three concepts:

1) current graphs
2) graphs with rotation
3) Kirchoff's law.

If $X(S)$, the chromatic number of S, the sphere, is four then consider M, any map on S. Without loss of generality we may take M to be a strict cellular decomposition of S; that is the closure of any k-cell is a closed k-cell, $k = 1,2$. In addition it may be assumed that M', the 1-skeleton of M, is a cubic graph. If M is four colorable then there exists a 2-chain c_2 in the chain group $C_2(M;\mathbb{K})$, where $\mathbb{K}$ is the Klein 4-group $(Z_2 \times Z_2)$, such that $\partial\ c_2\ \varepsilon\ C_1(M;\mathbb{K})$ is *totally non-trivial*; that is, *each* directed arc in M is assigned a non trivial element of $\mathbb{K}$ by the boundary homomorphism $\partial\colon C_2(M;\mathbb{K}) \to C_1(M;\mathbb{K})$.

In the language of Ringel and Youngs c_2 is a *coloring* of M and ∂C_2 is a *current graph.*

We also have a boundary homomorphism $\partial : C_1(M,\mathbb{K}) \to C_0(M,\mathbb{K})$, and it is well known that $\partial\partial$ is trivial, hence $\partial\partial c_2 = 0$. But this is simply another way of saying that *Kirchoff's current law holds at each vertex of* M°.

If a, b, c, 0 are the elements of $\mathbb{K}$ then $2a = 2b = 2c = 0$ and $a + b + c = 0$ provide complete information on the addition table. It is easy to see that given c_2 the currents assigned to three arcs at each vertex of M exhaust, without repetition, the elements a, b and c. This leads to the classical result that *if the four color conjecture is true then every planar cubic graph is (arc) three colorable.*

The converse is proved by considering any map M (with cubic 1-skeleton) and its dual D. A "potential change" is then designed in terms of a 3-coloring of M' (using a, b and c) as follows. Adjacent vertices of D have a potential change x if and only if the arc of M' crossed by the arc in D' joining the vertices carries current x. The potential change along any path in D' is defined in the usual fashion, and it is shown that the potential change along any closed path is zero - thus Kirchoff's potential law holds. This suffices to four color the vertices in D and thus *four color the* 2-cells *in* M.

No new results are obtained by those methods but, on complete exposition, all the standard equivalence theorems can be proved in a rather attractive way. Details will be forthcoming in a report on a symposium sponsored by the American Mathematical Association in April 196

ON THE 1-FACTORS OF n-CONNECTED GRAPHS

Joseph Zaks

University of Washington, Seattle, WA, U.S.A.

L. W. Beineke and M. D. Plummer have recently proved [1] that every n-connected graph with a 1-factor has at least n different 1-factors. The main purpose of this paper is to prove that every n-connected graph with a 1-factor has at least as many as $n(n-2)(n-4) \ldots 4.2$, or: $n(n-2)(n-4) \ldots 5.3$, 1-factors. The main lemma which is used here is that if a 2-connected graph G has a 1-factor, then G contains a vertex V (and even two such vertices), such that each edge of G, incident to V, belongs to some 1-factor of G.

1. <u>Introduction and statement of results</u>. We consider here finite graphs, without loops or double edges. A 1-factor of a graph G is a subgraph of G, which contains all the vertices of G, each one with valence 1.

Let $f(n)$ denote the maximal k for which the following is true:

"If G is an n-connected graph with a 1-factor then G has at least k (different) 1-factors."

One can now reformulate Theorem 1*of [1] simply as: $f(2) \geq 2$, and Theorem 2 of [1] becomes: $f(n) \geq n$, for each $n \geq 2$.

Our extension of [1] is the following

<u>Theorem 1</u>. *$f(n) \geq n(n-2)(n-4) \ldots 4.2$ for even n, while for odd n $f(n) = n(n-2)(n-4) \ldots 5.3$.*

Let $F(G)$ denote the subgraph of G, which is the union of all the 1-factors of G. A vertex V of G will be called *totally covered by* $F(G)$ if all the edges of G, incident to V, belong to $F(G)$.

---

* A simpler proof of this theorem was recently given by B. Grünbaum.

The following is our main idea:

Lemma 4. If G is a 2-connected graph with a 1-factor, then G contains a vertex V, which is totally covered by $F(G)$.

To the proof of this lemma we have the following

Corollary 1. If G is a 2-connected graph with a 1-factor, then G contains two vertices, which are totally covered by $F(G)$.

I would like to express my great appreciation to Professor Branko Grünbaum for his excellent encouragement.

2. Terminology. A graph G is connected if for each pair of its vertices V and W, G contains a $V - W$ path p: e.g., an ordered collection $(V_1; E_1; V_2; E_2; \ldots; E_k; V_{k+1})$ of vertices V_i and edges E_i of G, such that $V_1 = V$, $V_{k+1} = W$ and for all $1 \leq i \leq k$, E_i is the edge $(V_i;V_{i+1})$ of G.

A graph G is n-connected between two of its vertices V and W if G contains n $V - W$ paths $p_1, \ldots, p_n$ such that $p_i \cap p_j = \{V;W\}$ for all $i \neq j$, $1 \leq i; j \leq n$. G is *n-connected* if it is n-connected between each pair of its vertices.

It is well known that G is n-connected if it contains at least n+1 vertices, and the deletion of no set of n-1 or fewer vertices disconnects G.

Clearly, if G is n-connected, having a vertex V and an edge E, then both $G - V$ and $G - E$ are n-1 connected.

The following is well known:

Lemma 1. If G is a 2-connected graph, and $E = (V_1;V_2)$ is an edge of G, such that $G - E$ is only 1-connected, then $G - E$ is only 1-connected between V_1 and V_2.

3. With very little effort we can improve considerably the inequality $f(n) \geq n$, of [1], while using $f(2) \geq 2$, as follows:

Lemma 2. for each $n \geq 3$, $f(n) \geq f(n-1) + f(n-2)$.

Proof. Let G be an n-connected graph with a 1-factor F, $n \geq 3$. Since G is 2-connected, it has by [1] another 1-factor, say F_1. Let $E = (V_1;V_2)$ be an edge of G, such that $E \in F$ and $E \notin F_1$.

$G - E$ is $(n-1)$-connected, and has F_1 as a 1-factor, therefore $G - E$ has at least $f(n-1)$ 1-factors, each one of which is a 1-factor of G, which does not contain E. $G^* = G - \{V_1;V_2\}$ is $(n-2)$-connected, and has $F - E$ as a 1-factor, hence G^* has at least $f(n-2)$ 1-factors: Adding the edge E to a 1-factor of G^* yields a 1-factor of G; therefore G has at least $F(n-2)$ 1-factors, each one of which contains the edge E.

Clearly, these two collections of 1-factors of G are disjoint, therefore G has at least $f(n-1) + f(n-2)$ 1-factors.

This completes the proof of the Lemma.

4. Main Results. The main ideas of this paper are contained in the following two lemmas:

Lemma 3. If every n-connected graph with a 1-factor contains a vertex V, such that at least n edges of G, incident to V, belong to $F(G)$, then for each k, $k \geq 1$,

$$f(k) \geq k(k-2)(k-4) \ldots$$

Proof. The proof is by induction on k, as follows: for $k = 1$ this is the triviality that $f(1) \geq 1$; for $k = 2$ the claim is that $f(2) \geq 2$ which has been proven in [1]. To complete the proof it suffices to show that for each k, $k \geq 3$,

$$f(k) \geq k \cdot f(k-2).$$

Let G be a k-connected graph with a 1-factor. From the assumption of the lemma it follows that G contains vertices $V, V_1, \ldots, V_k$ and edges $E_i = (V,V_i)$, $1 \leq i \leq k$, such that each E_i belong to some 1-factor F_i of G.

Let $G_i = G - \{V;V_i\}$, for each $1 \leq i \leq k$. Then each graph G_i is $(k-2)$ connected, and has a 1-factor: $F_i - E_i$. Let $\tilde{F}_i$ be the collection of all the 1-factors of G_i; therefore, Card $\tilde{F}_i \geq f(k-2)$. Let $F_i^* = \{A \cup E_i | A \in \tilde{F}_i\}$, for each $1 \leq i \leq k$.

Clearly, each member of F_i^* is a 1-factor of G. Moreover, $F_i^* \cap F_j^* = \emptyset$ for $i \neq j$, $1 \leq i, j \leq k$, because each member of F_i^* contains the edge (V,V_i) and each member of F_j^* contains the edge (V,V_j). Therefore G has at least $k \cdot f(k-2)$ 1-factors, and the proof of Lemma 3 is completed.

In order to prove that the condition of Lemma 3 holds, we prefer to prove the following stronger claim:

Lemma 4. If G is a 2-connected graph with a 1-factor F, then G contains a vertex V which is totally covered by $F(G)$.

Proof. The proof is by induction on the number k of the edges of G. The induction begins with $k = 4$, in which case $G = F(G)$, hence all the vertices of G are totally covered by $F(G)$, because in this case G must be a (simple) cycle of length 4.

Let us assume that the assertion is true for all the appropriate graphs with less than k edges, and let G be a 2-connected graph with 1-factor, having k edges.

To prove that G has a vertex, which is totally covered by $F(G)$, we suppose, on the contrary, that G has no such a vertex.

Claim 1. If G has an edge E, which belongs to one 1-factor F^*, but does not belong to another 1-factor F^{**}, of G, then $G - E$ is only 1-connected.

Proof. Suppose an edge E of G belongs to F^*, but not to F^{**}, and $G - E$ is 2-connected. F^{**} is a 1-factor of $G - E$, and $G - E$ has k-1 edges, hence, by the inductive assumption, $G - E$ contains a vertex V, which is totally covered by $F(G - E)$. Now, $E \in F^*$ therefore $E \in F(G)$, and clearly $F(G - E) \subset F(G)$; therefore the vertex V of G is totally covered by $F(G)$; this contradicts our supposition that G contains no such a vertex, and hence completes the proof of Claim 1.

Now, G is 2-connected and has a 1-factor F, therefore by Theorem 1 of [1], G has another 1-factor F'. The subgraph $(F' - F) \cup (F - F')$ of G is not empty, and all of its vertices are of valence 2; therefore its connected components are (simple) cycles of length r, where r is even and $r \geqslant 4$.

Let $S = (V_1; E_1; V_2; E_2; \ldots; V_r; E_r; V_1)$ be one such component, where $E_{2i} \in F$ and $E_{2i+1} \in F'$, $1 \leqslant 2i, 2i+1 \leqslant r$.

For each $1 \leqslant i \leqslant r$, we have by Claim 1 that $G - E_i$ is only 1-connected, because each one of the edges E_i belongs to precisely one 1-factor, among F and F', of G.

Since no vertex of G is totally covered by $F(G)$, and clearly $S \subset F \cup F' \subset F(G)$, therefore each vertex V_i of S has valence $\geqslant 3$. Therefore there are edges $E_1^*, \ldots, E_r^*$ in G, such that $E_i^* \cap S = V_i$, for all $1 \leqslant i \leqslant r$. Let $E_i^* = (V_i, V_i^*)$, for all $1 \leqslant i \leqslant r$.

G is 2-connected, therefore $G - V_i$, for each $1 \leqslant i \leqslant r$, is (at least 1-) connected. Let α_i be a path in $G - V_i$, connecting the vertex V_i^* with some vertex V_i^{**} of $S - V_i$, such that V_i^{**} is the only vertex of S on

α_i $(1 \leq i \leq r)$. Clearly $V_i^{**} \neq V_i$.

Claim 2. V_i^{**} is not a neighbor, in S, of V_i; for all $1 \leq i \leq r$. (Remark that this is a contradiction if $r = 3$, but as it was mentioned earlier, $r \geq 4$).

Proof. Suppose, on the contrary, that for some i, $1 \leq i \leq r$, V_i^{**} is a neighbor, in S, of V_i; say $V_i^{**} = V_{i+1}$*. G is 2-connected, while $G - E_i$ is only 1-connected, hence by Lemma 1 $G - E_i$ is only 1-connected between V_i and V_{i+1}. However, $\alpha_i \cup E_i^*$ and $S - E_i$ are two disjoint $V_i - V_{i+1}$ paths in $G - E_i$. This is a contradiction, and therefore Claim 2 is true.

Claim 3. There is an index i for which V_i V_{i+1} V_i^{**} V_{i+1}^{**} are in cyclic order on S.

Proof. Let i be the index for which the directed arc from V_i to V_i^{**} on S contains the minimal number of vertices of S, among all such possible (directed) arcs. This minimal number is not zero, because V_i^{**} is different from V_i and from its two neighbors in S; if V_{i+1}^{**} would be between V_{i+1} and V_i^{**}, the arc from V_{i+1} to V_{i+1}^{**} on S would contain less vertices than possible, which is impossible. Therefore V_{i+1}^{**} is not between V_{i+1} and V_i^{**}, as required.

Claim 4. The index i, as given by Claim 3, is such that $G - E_i$ is 2-connected.

Proof. Suppose, on the contrary, that $G - E_i$ is not 2-connected. Since G is 2-connected, it follows, using Lemma 1, that $G - E_i$ is only 1-connected between V_i and V_{i+1}.

To prove that this is not the case, we will show that $G - E_i$ contains two $V_i - V_{i+1}$ paths, disjoint except for common end points, as follows:

If $\alpha_i \cap \alpha_{i+1} \neq \emptyset$, then $S - E_i$ is one such a path, while the other is contained in $\alpha_i \cup \alpha_{i+1} \cup E_i^* \cup E_{i+1}^*$.

If $\alpha_i \cap \alpha_{i+1} = \emptyset$, then $E_i^* \cup \alpha_i$ together with the part of S from V_{i+1} to V_i^{**} is one path, while the other one is $E_{i+1}^* \cup \alpha_{i+1}$, together with the part of S between V_{i+1}^{**} and V_i. This two parts of S are disjoint because V_i V_{i+1} V_i^{**} V_{i+1}^{**} are in cyclic order in S, as given by Claim 3.

The proof of Claim 4 is complete.

---

* $i + 1$, as an index of a vertex of S, is $i + 1$ (mod r).

Clearly, Claims 1 and 4 are in contradiction, obtained by assuming that no vertex of G is totally covered by $F(G)$. The proof of Lemma 4 is complete.

Proof of Theorem 1. Theorem 1 states that if an n-connected graph has a 1-factor, it has at least as many as $n(n-2)(n-4) \ldots$ 1-factors, $n \geq 1$.

Since the assertion is trivial for $n = 1$, while it becomes just Theorem 1 of [1] for $n = 2$, we will assume that $n \geq 3$.

Every n-connected graph G with a 1-factor has, by Lemma 1, a vertex V which is totally covered by $F(G)$, since G is 2-connected. The valence of V is at least n, therefore the condition of Lemma 3 holds, and the proof of Theorem 1 follows immediately from Lemma 3.

To show that equality holds in $F(n) \geq n(n-2)(n-4) \ldots 5.3$, for odd n, observe that the complete graph with n+1 vertices, for odd n, has precisely $n(n-2)(n-4) \ldots 5.3$ 1-factors.

5. Remarks. We extend our Lemma 4 as follows:

Corollary 1. If G is a 2-connected graph with a 1-factor, then G contain two vertices which are totally covered by $F(G)$.

Proof. The proof is similar to that of Lemma 4, but here we suppose that G has at most one vertex, which is totally covered by $F(G)$. Therefore, at most one vertex V_j of S can have valence 2, in which case we do not define a corresponding path α_j. The proof of Claim 2 is valid, but restricted to all i, $i \neq j$, and $1 \leq i \leq r$. Claim 3 becomes meaningful if $i \neq j$ or $i + 1 \neq j$, and the proof is as follows:

Starting with an index i, $i \neq j$, the arc from V_i to V_i^{**} in S contain V_j, or not. If it does not contain V_j, we proceed as follows: If V_{i+1}^{**} is not on the arc from V_i to V_i^{**}, then $V_i\ V_{i+1}\ V_i^{**}\ V_{i+1}^{**}$ are in cyclic order. If V_{i+1}^{**} is on the arc from V_i to V_i^{**}, we apply a similar argument to the arc from V_{i+1} to V_{i+1}^{**}, which is shorter than that from V_i to V_i^{**}; This process ends with the required index.

If, however, the arc from V_i to V_i^{**} in S contains V_j, we change notations as follows: for some t, $V_i^{**} = V_t$, we replace E_t by the last edge of α_i, if it is not that edge; we replace α_t by $\alpha_i \cup E_i$ less the last edge of α_i, if it is not that path; As a result, V_t^{**} becomes V_i, if it was not.

Using these notations, we observe that V_j is not a vertex of the arc from V_t to V_t^{**} in S, and the first part of the proof is applicable again.

This completes the proof of the corollary.

Theorem 1 gives the precise value of $f(n)$, for all the odd n, while for the even n it gives only a lower bound.

Since every (simple) cycle of even length has precisely 2 1-factors, it follows that $f(2) = 2$.

The party graph*) P_{2k} is defined as the complement of the graph consisting of k disjoint edges. P_6 is a 4-connected graph with precisely 8 1-factors, hence $f(4) \leqslant 8$, and using Theorem 1 it follows that $f(4) = 8$.

For even n, $n \geqslant 6$, we have the following:

$$48 \leqslant f(6) \leqslant 60$$
$$384 \leqslant f(8) \leqslant 544$$
$$3840 \leqslant f(10) \leqslant 6040, \text{ etc.}$$

where the upper bound for $f(n)$ is obtained using the graph P_{n+2}.

6. Remarks and Conjectures. A vertex V of a graph G is called totally covered by $M(G)$ if each edge of G, incident to V, belongs to some maximal matching of G. Let $g(n)$ be the maximal k for which the following is true: "If G is an n-connected graph with no 1-factors, then G has at least k different maximal matchings."

We have the following

Lemma 5. If all the vertices of a graph G are of valence $\geqslant k$, and G has no 1-factors, then G has $k+1$ vertices which are totally covered by $M(G)$.

This implies

Theorem 2. *For all* $n \geqslant 3$, $g(n) \geqslant n \cdot g(n-2) + \min\{g(n-1); f(n-1)\}$.

Proofs will be published somewhere else.

Conjecture 1. If G is an n-connected graph with a 1-factor, then G has n vertices which are totally covered by $F(G)$, $n \geqslant 3$.

Conjecture 2. For all $k \geqslant 3$, $f(2k) > 2^k \cdot k!$

---

*) P_{2k} "represents" a cocktail party of k couples, where everybody is talking to everybody else, except for each husband talking to his wife.

Problem. Determine $f(2k)$, for $k \geq 3$.

Conjecture 3. For every odd n, and for $n = 4$, there is a unique graph G_n which is n-connected and has precisely $f(n)$ 1-factors.

Assuming the last conjecture is true, let $f^*(n)$ be the maximal k for which the following is true: "If G is an n-connected graph with a 1-factor, and $G \neq G_n$, then G has at least k different 1-factors".

Conjecture 4. $f^*(3) = 4$, $f^*(4) = 10$ and $f^*(5) = 30$.

Let $f(n,v)$ be the maximal k for which the following is true: "If G is an n-connected graph with v vertices, and G has a 1-factor, then G has at least k different 1-factors".

Let $B(n,k)$ be the graph, obtained from the bipartite graph on n and $n + 2k$ vertices, by adding k disjoint edges with end points in the set of $n + 2k$ vertices. $B(n,k)$ is n-connected and has $n!$ 1-factors, therefore $f(n,v) \leq n!$, for all $v \geq 2n$.

Conjecture 5. $f(n,v) = n!$, for all $v \geq 2n$.

$B(n,1)$ is a counterexample to the following

Conjecture 6[*)]. If G is an n-connected graph with a 1-factor, and n is large enough, then G has two 1-factors which have no edge in common.

Reference

1. Beineke, L. W. and Plummer, M. D., On the 1-factors of a non-separable Graph, *J. Comb. Theory* 2 (3) (1967), pp. 285-289.

*) It was asked by W. T. Tutte, in private communication.

DECOMPOSITION OF THE COMPLETE DIRECTED GRAPH INTO FACTORS WITH GIVEN DIAMETERS

Štefan Znám

Bratislava, Czechoslovakia

We shall consider directed graphs without loops and multiple edges, but two edges ab and ba are allowed. The complete directed graph of this kind with n vertices (n is a cardinal number) is denoted by $\ll n \gg$. Our article is devoted to the decomposition of $\ll n \gg$ into factors with given diameters $d_1, \ldots, d_k$ (d_i are naturals or symbols ∞). In the article [1] the analogical problem for the non-directed case was studied.

We shall denote by the sumbol $E(d_1, \ldots, d_k)$ the smallest cardinal M for which the graph $\ll M \gg$ is decomposable into k factors of diameters $d_1, \ldots, d_k$. The case $k = 2$ was studied by E. Tomová, who proved that

$$E(d_1, d_2) = \begin{cases} d_2 + 1 & \text{if } 2 \leqq d_1 \leqq d_2 < \infty \\ \infty & \text{if } 1 = d_1 \leqq d_2 < \infty \\ d_1 + 1 & \text{if } 1 \leqq d_1 < d_2 = \infty \\ 2 & \text{if } \quad d_1 = d_2 = \infty \end{cases}$$

The importance of the number $E(d_1, d_2)$ is evident from the following theorem:

<u>Theorem</u>. *If the complete graph $\ll n \gg$ can be decomposed into two factors with given diameters d_1 and d_2, then for any cardinal $N > n$ the complete graph $\ll N \gg$ can be also decomposed into two factors with diameters d_1 and d_2.*

In the non-directed case an analogical theorem was true for decomposition into any number of factors, however in the directed case it does not hold, as will be shown in the following.

We shall need three lemmas:

Lemma 1. Any directed graph of diameter k with $k+2$ vertices contains at least $k+4$ edges (Proof see [4]).

Lemma 2. The graph $\ll k+2 \gg$ cannot be decomposed into k factors of diameter k for any $k > 3$.

Proof. Lemma 2 immediately follows from Lemma 1.

Lemma 3. If k is even, $k > 3$ or if $k = 21$, then $\ll k+1 \gg$ can be decomposed into k factors of diameters k.

Proof. For the case k even, $k > 3$ the assertion follows from the well-known fact that any undirected complete graph with an odd number of edges can be decomposed into Hamiltonian cycles. Further, Mendelsohn proved that $\ll 21 \gg$ can be decomposed into 21 directed hamiltonian paths, hence $\ll 22 \gg$ can be decomposed into 21 Hamiltonian cycles.

From the lemmas 2 and 3 it follows that for k even, $k > 3$ and for $k = 21$, the graph $\ll k+1 \gg$ can be, but the graph $\ll k+2 \gg$ cannot be decomposed into k factors of diameter k, and hence the above mentioned theorem does not hold for these k. For the remaining k's the problem is open. The most interesting case is $k = 3$.

References

1. Bosák, J., Rosa, A., Znám, Š., On decomposition of complete graphs into factors with given diameters, *Theory of graphs, Proc. Coll. Tihany* 1966.

2. Mendelsohn, N.S., Hamiltonian decomposition of the complete directed n-graph, *ibidem*.

3. Tomová, E., On decomposition of a complete directed graph into two factors with given diameters, *Mat. Časop.* - to appear.

4. Znám, Š., Decomposition of complete directed graphs into factors with given diameters, *Mat. Časop.* - to appear.

Problem Session of the Conference

There have been 3 problem sessions at the C.I.C.O.C.S.A.T.A. We include here these problems and several others submitted by participants of the conference. If one of these problems was solved during the conference the solution appears immediately behind the problem.

Problem 1. Branko Grünbaum

Valence Sequences

A sequence $V = (v_1,\ldots,v_n)$ of integers is a *valence sequence* provided there exists a graph G with n vertices such that the ith vertex of G has valence v_i : G is said to *realize* V. Necessary and sufficient conditions for a sequence V to be a valence sequence were given by Havel, Erdös-Gallai (see also Harary), and Fulkerson-Hoffman-McAndrew.

(1) *Problem.* Characterize valence sequences of planar graphs.

It is known (Hakimi) that if $v_1 \geq v_2 \geq \ldots \geq v_n$, then

$$\sum_{i=1}^{k} v_i \leq 6\ (k-2) + \sum_{i=k+1}^{n} v_i \qquad \text{for } k = 3,4,\ldots,n$$

is a necessary condition. However, this condition is not sufficient, as shown by the examples $V = (5,4,4,4,4,3)$ (Böttger-Harders), $V = (4,4,4,4,4,2)$, etc.

References

1. Erdös, P. and Gallai, T., Graphen mit Punkten vorgeschriebenen Grades, (Hungarian), *Matem. Lapik,* 11 (1960), 264-274.

2. Havel, V., Poznámka o existenci konecných grafu, *Casop. Pest. Mat.,* 80 (1955), 477-480.

3. Harary, F., Graph Theory, Addison-Wesley, Reading 1969.

4. Böttger, G. and Harders, H., Note on a problem by S.L. Hakimi concerning planar graphs wihtout parallel elements, *J. SIAM,* 12 (1964), 838-839.

5. Fulkerson, D.R., Hoffman, A.J., and McAndrew, M.H., Some properties of graphs with multiple edges, *Canad. J. Math.,* 17 (1965), 166-177.

Problem 2. Branko Grünbaum

1-factors

If $V = (v_1,\ldots,v_n)$ may be realized by a graph G which has a 1-factor, then clearly $(v_1-1, v_2-1, \ldots, v_n-1)$ is also a valence sequence. We venture

(1) *Conjecture.* A sequence $(v_1,\ldots,v_n)$ may be realized by a graph which has a 1-factor if and only if n is even and both $(v_1,\ldots,v_n)$ and $(v_1-1, v_2-1,\ldots,v_n-1)$ are valence sequences.

(2) *Problem.* Characterize those valence sequences which have the property that every graph realizing them has a 1-factor. (Examples of such sequences: Valence sequences of complete graphs with an even number of vertices: (3,3,3,3,3,3) : (3,3,3,3,3,3,3,3) : (4,4,3,3,3,3) : etc.)

References

1. Hakimi, S.L., On the realizability of a set of integers as degrees of the vertices of a linear graph II, *J. SIAM*, 11 (1963), 135-147.

2. Larman, D.G. and Mani, P., On the existence of certain configuration within graphs and the 1-skeletons of polytopes, *Proc. London Math. Soc.*, (to appear).

Problem 3. Branko Grünbaum

Connectedness

Let $f(n)$ denote the least integer k with the property: For every k-connected graph G and for every $2n$ distinct vertices $A_1,\ldots,A_n$, $B_1,\ldots,B_n$ of G, there exists a family of n disjoint paths P_i such that P_i has endpoints A_i and B_i. It is known that $f(1) = 1$, $f(2) = 6$ (Mani; Jung [2]), and $2n-1 \leqslant f(n) \leqslant 3n.2^{\binom{3n}{2}}$ for $n \geqslant 3$ (see Watkins and Larman-Mani).

M. Rosenfeld has made the following

(1) *Conjecture.* $f(n) = \binom{2n}{2}$ for all n.

References.

1. Fulkerson, D.R., Hoffman, A.J., and McAndrew, M.H., Some properties of graphs with multiple edges, *Canad. J. Math.*, 17 (1965), 166-177.

2. Mani, D., Bridges in 6-connected graphs. (To appear).

3. Jung, H.A., A variation of n-connectedness. (To appear).

4. Larman, D.G. and Mani, P., On the existence of certain configurations within graphs and the 1-skeletons of polytopes. *Proc. London Math. Soc.* (to appear).

Problem 4. Branko Grünbaum

Connectedness

Let $g(n)$ be the least integer k with the property: Every k-connected graph contains a subdivision of the complete graph with n vertices. It is known that

$$g(n) = n-1 \quad \text{for } n = 2,3,4.$$
$$6 \leqslant g(5) \leqslant 256$$
$$\frac{1}{8}(n^2-1) \leqslant g(n) \leqslant (n-1)\cdot 2^{\binom{n-1}{2}} \quad \text{for } n \geqslant 5.$$

(Dirac, Mader, Jung [1]).

(1) *Conjecture.* $g(5) = 6$.

References

1. Dirac, G.A., In abstrakten Graphen vorhandene vollständige 4-Graphen und ihre Unterteilungen, *Math. Nachr.*, 22 (1960), 61-85.

2. Mader, W., Homomorphieeigenschaften und mittlere Kantendichte von Graphen, *Math. Ann.*, 174 (1967), 265-268.

3. Jung, H.A. [1], Zusammenzüge und Unterteilungen von Graphen, *Math. Nachr.*, 35 (1967), 241-267.

Problem 5. Branko Grünbaum

A. Kotzig (private communication) noted that the solution of (P 2.2) is trivially affirmative if (loopless) multigraphs are allowed. For such multigraphs he also completely solved problem (P. 31.1); the solution is somewhat involved for small n, but if the number of vertices is at least 10, then $(1,1,\ldots,1)$ is the only valence sequence for which every realization by a (loopless) multigraph has a 1-factor.

References

1. Dirac, G.A., In abstrakten Graphen vorhandene vollständige 4-Graphen und ihre Unterteilungen, *Math. Nachr.*, 22 (1960), 61-85.

2. Mader, W., Homomorphieeigenschaften und mittlere Kantendichte von Graphen, *Math. Ann.*, 174 (1967), 265-268.

3. Jung, H.A. [1], Zusammenzüge und Unterteilungen von Graphen, *Math. Nachr.*, 35 (1967), 241-267.

Problem 6. A. Kotzig

Let G be a graph which has the following properties:

(1) G does not contain any linear factor;

(2) If we add to G one arbitrary edge (not belonging to G) we obtain always a graph containing at least one linear factor.

Let $n, k_1, k_2, \ldots, k_{n+2}$ be natural numbers or 0 and let F be the graph constructed as follows:

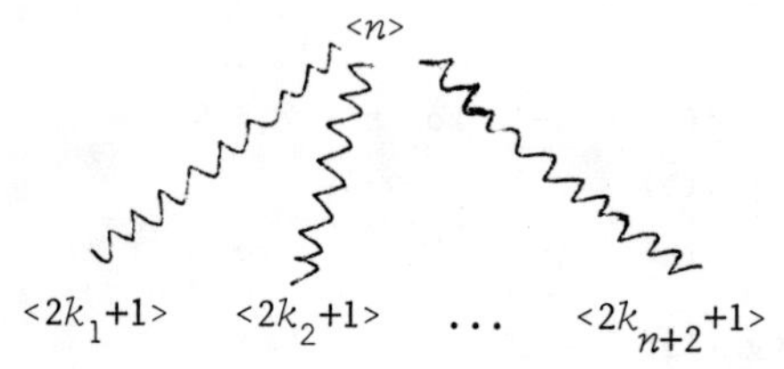

where: $\langle m\rangle$ is the complete graph with m vertices; $\langle 0\rangle$ means the empty graph and $G_1 \leftrightsquigarrow G_2$ means that each vertex from the graph G_1 is joined by an edge with each vertex from the graph G_2.

It is clear that F has the properties (1) and (2).

<u>Conjecture</u>. Every graph with the properties (1), (2) can be constructed by the described way.

Problem 7. A. Rosa

An L_n-factor F is a nearly linear factor of the complete graph $\langle 2n-1\rangle$ (all vertices of F but one have degree one, one vertex is isolated in F).

Let $v, v_1, \ldots, v_{2n-1}$ be the vertices of $\langle 2n-1\rangle$. Define the distance ρ_{ij} between two vertices v_i, v_j (the length of the edge $[v_i, v_j]$) as follow $\rho_{ij} = \min(|i-j|, 2n-1-|i-j|)$. Two edges $[v_i, v_j]$, $[v_m, v_n]$ are said to be parallel if $\rho_{im} = \rho_{jn}$. An edge $[v_i, v_j]$ and a vertex v_m are parallel if $\rho_{im} = \rho_{jm}$. A perfect L_n-factor F is an L_n-factor satisfying the following conditions:

(1) No two edges of F have the same length;

(2) No two edges of F are parallel;

(3) There is no edge in F parallel to the isolated vertex in F.

<u>Conjecture</u>. A perfect L_n-factor exists if and only if $n \geqslant 4$, $n \neq 5$.

Problem 8. P. Erdös

Let $G(n; \left[\frac{n^2}{4}\right] + k)$ be a graph of n vertices and $\left[\frac{n^2}{4}\right] + k$ edges. I can prove that if $k < cn$ then our graph contains k edge-disjoint triangles using the method of my paper "On a theorem of Rademacher-Turán, Illinois Journal of Math, 6 (1962), 122-127.

It seems that the result remains true for very much larger values of k, perhaps for all $k < cn^2$ where c is a sufficiently small constant.

Problem 9. P. Erdös

Let G be a k-chromatic graph which contains no $\langle k-r \rangle$. Dirac and I thought that then G contains at least $k+2r$ vertices. This result certainly holds if r or $k-r$ are very small. Chvatal disproved our conjecture in fact he showed that it is false even with $k+r+r^{\frac{1}{2}+\varepsilon}$.

It seems that our conjecture remains true if $r < ck$ where c is a sufficiently small constant.

Solution of Problem 9. V. Chvátal

On a problem of Erdös

P. Erdös asked the following question.

Must every k-chromatic graph containing no complete subgraph with $k - r + 1$ vertices contain at least $k + 2r$ vertices?

We shall show that the answer is negative.

Let us denote by $f(n,m)$ the least integer f such that every graph with f vertices contains either n independent vertices or a complete subgraph with m vertices. Erdös [1] has proved that

$$f(3,m) > c\left(\frac{m}{\log m}\right)^2 \tag{1}$$

for an absolute constant $c > 0$. Particularly, there is an integer m_0 such that

$$f(3,m) > 4m - 5 \tag{2}$$

whenever $m \geq m_0$. Given an integer m satisfying (2), set $r = m - 1$ and $k = 2r$. Then (2) becomes $f(3,k-r+1) > k+2r-1 = 2k-1$ and there is a graph G with $k + 2r - 1$ vertices containing neither a complete graph with $k - r + 1$ vertices nor three independent vertices. The last fact implies that the chromatic number of G is not less than k (since there are no three vertices of the same color and the total number of vertices is $2k - 1$) and the conjecture is disproved. The smallest counterexample obtained in this way is a 16-chromatic graph with 31 vertices containing no complete subgraph with 9 vertices. Its existence is justified by $f(3,9) \geq 36$ (see [2]).

Similarly, the answer is also negative whenever $k + 2$ is replaced by $k + r + r^{\frac{1}{2}+\varepsilon}$, $\varepsilon > 0$ a constant.

Without any loss of generality, we may assume $\varepsilon < 1$. Find m so large that

$$\tfrac{1}{2}\left[c\left(\frac{m}{\log m}\right)^2\right] - m > m^{2-\varepsilon}$$

and set $k = \left[\tfrac{1}{2}\left[c\left(\frac{m}{\log m}\right)^2\right]\right]$, $r = k - m + 1$. By (1), there exists a graph with $\left[c\left(\frac{m}{\log m}\right)^2\right]$ vertices containing neither a complete subgraph with $m = k - r + 1$ vertices nor three independent vertices. Clearly, its chromatic number is not less than $\left[\tfrac{1}{2}\left[c\left(\frac{m}{\log m}\right)^2\right]\right] = k$. However, one has

$$\begin{aligned} r^{\frac{1}{2}+\varepsilon} &= \left(\left[\tfrac{1}{2}\left[c\left(\frac{m}{\log m}\right)^2\right]\right] - m + 1\right)^{\frac{1}{2}+\varepsilon} \\ &\geq \left[\tfrac{1}{2}\left[c\left(\frac{m}{\log m}\right)^2\right] - m\right)^{\frac{1}{2}+\varepsilon} \\ &> m^{(2-\varepsilon)(\frac{1}{2}+\varepsilon)} > m \end{aligned}$$

or

$$\begin{aligned} \left[c\left(\frac{m}{\log m}\right)^2\right] &< \left[c\left(\frac{m}{\log m}\right)^2\right] + r^{\frac{1}{2}+\varepsilon} - m \\ &\leq 2k + 1 - m + r^{\frac{1}{2}+\varepsilon} \\ &= k + r + r^{\frac{1}{2}+\varepsilon} \end{aligned}$$

References

1. Erdös, P., Graph theory and probability II, *Canad. J. Math.*, 13 (1961), 346-352.

2. Kalbfleisch, J.G., Upper bounds for some Ramsey numbers, *J. Combinatorial Th.*, 2 (1967), 35-42.

Problem 10. László Lovász

Let G be an arbitrary finite, $(a+b-1)$-chromatic graph, which is not a clique $(a,b\geq 2)$. Does G always contain two vertex-disjoint subgraphs of chromatic number a and b, respectively? (Even the case $a = 2$ is unsolved).

Problem 11. László Lovász

Let us construct a finite, connected, undirected graph which is symmetric and has no simple path containing all vertices. A graph is called symmetric, if for any two vertices x,y it has an automorphism mapping x onto y.

Problem 12. J. Sedláček

Let G be a finite connected graph without loops and multiple edges. Define a graph $c(G)$ the vertices of which are all the spanning trees of G. Let two vertices T_1, T_2 of $c(G)$ be joined by an edge if the tree T_2 can be derived from T_1 by a singular cyclic interchange (O. Ore: Theory of graphs, p. 104). Prove or disprove the following conjecture: If G contains at least one circuit then $c(G)$ has a Hamilton circuit.

Problem 13. Michael F. Capobianco

Associate with a graph having n vertices an n-tuple, $(x_0,x_1,\ldots,x_{n-1})$ where x_0 is the number of pairs of vertices which have no paths between them, and x_i is the number of pairs which have paths of length i between them, for $i = 1,2,\ldots,n-1$. We may call this n-tuple the *path length distribution* (PLD) of the graph.

Two isomorphic graphs have the same PLD, but the converse is not true. The problem I pose is that of realizability, i.e. what conditions must $(x_0,x_1,\ldots,x_{n-1})$ satisfy in order to be the PLD of some graph?

Problem 14. M.R. Williams

There exists a class of graphs, G, such that each edge of a graph $G(G \in G)$ is part of at least $n-2$ edge circuits of length three (i.e. triangles for some n. G may be divided into two subclasses:

1) those graphs which contain a K_n (a complete graph on n vertices)

2) those graphs which do not contain a K_n.

For example:

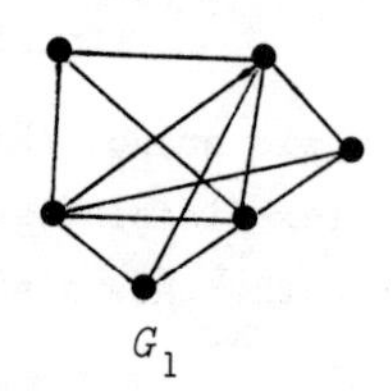

G_1

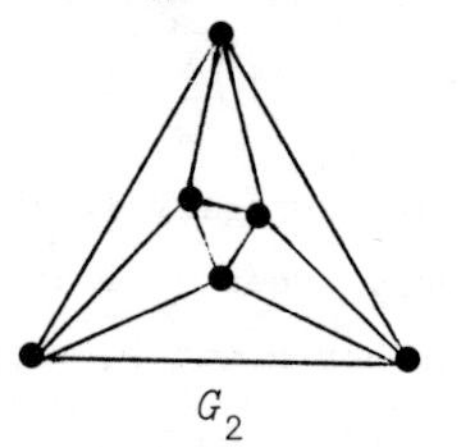

G_2

if $n = 4$ then each edge in the graphs G_1 and G_2 is a member of at least $n-2$ edge circuits of length three. G_1 belongs to subclass 1 because it contains three examples of a K_4 while G_2 belongs to subclass 2 because it does not contain a K_4.

Problem. Given a graph G ($G \in G$) devise a nonrecursive algorithm which, given a specific n, will determine whether G is a member of subclass 1 or subclass 2 of G. This algorithm should not resort to checking of all possible combinations of n vertices in the graph.

Problem 15. R.L. Graham

What is the maximum area of the convex hull of a set of n points in the plane given that no distance between any pair of points exceeds 1? For *odd* n, the optimal configurations are almost certainly the n vertices of the regular n-gon of diameter 1. For *even* n, the situation is more difficult. For $n = 4$, the vertices of a square of diameter 1 form an optimal configuration although it is not unique. For n even and ≥ 6, the optimal configurations are *not* the vertices of the regular n-gon of diameter 1. I have determined the unique optimal configuration for $n = 6$. It doesn't seem to be particularly nice, requiring, for example, a root of an irreducible 10th degree polynomial.

We can ask the same question in higher dimensions, the first interesting case being the maximum volume of the convex hull of a set of 5 points in 3-space which has diameter 1.

Problem 16. R.L. Graham

Suppose the plane is covered by a family of sets of diameter ≤ 1. We can assume without loss of generality that the sets are convex. How small can the average area of the sets be? (The average is defined in the usual manner for such packing and covering problems). It is conjectured that the average area is always $\geq \frac{3\sqrt{3}}{8}$. This bound can be attained by regular hexagons of diameter 1 and *also* by an uncountable number of pairs of irregular hexagons of diameter 1.

Problem 17. R.L. Graham

Suppose the finite subsets of a countable set S are split into r classes. Is it true that we can always find an infinite family of disjoint finite subsets of S, all of whose finite (nonempty) unions are in one class?

This is true if we just ask for a finite number of subsets of S.

Problem 18. E.N. Gilbert, H.O. Pollak, R.L. Graham

For a finite set S of points in the plane, let $\lambda(S)$ denote the minimal length of any tree having S as its set of vertices (where the length of a tree is the sum of its edge lengths). Is it true that

$$\frac{\lambda(T)}{\lambda(S)} \leq \frac{2}{\sqrt{3}} \quad \text{for} \quad T \subseteq S?$$

One example (there are others) which achieves the bound is given by: S = vertices + centroid of an equilateral triangle, T = vertices of the triangle.

Problem 19. R.L. Graham and S. Silverman

Let L denote a finite set of lines in the plane. For a fixed integer r, call a finite set S of m points *admissible* if any line parallel to a line of L is either disjoint from S or intersects S in exactly r points. Is it clear that $m \equiv 0 \pmod{r}$ is necessary. I can show that if $m \equiv 0 \pmod{r}$ and m is sufficiently large then an admissible configuration of m points exists. What is (a reasonable bound on) the least integer n for which an admissible configuration of nr points does not exist? In particular, for L = set of 4 lines with slopes $0,\infty,1,-1$ and $r = 2$, is there an admissible configuration of 14 points? (This is the only undetermined value for this L and r).

Problem 20. R.L. Graham and S. Silverman

A set C of edges of a graph G is said to be a *cutset* of G if the removal of the edges C from G disconnects G. C is said to be a *simple* cutset if no two edges of C have a common vertex. Call G *primitive* if G has no simple cutset but every proper subgraph of G has a simple cutset (possibly empty). What are the primitive graphs?

Some examples are given as follows. Let $T(G)$ denote the class of graphs which is closed under the operation

and which contains G. Then $T(X)$ consists entirely of primitive graphs when X is either of the graphs:

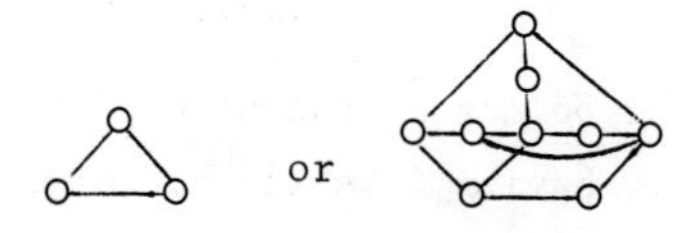

Are these primitive graphs with no cycle of length ≤ 4? $\leq k$? If $e(n)$ denotes the maximum number of edges a primitive graph on n vertices can have, is $e(n) = O(n)$? I can only show $e(n) < (\frac{1}{2}+\varepsilon)n \log n/\log 2$.

Problem 21. Branko Grünbaum

Vectors

Let $v(n)$ denote the least integer v with the property: Every set S of n points in general position in the plane determines at least v different vectors (having endpoints in S). It is not hard to see $v(2) = 3$, $v(3) = 7$, $v(4) = 9$.

(1) *Problem.* Determine $v(n)$.

One may easily show that $v(2^k) \leq 3^k$, and for small k it is probable that equality holds. However, L. Danzer (private communication) has found that $v(2^k) < 3^k$ for sufficiently large k.

Problem 22. Branko Grünbaum

Polytopes

The famous theorem of Steinitz (see Grünbaum or Barnette-Grünbaum [2] for this formulation) asserts that a graph G is isomorphic to the graph determined by the vertices and edges of some 3-dimensional (convex) polytope P if and only if G is 3-connected and planar.

For a given 3-connected planar G, there is quite a lot of freedom in the choice of the shape of P; as a matter of fact, the freedom seems to be sufficient to justify the following

(1) *Conjecture.* Let G be a 3-connected planar graph, let G be a simple circuit in G containing n vertices, and let F be a given convex n-gon in a plane L. Then there exists a 3-polytope P such that

(i) its vertices and edges determine a graph isomorphic to G;

(ii) the orthogonal projection π of E^3 onto L maps P onto F;

(iii) the inverse image $\pi^{-1}(bd\ F)$ of the boundary of F intersects P precisely in the vertices and edges that correspond to C.

The conjecture is known to be true in two special situations:

(1) If C is a non-separating circuit of G (see Barnette-Grünbaum [1]);

(2) If no specific shape is prescribed for the n-gon F (see Barnette).

References

1. Steinitz, E., Polyeder und Raumeinteilungen, *Enzykl. Math. Wiss.*, 3 (1922), (Geometrie), Part 3AB12, 1-139.

2. Barnette, D.W. and Grünbaum, B. [2], On Steinitz's theorem concerning convex 3-polytopes and on some properties of planar graphs. "The many facets of graph theory", edited by G. Chartrand and S.F. Kapoor. Springer 1969 (to appear).

3. Barnette, D.W., Projection of 3-polytopes. (To appear).

Problem 23. Branko Grünbaum

Projective plane

Consider n lines in general position in the projective plane, and the convex polygons into which they partition the projective plane. For $n = 3$, 6, or 10 it is possible to place the n lines so that none of the polygons is a quadrangle.

(1) *Conjecture.* If $n \neq 3,6,10$, every n lines in general position in the projective plane determine some quadrangle.

References

1. Barnette, D.W. and Grünbaum, B. [1], Preassigning the shape of a face, *Pacific J. Math.* (to appear).

2. Grünbaum, B., Convex polytopes, Interscience, New York 1967.

Problem 24. E. Jucovič

(a) What is the minimum number of horocycles determined by n points in the hyperbolic plane?

(b) What is the minimum number of ordinary horocycles (= incident with exactly 2 of the given points) determined by n points?

(c) What is the maximum number $f_m(n)$, $n > 2$, of horocycles incident with exactly m from $n > m$ points?

Problem 25. L. Moser

Given the r^n points $(x_1,\ldots,x_n)$, where the x_i's are positive integers $\leq r$, estimate $f(n,r)$ = the maximum number of these points with no r collinear. I offer $10. for proof or disproof that $f(n,3) = 0(3^n)$. Is $f(n,r) = 0(r^n)$ for every fixed r.

Problem 26. J. Smith

Croft has asked whether there exists a set of measure zero in the plane all of whose translations (or all of whsoe rotations about finite centers) contain infinitely many lattice points. A negative answer in these cases (and for several other sets of motions such as all reflections or all rotations of $\pi/2$) is contained in the following theorem.

Theorem. *Let G be a locally compact group of isometries, containing the translations, of Euclidean n-space E (topology of uniform convergence on compacts). Let $S \subset E$ have zero (Lebesque) measure, $T \subset E$ be countable, and $A \subset G$ have positive (left Haar) measure. Then $g(T) \cap S = \phi$ for almost all $g \in A$ (in sense of measure).*

Sketch of proof. Since G includes translations, a patching argument shows that for any $x \in E$ the (left Haar) measure of $\{g \in G | g(x) \in U\}$ is proportional to the (Lebesque) measure of $U \subset E$. Hence if S is of measure zero and T countable the measure of $\{g | g(T) \cap S \neq \phi\} =$
$= \bigcup_{x \in T} \{g | g(x) \in S\}$ is zero. Hence if A has positive measure, almost all of A is outside this set.

Problem 27. E.R. Berlekamp

A *complete, balanced Howell rotation* for a bridge tournament on n partnerships and n-1 boards may be defined by two $(n-1) \times n$ matrices.

The columns of each matrix represent the partnerships and the rows represent the boards. One matrix contains integers which specify the round numbers on which the various partnerships play the various boards; the other matrix contains signs which specify the direction in which the various partnerships play the various board. The matrices must satisfy 3 conditions:

i) Annexing an all-plus row to the signs gives a Hadamard matrix.

ii) Every column contains every number once.

iii) Numbers in any pair of columns agree in a unique row, and the corresponding signs are opposite.

Example with $n = 8$:

0-	0+	1+	2+	4-	4+	2-	1-
3-	4-	3+	4+	5+	0-	0+	5-
6-	1-	0-	6+	0+	1+	3-	3+
2-	6+	4-	3-	2+	3+	4+	6-
5-	2-	2+	0-	6-	5+	6+	0+
1-	3+	5-	5+	3-	2-	1+	2+
4-	5+	6+	1-	1+	6-	5-	4+

In a paper which has been submitted to the American Mathematical Monthly, Berlekamp & Huang have shown that such tournaments exist if $n \equiv 0 \bmod 4$ and $n - 1$ is a prime power greater than 3. It is known that no such tournament exists with $n = 4$, but all other values of $n \equiv 0 \bmod 4$ are unsolved.

Problem 28 K.A. Bush

Very little is known about symmetric, orthogonal matrices. In particular we ask whether Hadamard matrices of order $4t$ exist which are symmetric and admit a partitioning into submatrices of order $2t^2$ such that the submatrices on the principal diagonal are J (all entries +), and the off-diagonal submatrices have row and column sum zero. If there is no such matrix, then there is no finite geometry of order $2t$.

Problem 29. K.A. Bush

A is a 0-1 matrix of order $n \cdot k$, and A is symmetric. Each row contains u zeros. Subsidiary conditions such as $a_{ii} = 1$ may be imposed. Call this set of conditions C. With these conditions, can a permutation matrix P be found such that for all sufficiently large

$$PAp^t = \begin{pmatrix} J & & \\ & J & \\ & & \ddots \\ & & & J \end{pmatrix}$$

where each J is $N \times n$ and there are k of them? In particular if $k = 3$, $u = 4$, and C is the set of conditions

(i) $a_{ii} = 1$

(ii) for any triple i,j,k is two of the $a_{i,j,k}$ are zero, then so is third (e.g. if $a_{ij} = a_{ik} = 0$, then $a_{jk} = 0$). Then if P can be found, it is possible to color a planar map in 4 colors.

Problem 30. Jean Doyen

Let $C(n)$ denote the number of non isomorphic cyclic Steiner triple systems of order n (such a system exists for every $n \equiv 1$ or $3 (\text{mod } 6)$ except for $n = 9$). It is known that

$C(1) = C(3) = C(7) = 1$, $C(9) = 0$, $C(13) = 1$,
$C(15) = 2$, $C(19) = 4$, $C(21) = 7$, $C(25) = 12$,
$C(27) = 8$, $C(31) = 80$, $C(33) = 84$.
$C(37) = 820$, $C(43) = 9380$, and that
$C(n) \gtrsim 2^{k-1}/k$ whenever $n = 6k + 1$ is a primer number. Is it true that $C(n)$ goes to infinity with n, where $n \equiv 1$ or $3 (\text{mod } 6)$?

Problem 31. Branko Grünbaum

Tournaments

It is well known (Rédei, Szele, Moon) that every n-tournament (= oriented complete graph with n nodes) has an odd number of different Hamiltonian paths (= spanning paths). The n-tournaments with precisely one Hamiltonian path are just those corresponding to totally ordered sets. Some information concerning $h(n)$, the maximal number of different Hamiltonian paths possible in an n-tournament, is available, but even Szele's conjecture that $\lim_{n \to \infty} (h(n)/n!)^{1/n} = \frac{1}{2}$ is still open.

Possibly easier to decide is

(1) *Conjecture.* For each odd integer h with $1 \leq h \leq h(n)$, there exists an n-tournament with precisely h different Hamiltonian paths.

Douglas has recently characterized those n-tournaments which have a single Hamiltonian circuit (= spanning cycle). However, the problem of determining the maximal number $h_o(n)$ of different Hamiltonian circuits possible in an n-tournament is still open, and so is the characterization of the tournaments having $h_o(n)$ different Hamiltonian circuits.

In analogy to the above we have

(2) *Conjecture.* For every integer h with $0 \leq h \leq h_o(n)$ there exists an n-tournament with precisely h Hamiltonian circuits.

References

1. Douglas, R.J., Tournaments that admit exactly one Hamiltonian circuit. (Preprint, 1969).

2. Moon, J.W., Topics on tournaments, Holt, Rinehart and Winston, New York, 1968.

3. Szele, T., Kombinatorische Untersuchungen über gerichtete vollständige Graphen. (Hungarian) *Mat. Fiz. Lapik,* 50 (1943), 223-256. German translation in *Publ. Math. Debrecen,* 13 (1966), 145-168.

4. Rédei, L., Ein kombinatorischer Satz., *Acta Math. (Szeged).,* 7 (1934), 39-43.

Problem 32. J. Schönheim

Given a $n \times n$ chasseboard, coloured with n colours, such that each colour occurs in each row, in each column and in the main diagonal. A *super rook* standing on colour x is allowed to move to any place of his row, of his column or having colour x.

How many super rooks placed on the main diagonal are sufficient to attack every off main diagonal place:

(a) by any colouring

(b) by a colouring which is symmetrical in respect to the main diagonal?

Problem 33. J. Schönheim

Let $Q_1, Q_2, \ldots, Q_m$ be a set of quasigroups, each of order n none of which is isotopic to a group. Determine the maximum value of m so that the Cayley tables of the Q_i form a family of mutually orthogonal latin squares.

Conjecture. $m < n - 1$.

Problem 34. P. Erdös and L. Moser

Find bounds for $f(n)$ = the least number of subsets of a set A of n elements such that every subset of A is the union of two of the $f(n)$ subsets. It is easy to prove that

$$\sqrt{2} \cdot 2^n < f(2n) \leq 2 \cdot 2^n.$$

We offer \$25. deciding (with proof) whether $f(2n)$ is > or < $(1.75)2^n$ for sufficiently large n.

Problem 35. P. Erdös and L. Moser

Find bounds on $f(n)$ = the largest number of subsets $A_1, A_2, \ldots, A_{f(n)}$ of a set of n elements such that the $\binom{f(n)}{2}$ sets $A_i \cup A_j$, $1 \leq i < j \leq f(n)$, are distinct. We can prove that for large n

$$(1+\varepsilon_1)^n < f(n) < (1+\varepsilon_2)^n,$$

where $0 < \varepsilon_1 < \varepsilon_2 < 1$, and offer \$25. for finding $\varepsilon_1, \varepsilon_2$ with $\frac{\varepsilon_2}{\varepsilon_1} < 1.01$.

Problem 36. Joel Spencer

Define $N(n,k)$ as the minimal size of a family $R_1, \ldots, R_s$ of simple orders on $\{1,2,\ldots,n\}$ having the property that for all $x_1, \ldots, x_k \in \{1, \ldots$ we have $x_2, \ldots, x_k \; R_i \; x_1$ for some i. [This notion is due to Duchnik - 19 Thus $N(n,2) = 2$ by taking R_2 as the reverse of R_1. Fixing k find asymptotic bounds on $N(n,k)$. I can show

$$3 + [\log_2 \log_2 (n-1)] \leq N(n,3) \leq 4 + 2[\log_2 \log_2 (n-1)]$$

$$\log_2 \log_2 n \leq N(n,k) \leq C_k (\log \log n)^{k-2}$$

Problem 37. J. Moon and L. Moser

Given a square matrix A of 1's and -1's, you are permitted to multiply any of its rows, and any of its columns, by -1, to obtain anoth matrix $B = (b_{ij})$ of 1's and -1's. How small can $\left|\sum_{i,j=1}^{n} b_{ij}\right|$ be made?

We offer \$10. for proof or disproof that it can always be made ≤ 2.

Problem 38. H. Abbott and L. Moser

Find bounds for $f(n)$ = the number of Hamiltonian circuits on the n dimensional cube. It is known that

$$c^{2^n} < f(n) \leq \binom{n}{2}^{2^n},$$

where $c > 1$ is constant. We conjecture that, for large n, $f(n)$ is approximately $(n/e)^2$, and offer \$10. for proof, or disproof, that $f(n) > (kn)^2$, where $0 < k \leq \frac{1}{2}$.

Problem 39. R.L. Graham

Is it true that A must always contain a monotone arithmetic progression of length 4? It is easy to see that length 3 must occur; it is slightly harder (but possible) to construct a permutation A which has no progression of length 5.

Problem 40. R.L. Graham

Let $A = (a_1, a_2, \ldots)$ be a permutation of the positive integers (i.e., for any n there is a k such that $a_k = n$).

We say that $a_{i_1}, a_{i_2}, \ldots, a_{i_k}$ is a *monotone arithmetic progression of length* k if:

(i) the a_{i_j} form an arithmetic progression

(ii) $i_1 < i_2 < \ldots < i_k$

(iii) either $a_{i_1} < a_{i_2} < \ldots < a_{i_k}$ or $a_{i_1} > a_{i_2} > \ldots > a_{i_k}$.

Problem 41. K. Leeb

Many problems in combinatorics arise from classifications of sub- and quotientobjects of sets into equivalence types. One could mention Ramsey, Rado-Harzheim, Hall, combinatorial designs. All the concepts involved allow an immediate extension to categories, where sometimes ordering has to be managed by commutative diagrams. Rota's conjecture is only a rather poor example in the sense that vector spaces are still characterized by a number. One could for example ask for categories of relational systems satisfying some of the above mentioned theorems or having such structures. We do not have to deal with partial orders separately. But for illustration we could mention that there are "geometries" over powers of the natural numbers (cardinally ordered).

Problem 42. K. Leeb

Most known results about well-quasi-order deal with proper classes which are ordered by the existence of morphisms. This forgets some of the available structure and hence we define the following statement about categories: Let D be a finite category, A and B be subcategories of E. Every finite sequence of E-diagrams over A allows an infinite subsequence such that any two neighbored diagrams can be linked by a natural transformation over B. Example: Let D be $\cdot \to \cdot$, A and B be any of the categories of 1-1 and onto-functions of finite sets, then the above statement holds.

Problem 43. Walter Taylor

It is known that for some cardinal $\aleph$, the following is true: every graph of chromatic number $\aleph$ is elementarily equivalent to graphs of arbitrarily high chromatic number. It is also known that $\aleph_0$ does not enjoy the above property. The problem is to find the least such $\aleph$; in particular, is the above statement true of $\aleph_1$?

Such would be the case if the following problem had a positive solution. Let $G = G(k,n)$ be the graph constructed by Erdös and Hajnal (mich. Math. J. II (1964), p. 118), which has chromatic number and no odd circuits of length $\leqslant 2k + 1$. Is it true that given a graph G of chromatic number $\aleph_1$, there is a k such that for all n G contains a homomorphic image of $G(k,n)$?

A solution of either or both of these problems would have applications in model theory.

Problem 44. J.L. Selfridge

Pack n unit circles into a convex region having minimum area. Conjecture. if $n = 3k^2 + 3k + 1$ the region is a regular hexagon with rounded corners. For similar problems, such as minimizing the perimeter, or minimizing the maximal distance between centres, it seems likely that the shape is roughly circular as n gets large.